Lecture Notes of the Institute for Computer Sciences, Social Informatics and Telecommunications Engineering 333

More information about this series at http://www.springer.com/series/8197

Xiaolin Jiang · Peng Li (Eds.)

Green Energy and Networking

7th EAI International Conference, GreeNets 2020
Harbin, China, June 27–28, 2020
Proceedings

 Springer

Editors
Xiaolin Jiang
Heilongjiang University of Science
and Technology
Harbin, China

Peng Li
Dalian Polytehnic University
Dalian, China

ISSN 1867-8211   ISSN 1867-822X   (electronic)
Lecture Notes of the Institute for Computer Sciences, Social Informatics
and Telecommunications Engineering
ISBN 978-3-030-62482-8   ISBN 978-3-030-62483-5   (eBook)
https://doi.org/10.1007/978-3-030-62483-5

This Springer imprint is published by the registered company Springer Nature Switzerland AG
The registered company address is: Gewerbestrasse 11, 6330 Cham, Switzerland

Preface

We are delighted to present the proceedings of the 7th edition of the 2020 European Alliance for Innovation (EAI) International Conference on Green Energy and Networking (GreeNets 2020). This conference aims at establishing a multidisciplinary scientific meeting to discuss complex societal, technological, and economic problems of green communication and green IoT for researchers, developers, and practitioners from around the world. All of the topics related to these subjects were addressed during the GreeNets 2020 conference.

The technical program of GreeNets 2020 consisted of 35 full papers in oral presentation sessions at the main conference tracks. The conference tracks were: Track 1 – Green Communication; Track 2 – Green Energy; Track 3 – Green Networking. Aside from the high-quality technical paper presentations, the technical program also featured two keynote speeches. The two keynote speeches were Dr. Weixing Li from Dalian University of Technology of China and Dr. Fan-Yi Meng from Harbin Institute of Technology, China.

It was a great pleasure to work with the excellent organizing team of the EAI, which was absolutely essential for the success of the GreeNets 2020 conference. In particular, the peer-review process of technical papers was completed by the Technical Program Committee, which made for a high-quality technical program. We are also grateful to all the authors who submitted their papers to the GreeNets 2020 conference.

We strongly believe that the GreeNets 2020 conference provided a good forum for all researchers, developers, and practitioners to discuss all science and technology aspects related to green energy and networking. We also expect that the future GreeNets conferences will be as successful and stimulating, as indicated by the contributions presented in this volume.

August 2020

Xiaolin Jiang
Peng Li

Organization

Steering Committee

Imrich Chlamtac   University of Trento, Italy

Organizing Committee

General Chairs

Peng Li   Dalian Polytechnic University, China
Jinxian Zhao   Heilongjiang University of Science and Technology, China
Zidian Xie   Heilongjiang University of Science and Technology, China

TPC Chairs

Xiaolin Jiang   Heilongjiang University of Science and Technology, China
Qingjiang Yang   Heilongjiang University of Science and Technology, China
Xunwen Su   Heilongjiang University of Science and Technology, China

Web Chairs

Huadong Sun   Harbin University of Commerce, China
Juan Wang   Heilongjiang University of Science and Technology, China

Publicity and Social Media Chair and Co-chair

Yiqi Liu   Northeast Forestry University, China
Fugang Liu   Heilongjiang University of Science and Technology, China

Workshops Chairs

Aili Wang   Harbin University of Science and Technology, China
Yaoqun Huang   Heilongjiang University of Science and Technology, China

Sponsorship and Exhibits Chairs

Hongquan Zhang	Heilongjiang University of Science and Technology, China
Huadong Sun	Harbin University of Commerce, China
Juan Wang	Heilongjiang University of Science and Technology, China

Publications Chair

Yannan Yu	Heilongjiang University of Science and Technology, China

Posters and PhD Track Chair

Xianhui Zhu	Heilongjiang University of Science and Technology, China

Local Chair

Weiguang Zhao	Heilongjiang University of Science and Technology, China

Track Chairs

Aili Wang	Harbin University of Science and Technology, China
Zhixin Zhao	Lingnan Normal University, China
Huadong Sun	Harbin University of Commerce, China
Mingyuan Ren	Harbin University of Science and Technology, China
Yiqi Liu	Northeast Forestry University, China

Technical Program Committee

Fanyi Meng	Harbin Institute of Technology, China
Weixing Li	Dalian University of Technology, China
Zhengyu Tang	Heilongjiang University of Science and Technology, China
Wenxiang Zhang	Wuzhou University, China
Susu Qu	Heilongjiang University of Science and Technology, China

Contents

Green Communications

Green Networking

Green Energy

Research on Sub-synchronous Oscillation in Wind-HVDC-Thermal System

Yuming Pei, Xunwen Su[⊠], Hanqing Cui, and Rongbo Ma

Heilongjiang University of Science and Technology, Harbin 150022, China
suxunwen@163.com

Abstract. To comply with the characteristics of energy and source-load inverse distributed, the thermal-wind bundling system has been used in the grid of our country, such as Hami-Zhengzhou UHVDC channel and Jiuquan wind farm base power delivery channel. But it can induce some stability problem like sub-synchronous oscillation, especially in some limit operation state. This paper build a wind-thermal bundling system with HVDC in PSCAD/EMTDC simulation software, then by means of changing DC line parameter and manual putting 3-phase fault into inverter side PCC three aspects to explore the factor that entails SSO phenomenon. In conclusion, parameter mismatch and DC fault and so on factor can induce SSO phenomenon.

Keywords: Thermal-HVDC-wind farm system · SSO factor · Common coupled point

1 Introduction

Wind-Thermal bundling system operates with a method of bundling wind power and thermal power, and realizing power delivery through power transmission lines. This method can not only meet the needs of inter-regional power transmission and the construction of a strong grid, but also increase the use and development of renewable energy sources such as wind power. The wind-thermal bundling system can improve the power transmission capacity and reduce the cost of power supply, but it also brings some problems to the stable operation of the power grid. In the initial operation period of the wind and fire bundling energy base, the system is easy to run near the limit state, and the system stability cannot be guaranteed. At this time, it is particularly important to study the related technical issues such as the sub-synchronous oscillation caused by the transient stability and disturbance of the wind-fired AC/DC delivery system, and the coordination between the power supply and the line [1, 2].

Sub-synchronous oscillation (SSO) is an abnormal electromagnetic and mechanical oscillation phenomenon that occurs when the operating system's operating equilibrium point is disturbed or multiple natural oscillation frequencies for significant energy exchange [3]. The sub-synchronous oscillation belongs to the oscillation instability of

X. Jiang and P. Li (Eds.): GreeNets 2020, LNICST 333, pp. 3–13, 2020.
https://doi.org/10.1007/978-3-030-62483-5_1

the system. It is caused by a special electromechanical coupling effect in the power system. The biggest harm is that the serious electromechanical coupling effect may directly cause the serious damage to the shaft system of the large steam turbine generator set, which cause major accidents to endanger the safe operation of the power system. Wind-fired bundling via a DC transmission system is a new system structure that has emerged with the rapid development of wind power in recent years.

At present, most of the researches on sub-synchronous oscillation are focused on the single-synchronous generating unit, single wind-powered unit via DC transmission system and wind-fired bundling via series-supply delivery system, and the sub-synchronous oscillation of wind-fired bundling via DC-supply system Little research has been done [4]. With the increase of installed wind power capacity and the development of DC transmission projects, the possibility of the system structure increasing. In actual engineering, the problem of sub-synchronous oscillation does exist in the wind-fired bundled DC delivery system. On July 1st, 2015, the Hami region of Xinjiang, China, after the direct-drive wind farm was connected, a sub-synchronous oscillation accident occurred in the power system, which caused the torsional vibration protection of three thermal power units in the thermal power plant 300 km away from the wind farm to successively cause Power loss is 1280 MW. In July 2015, a thermal power plant accident caused by a synchronous synchronization occurred at a thermal power plant near the transmission terminal of the Harbin-Zhengzhou DC transmission was related to the large-scale wind power transmission at the transmission-side AC system. Therefore, it is of great significance to study the problem of sub-synchronous oscillation of wind and fire bundling via a DC output system.

2   Principe of Sub-synchronous Oscillation

Sub-synchronous oscillation (SSO) is a type of system stability problem that belongs to the dynamic process of electromechanics. Since the Mohave power station in the United States has experienced two times of generator shaft failure due to series compensation, this phenomenon has raised worldwide electrical workers. Note that the issue of sub-synchronous oscillation has also become a research hotspot in power system stability [5].

According to the different generation mechanisms, the mainstream problem of sub-synchronous oscillation is divided into two categories, namely: sub-synchronous resonance (SSR) problems and sub-synchronous control interaction (SSCI) problems. SSR problems can be divided into induction generator effect problems according to their causes and effects. SSR problems can be divided into three aspects according to their causes and effects: induction generator effect (IGE), sub-synchronous torsional interaction (SSTI), and sub-synchronous torque amplification (SSTA) [6].

IGE is because the rotor speed is higher than the sub-synchronous rotating magnetic field speed generated by the sub-synchronous current component on the stator side. From the perspective of the stator, the equivalent resistance of the rotor to the sub-synchronous stator current has a negative resistance characteristic. Effect, when this negative resistance is large enough, it can cause self-excited oscillation of the electrical system.

SSTI is a kind of electromechanical coupling self-excited oscillation phenomenon caused by the electrical system's electrical resonance frequency being complementary to the natural torsional vibration frequency of the generator shaft system. When the frequency generated on the generator rotor is equal to the natural torsional vibration frequency of the generator shaft system, it will A complementary secondary synchronization frequency voltage component is induced on the stator side. If the voltage component frequency is close to the electrical resonance frequency of the system at this time, the secondary synchronization torque generated by the rotor will be maintained. When the secondary synchronization torque is greater than or When it is equal to the mechanical damping torque of the generator, this trend will show increasing oscillation and form a continuous unstable process.

In the case of large disturbances such as faults, machine cuts, non-synchronous grid connection, and frequent operation of line switches, SSTA is due to the mutual increase of electromechanical oscillations, which causes mutual interactions on a certain natural frequency or frequencies on the shaft system and may cause severe damage in the first torsional vibration period [7].

3 Modeling of Wind-HVDC-Thermal System

3.1 Doubly-Fed Induction Generator (DFIG) Model

According to Aerodynamics, a wind turbine can transfer the wind power to the mechanical energy P_W as shown in Eq. 1. This Equation illustrates that the input mechanical power P_W is relevant to the air density ρ, blade radius R, wind speed V_W and power coefficient C_P [8].

$$P_W = \frac{1}{2} C_P \rho \pi R^2 V_W^3 \tag{1}$$

The power coefficient C_P as shown in the Eq. 2 could accurately demonstrates the power losses on the aspects of wind power utilization. And the variable value C_P is relative to one variation, the turbine angular velocity ω_W multiply by the blade radius R and divided by wind speed V_W, which we called tip speed ratio λ.

$$C_P = 0.22 \left(116 \left(\frac{1}{\lambda} - 0.035 \right) - 5 \right) e^{-12.5 \left(\frac{1}{\lambda} - 0.035 \right)} \tag{2}$$

We can make a derivation of input mechanical torque T_W demonstrated in the Eq. 3 by combining Eq. 1 and 2.

$$T_W = \frac{P_W}{\omega R} = \frac{1}{2} \frac{C_P \rho \pi R^3 V_W^2}{\lambda} \tag{3}$$

The drive mechanism in the turbine could be divided into 3 aspects: the hub, the transmission shaft and the gear box. On the static state, it could be illustrated by the Eq. 4 of the drive mechanism model. In Eq. 4, τ_W is the time constant of the entire mechanism.

$$\frac{dT_M}{dt} = \frac{1}{\tau_W} (T_W - T_M) \tag{4}$$

The turbine and the generator is connected by a speed-variable gear box, and there exist a N-times relationship of the angular velocity between the generator ω_G and turbine ω_W, which is shown in the Eq. 5.

$$\omega_G = N\omega_W \tag{5}$$

3.2 High Voltage Directive Current (HVDC) Model

At present, quasi steady state (QSS) models are used to model the dynamic problems caused by DC transmission. The converter uses a steady-state mathematical model and the related control system uses a dynamic model to describe transient process in the QSS model [9]. Figure 1 shows the converter part of the HVDC system. For this system, the AC system voltage E, the rectifier-side and inverter-side trigger angles α, β, and the commutation angle γ are known. U_{d0} is the converter-side equivalent voltage. X_T is the equivalent reactance of converter loss and transformer leakage reactance (Fig. 2). The steady-state model of the rectifier side and the inverter side and the dynamic model of the dc line are respectively shown in Eqs. (6)–(8).

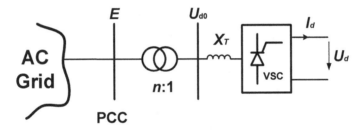

Fig. 1. HVDC converter model

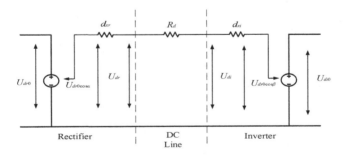

Fig. 2. QSS equivalent model

$$\begin{cases} U_{dr0} = \frac{3\sqrt{2}E_r}{n\pi} \\ \Delta U_{dr} = d_{xr}I_{dr} = \frac{1}{2}U_{dr0}(\cos\alpha - \cos(\alpha + \gamma)) \\ U_{dr} = U_{dr0}\cos\alpha - d_{xr}I_{dr} \end{cases} \tag{6}$$

$$\begin{cases} U_{di0} = \frac{3\sqrt{2}E_i}{n\pi} \\ d_{xi} = \frac{3X_T}{\pi} \\ U_{di} = U_{di0}\cos\beta + d_{xi}I_{di} \end{cases} \tag{7}$$

$$\begin{cases} C\frac{dU_{dr}}{dt} = I_{dr} - I_d \\ C\frac{dU_{di}}{dt} = I_{di} + I_d \\ L_d\frac{dI_d}{dt} = U_{dr} - U_{di} - R_dI_d \end{cases} \tag{8}$$

3.3 Thermal-HVDC-Wind Farm System

Figure 3 shows the schematic diagram of thermal-HVDC- wind farm system. The power frequency of the whole system is 60 Hz.The thermal power unit adopts the IEEE first standard model. The thermal power unit adopts the four-masses (LPA-LPB-GEN-EXT) block model, the rated voltage is 7.97 kV meanwhile the rated current is 3.136 kA, and the rated power is 892.4 MV.A, and is fed into the HVDC system through 13.8/62.5 kV boost transformer.

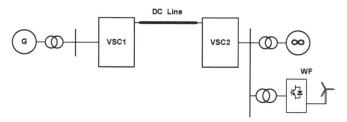

Fig. 3. Schematic graph of thermal-HVDC-wind farm system

The HVDC system transmits 115 kV DC voltage and 75 MW rated power, and the length of the DC line is 50 km. In order to ensure system voltage stability and stable power transmission, the rectifier side adopts the fixed active power control strategy shown in Fig. 4(a), and the inverter side adopts the fixed DC power control strategy shown in Fig. 4(b). The inverter is connected to a 110 kV power grid. The final equivalent of the power grid is an infinite system with a rated voltage of 110 kV and a rated power of 100 MV.A.

The wind farm is a N identical 2 MW DFIG wind turbines of the same model, with a rated voltage of 0.69 kV, which is connected to the inverter-side common coupling point (PCC) after 0.69/110 kV step-up transformer. The rotor side of the inverter adopts stator flux-oriented vector control, and decoupling control of active and reactive power is achieved through coordinate transformation. The stator side adopts the grid voltage-oriented vector control strategy to achieve stable output of the generator port voltage according to the grid current and provide a trigger angle for the rotor side flux control (Fig 5).

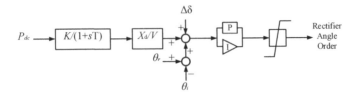

(a) Rectifier control strategy

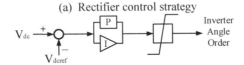

(b) Inverter control strategy

Fig. 4. HVDC system control strategy

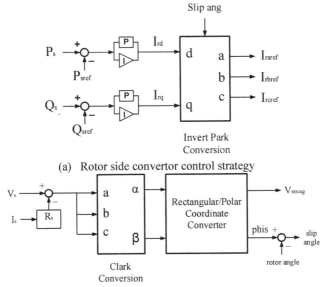

(a) Rotor side convertor control strategy

(b) Stator side convertor control strategy

Fig. 5. Wind farm converter control strategy

4 Simulation

This paper uses PSCAD/EMTDC simulation software to build a model of wind-thermal bundle system with HVDC as shown in Fig. 3. By changing the related parameters of the DC transmission system, and the PCC fault on the inverter side, the factor and influence of SSO is explored in such wind-thermal system with HVDC.

4.1 Infects of the DC Line Parameters

To explore the effect of SSO by changing DC line parameter, the experience will change the length from 50 km to 200 km. It is shown that when the line span 50-km long, the system will operate at static state as demonstrated in Fig. 6.

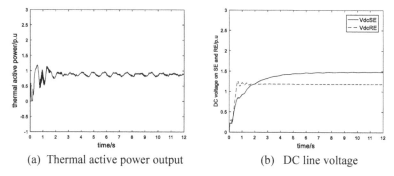

(a) Thermal active power output   (b) DC line voltage

Fig. 6. Static state power and DC voltage characteristic

When the length become longer, the stability of the system may changed. At the length of 125 km, the active power of the thermal generator suffered a variation which was beginning at the time of 3 s and maintaining 1 s, then the power operated with small scale vibration as is shown in Fig. 7.(a). So did the DC voltage both on the rectifier side and on the inverter side in Fig. 7.(b), but after a 1 s period variation it finally operated in a new static state. So in reality, if the shaft is robust enough that can bear the first period torque amplification, the system could finally operates in a stable state.

However, there is a severe damage at the length of 150 km as is vividly demonstrated in Fig. 8(a)–(d). There is a variation at the time of 3 s and maintaining 1 s, rectifier side DC voltage V_{dcR} dropped to a very low level. About 3.5 s later, second variation occurs and it induced a severe unstable state: Torque LPA-LPB and Torque LPB-GEN amplified at times and maintaining till the end of the simulation.

When the length span 200 km, a variation about 3–4 s occurs in the system. And the DC voltage finally emerged a trend to get close to a new stable state as shown in Fig. 9.(b). In the new state, DC line voltage dropped at the sending end.

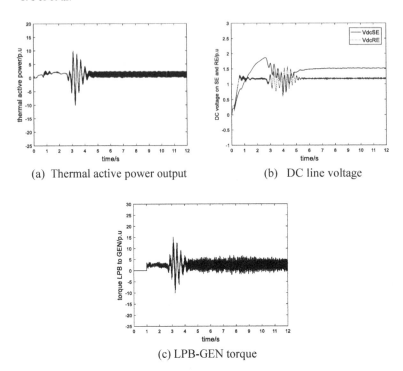

(a) Thermal active power output

(b) DC line voltage

(c) LPB-GEN torque

Fig. 7. Power, DC voltage and torque characteristic when L = 125 km

In conclusion, the mismatch of the DC line parameter may enduce SSO phenomenon in the system. A matched parameter is essential in the period of the system modeling and operating.

4.2 Infects of the Inverter Side PCC Fault

In this section, a single line short circuit, 3-phase fault will be manually put into at the 6 s, and the fault will sustain 1 s. When the single line to ground short circuit was putting into the PCC, it is shown that there is a small disturbance on generator output power and torque. There is a small transient voltage drop at inverter side DC line, while rectifier side shows no influence when the fault was put into operation (Fig. 10).

When the 3-phase short circuit was putting into the PCC, it could be seen that active power of thermal generator and torque amplified swiftly, meanwhile DC voltage at the inverter side dropped to a very low standard. But 2 s later, the fault was cleared and the system recovered to a new steady state. And if the fault time sustained longer, such transient process would be longer (Fig. 11).

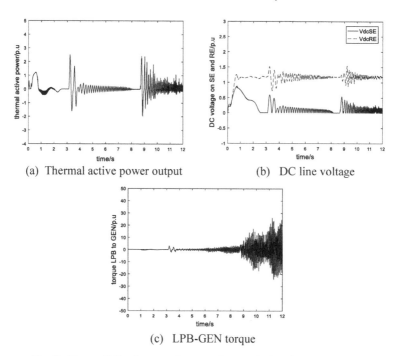

(a) Thermal active power output   (b) DC line voltage

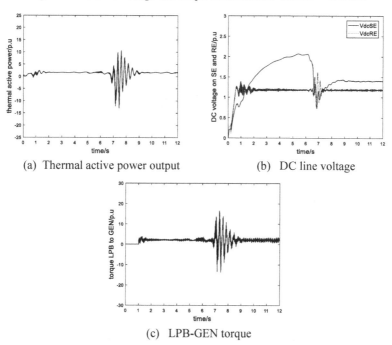

(c) LPB-GEN torque

Fig. 8. Power, DC voltage and torque characteristic when L = 150 km

(a) Thermal active power output   (b) DC line voltage

(c) LPB-GEN torque

Fig. 9. Power, DC voltage and torque characteristic when L = 200 km

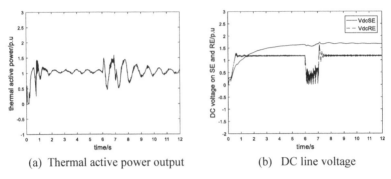

(a) Thermal active power output   (b) DC line voltage

Fig. 10. Power and DC voltage graph when single line to ground short curcuit occurs

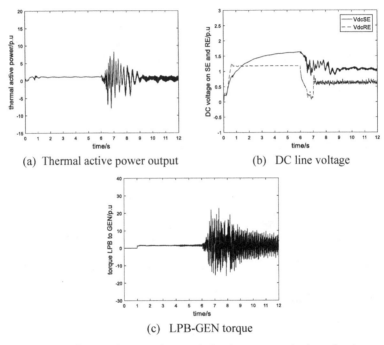

(a) Thermal active power output   (b) DC line voltage

(c) LPB-GEN torque

Fig. 11. Power, DC voltage and torque characteristic when symmetric short circuit occurs in the system

5 Conclusion

(1) A mismatched DC line parameter may caused SSO in the thermal-HVDC-wind farm system. Such torsional interaction influence is stronger in the thermal system than in the DFIG wind farm. And the mainly factor of such SSO is power shortage, which entails DC line voltage unstable and thermal torque amplification. This SSO phenomenon should pay more attention on the first oscillaton period.

(2) Faults appears at inverter-side would induce a voltage drop at inverter DC line side, while the degree of the generator output power, torsional torque on the shaft and rectifier-side DC voltage will mostly depend on the fault type on inverter-side PCC and time scale of the fault.

Acknowledgment. This work was supported by the National Science Foundation of China under Grant(51677057), Local University Support plan for R&D, Cultivation and Transformation of Scientific and Technological Achievements by Heilongjiang Educational Commission(TSTAU-R2018005) and Key Laboratory of Modern Power System Simulation and Control & Renewable Energy Technology, Ministry of Education(MPSS2019-05).

References

1. Dong, X., Yang, W., Wu, L.: Subsynchronous oscillation characteristic study of wind-termal power bundling and EHV AC-DC hybrid transmission system. In: International Conference on Power Technology, pp. 1995–2000 (2018)
2. Hu, J., Huang, Y., Wang, D.: Modeling of grid-connected DFIG-based wind turbines for DC-LINK voltage stability analysis. IEEE Trans. Sustain. Energ. **6**(4), 1–12 (2015)
3. IEEE SSR Working Group.: First benchmark model for computer simulation of subsynchronous resonance. IEEE Trans. Power Apparatus Syst. **96**(5), 1565–1572 (1977)
4. IEEE SSR Working Group.: Second benchmark model for computer simulation of subsynchronous resonance. IEEE Trans. Power Apparatus Syst. **104**(5), 1057–1066 (1985)
5. Bahrman, M., Larsen, E.V., Piwko, R.J., Patel, H.S.: Experience with HVDC-turbine-generator torsional interaction at square butte. IEEE Trans. **99**, 966–975 (1980)
6. Lyu, J., Cai, X., Amin, M., Molinas, M.: Sub-synchronous oscillation mechanism and its suppression in MMC-based HVDC connected wind farms. IET Gener. Transm. Distrib. **12**(4), 1021–1029 (2018)
7. Song, Y., Breitholtz, C.: Nyquist stability analysis of an AC-grid connected VSC-HVDC system using a distributed parameter DC-cable model. IEEE Trans. Power Delivery **31**(2), 898–907 (2016)
8. Hu, J., Huang, Y., Wang, D.: Modeling of grid-connected DFIG-based wind turbines for DC-LINK voltage stability analysis. IEEE Trans. Sustain. Energy **6**(4), 1–12 (2015)
9. Stamatiou, G., Bongiorno, M., Song, Y., Breitholtz, C.: Analytical derivation of poorly-damped eigenvalues in two-terminal VSC-HVDC systems. In: EPE 2017 ECCE Europe (2017)

Research on SSO Influence with Consideration of Topology Structure Based on CTC Analysis Method

Hanqing Cui[✉], Xunwen Su, Yuming Pei, and Shiyan Zhao

Heilongjiang University of Science and Technology, Harbin 150022, China
969426818@qq.com

Abstract. In order to study the phenomenon of sub-synchronous oscillation (SS0) in thermal power plants considering different topologies, based on the IEEE first standard model, this paper uses the PSCAD simulation platform to build models of different topological structures of thermal power plants. The complex torque coefficient (CTC) analysis method is used to obtain the electrical damping (De) in each topology. The analysis shows that the topology will affect the sub-synchronous oscillation, if the system topology is more complex, the system's ability to withstand the sub-synchronous oscillation will be stronger, and it is verified by time-domain simulation..

Keywords: Sub-synchronous oscillation · Complex torque coefficient · Topology · Electrical damping

1 Introduction

Sub-synchronous oscillation is an abnormal operating state in the power system. When sub-synchronous oscillation occurs in the system, it will produce continuous or even increased oscillations in the related variables of the mechanical system's shafting and electrical system. In severe cases, it can even cause The shaft system is damaged or even broken, becoming a major threat to the stable operation of the power system [1]. Since 1970 and 1971, sub-synchronous oscillation accidents occurred in the Mohave power plant in the United States, resulting in damage to the shaft system. The issue of sub-synchronous oscillation began to receive widespread attention [2, 3]. Because the sub-synchronous oscillation is very harmful, the analysis and research of the sub-synchronous oscillation phenomenon is of great significance to the stability of the system.

Commonly used sub-synchronous oscillation analysis methods are: frequency scanning method, complex torque coefficient analysis method, eigenvalue analysis method and time domain simulation method [4]. The frequency scanning method can screen out system operating conditions with potential sub-synchronous oscillation threats, but the frequency scanning method results are not accurate and can only be used as a preliminary screening method to identify units that may have SSO problems. The eigenvalue analysis method needs to solve the eigenvalues of the system coefficient matrix. As the

X. Jiang and P. Li (Eds.): GreeNets 2020, LNICST 333, pp. 14–23, 2020.
https://doi.org/10.1007/978-3-030-62483-5_2

complexity of the system increases, the matrix dimension becomes larger and larger, and it has the problem of dimensional disaster. The complex torque coefficient analysis method can get the full picture of the electrical damping coefficient changing with frequency, and can also take into account the effects of dynamic processes and operating conditions of various control systems on sub-synchronous oscillation [5, 6].

The topology of the power grid will have a direct impact on the stability of the power system. A reasonable grid structure can improve its reliability. Therefore, it is particularly important to analyze and study the influence of network topology on sub-synchronous oscillation of power systems [7].

In this paper, simulation models of different topologies are built on the PSCAD electromagnetic simulation platform. The complex torque coefficient analysis method is used to analyze the electrical damping under different topologies. The influence of the topology on sub-synchronous oscillation is analyzed,and verify with time domain simulation.

2 Complex Torque Coefficient Analysis Method

2.1 Principle of Complex Torque Coefficient Analysis Method

The complex torque coefficient analysis method is to divide the system into two parts, namely the electrical part and the mechanical part. Consider the two subsystems separately and add small disturbances, then we can obtain the electric complex torque coefficient $K_E(j\zeta)$ and the mechanical complex torque coefficient $K_M(j\zeta)$, and write them in plural, when the sum of the real parts is equal to zero, and the sum of the imaginary parts is less than zero, the system will oscillate sub-synchronously.

The specific method of the complex torque coefficient method is: apply a forced small amplitude oscillation with a frequency f to the relative angle of the rotor of a generator set in the system, that is,

$$\Delta\delta = \Delta\delta_m e^{j\zeta t} \tag{1}$$

Where $\zeta = 2\pi f$, $\Delta\delta_m$ is the amplitude of the oscillation. The electrical complex torque increment ΔT_e and the mechanical complex torque increment ΔT_M in the generator electrical system and mechanical system response caused by this small amplitude oscillation $\Delta\delta$ can be obtained through calculation, respectively, and define the equivalent electrical complex torque coefficient $K_E(j\zeta) = \Delta T_e/\Delta\delta$, the equivalent effective mechanical complex torque coefficient $K_M(j\zeta) = \Delta T_M/\Delta\delta$. Usually $K_E(j\zeta)$ and $K_M(j\zeta)$ are plural.

$$K_E(j\zeta) = K_e(\zeta) + j\zeta D_e(\zeta) \tag{2}$$

$$K_M(j\zeta) = K_m(\zeta) + j\zeta D_m(\zeta) \tag{3}$$

Where K_e, D_e are called electrical elastic coefficient and damping coefficient respectively; K_m, D_m are called mechanical elastic coefficient and damping coefficient, respectively, they are functions of frequency ζ, D_e and D_m indicate the damping characteristics of the system at different frequencies. By comparing these coefficients, we can analyze the oscillation characteristics of the system at frequency $f = \zeta/2\pi$. The discriminant formula for unstable sub-synchronous oscillation is as follows:

$$[D_m(\zeta) + D_e(\zeta)]_{K_m(\zeta)+K_e(\zeta)=0} < 0 \tag{4}$$

When the frequency f changes from 0 to 60 Hz, the characteristic curves of the system's coefficients $K_m(\zeta), D_m(\zeta), K_e(\zeta), D_e(\zeta)$ as a function of frequency can be obtained. In fact, it is usually only necessary to scan the coefficients within a narrow frequency range near the natural torsional oscillation frequency of the shaft system to determine whether the system will have unstable sub-synchronous oscillations.

3   Model Establishment

3.1   Establishment of Shafting Model

In this paper, a six-mass block model is adopted. The generator shaft system is a six-mass block model. The high-pressure cylinder HP, the medium-pressure cylinder, the two low-pressure cylinders LPA, LPB, the generator GEN and the exciter EXC are connected in sequence through a massless spring. Figure 1 is the schematic diagram of the model.

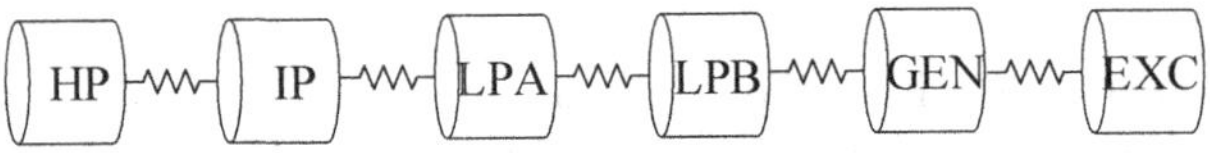

Fig. 1. Shaft structure diagram

The equation of the shafting system is Eq. (4) and Eq. (5)

$$\frac{d\delta_i}{dt} = \omega_i - \omega_0, \quad i = 1, 2, \cdots, 6 \tag{5}$$

Where δ_i is the electrical angular displacement of the i-th mass with respect to the synchronous rotation reference cycle, and ω_i is the electrical angular velocity of the i-th mass.

T_{Ji} Is the inertia time constant of the i-th mass, T_{mi} is the original torque on the i-th mass, D_{ii} is the self-damping of the i-th mass, and $D_{i,i+1}$ is between the i-th and i + 1-th masses Mutual damping.

$$\begin{cases} T_{J1}\frac{d\omega_1}{dt} = T_{m1} - D_{11}\omega_1 - D_{12}(\omega_1 - \omega_2) - K_{12}(\delta_1 - \delta_2) \\ T_{J2}\frac{d\omega_2}{dt} = T_{m2} - D_{22}\omega_2 - D_{12}(\omega_2 - \omega_1) - D_{23}(\omega_2 - \omega_3) \\ \qquad\quad - K_{12}(\delta_2 - \delta_1) - K_{23}(\delta_2 - \delta_3) \\ T_{J3}\frac{d\omega_3}{dt} = T_{m3} - D_{33}\omega_3 - D_{23}(\omega_3 - \omega_2) - D_{34}(\omega_3 - \omega_4) \\ \qquad\quad - K_{23}(\delta_3 - \delta_2) - K_{34}(\delta_3 - \delta_4) \\ T_{J4}\frac{d\omega_4}{dt} = T_{m4} - D_{44}\omega_4 - D_{34}(\omega_4 - \omega_3) - D_{45}(\omega_4 - \omega_5) \\ \qquad\quad - K_{34}(\delta_4 - \delta_3) - K_{45}(\delta_4 - \delta_5) \\ T_{J5}\frac{d\omega_5}{dt} = T_{m5} - D_{55}\omega_5 - D_{45}(\omega_5 - \omega_4) - D_{56}(\omega_5 - \omega_6) \\ \qquad\quad - K_{45}(\delta_5 - \delta_4) - K_{56}(\delta_5 - \delta_6) \\ T_{J6}\frac{d\omega_6}{dt} = -T_{m6} - D_{66}\omega_6 - D_{56}(\omega_6 - \omega_5) - K_{56}(\delta_6 - \delta_5) \end{cases} \quad (6)$$

Linearize Eqs. (5) and (6) at the operating point to obtain Eq. (7).

$$\begin{cases} T_{Ji}P\Delta\omega_i = \Delta T_i - D_{ii}\Delta\omega_i - D_{i-1,i}(\Delta\omega_i - \Delta\omega_{i-1}) - D_{i,i+1}(\Delta\omega_i - \Delta\omega_{i+1}) \\ \qquad\quad - K_{i-1,i}(\Delta\delta_i - \Delta\delta_{i-1}) - K_{i,i+1}(\Delta\delta_i - \Delta\delta_{i+1}) \\ P\Delta\delta_i = \omega_b\Delta\omega_i \end{cases} \quad (7)$$

$$i = 1, 2, \cdots, 6$$

3.2 Establishment of Disturbance Model

The addition of disturbances should not destroy the linearity of the system, so the amplitude of the disturbances is chosen as 0.01 pu. Due to the complexity of the system, adding only one disturbance per simulation will cause too low efficiency, but it will add disturbances of multiple frequencies at the same time. The superimposed perturbation signal will have spikes, and the amplitude is large enough to destroy the linearization of the system, so a nonlinear lag phase is added to the perturbation of each frequency [9]. In summary, the disturbance model established is as follows:

$$N = \sum_{k=5}^{60} T_k \cos(2\pi k f_0 + (k f_0)^5) \quad (8)$$

Since the sub-synchronous oscillation frequency range is 5–60 Hz, the disturbance frequency f_0 is taken as 5–60 Hz. Set up the disturbance model in PSCAD/EMTDC as shown below (Fig. 2).

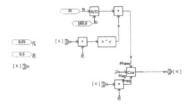

Fig. 2. Disturbance model

4   Simulation Analysis

In the PSCAD/EMTDC simulation platform, this paper combines the IEEE standard model and a 39-node system to build a simulation model to study the influence of the topology on the sub-synchronous oscillation of the system with series compensation. Because mechanical damping D_m is generally a small positive value, electrical damping D_e is the main factor affecting sub-synchronous oscillation. Therefore, electrical damping coefficients are often used to judge the factors that affect the system's sub-synchronous oscillation [10].

4.1   Topology 1: 39 Asynchronous System Connected to the Generator's Far End

The 39-node system is connected to the side of the series compensation capacitor, that is the end away from the steam turbine. Its topology is shown in Fig. 3.

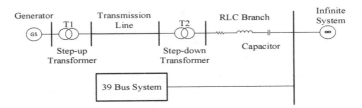

Fig. 3. Schematic of topology 1

The complex torque coefficient analysis method can be used to obtain the De value in this topology as shown in Fig. 4. The electrical damping coefficient has a large negative value around 10 Hz, which is about −8000pu. This shows that the system has a higher risk of sub-synchronous oscillation.

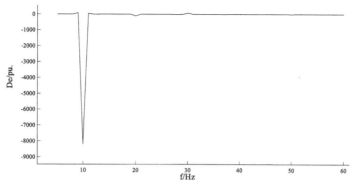

Fig. 4. De for topology 1

4.2 Topology 2: 39 Asynchronous System Connected to the Generator's Near End

The 39-node system is connected to the primary bus of the step-down transformer. Compared with the topology, a 39-node system is closer to the turbine. Its topology is shown in Fig. 5.

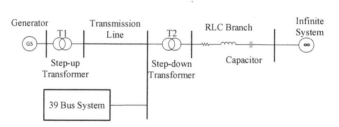

Fig. 5. Schematic of topology 2

The complex torque coefficient analysis method can be used to obtain the De value in this topology as shown in Fig. 6. The electrical damping coefficient has a large negative value of about −70pu around 24 Hz. It shows that the system has the risk of sub-synchronous oscillation.

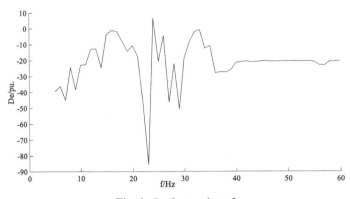

Fig. 6. De for topology 2

4.3 Topology 3: 39-Node System in Parallel with Step-Down Transformer and Inductor at Both Ends

The 39-node system is connected to the IEEE first standard model as shown in Fig. 7. It can be found that the topology of this model is more complicated than the above two models.

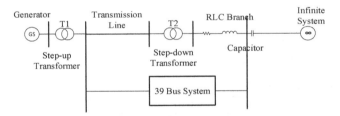

Fig. 7. Schematic of topology 3

Using the complex torque coefficient analysis method, the value of De in this topology can be obtained as shown in Fig. 8. The electrical damping coefficient has a large negative value around 10 Hz, which is about −1500 pu. This indicates that the system is at risk of sub-synchronous oscillation. The frequency at which the De value peaks is similar to that of Topology 1 at about 10 Hz, but its negative value is much smaller than that of Topology 1, and larger than the DE value of Topology 2.

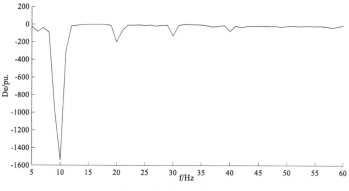

Fig. 8. De for topology 3

4.4 Time Domain Simulation to Verify

In the PSCAD/EMTDC electromagnetic simulation software, the time domain simulation is performed on the models of the above three topologies, and a three-phase short-circuit fault is added in 1.5 s, and the fault lasts for 0.075 s. Figure 9, 10 and 11 shows the torque between the two low-pressure cylinders LPA and PLB in each model.

It can be found that the three topologies have sub-synchronous oscillations, which are consistent with the results obtained by the complex torque coefficient method in the previous chapter. Among them, topology 1 oscillates most intensely, topology 2 has a tendency to converge, and topology 3 has For small amplitude equal amplitude oscillations. It can be seen that the complex torque coefficient method can effectively analyze the risk of sub-synchronous oscillations in the system, but the magnitude of De and the severity of the oscillations are not directly related. The topology structure will affect the sub-synchronous oscillation. The structure of topology 3 is more complicated, but the system is relatively more stable.

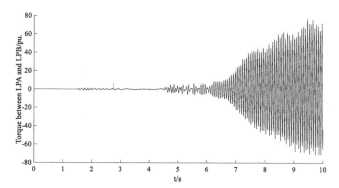

Fig. 9. Torque between LPA and LPB for topology 1

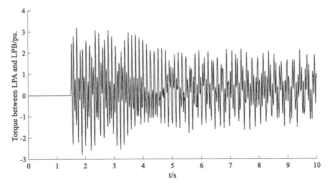

Fig. 10. Torque between LPA and LPB for topology 2

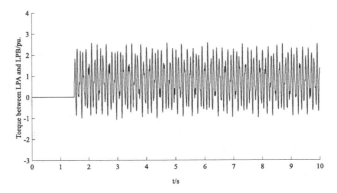

Fig. 11. Torque between LPA and LPB for topology 3

5 Conclusion

Based on the IEEE first standard model, this paper uses the PSCAD/EMTDC simulation platform to build models of different topologies. The complex torque coefficient analysis method is used to obtain the electrical damping of each system, and the risk of sub-synchronous oscillation in each system is evaluated. And verified by time domain simulation, the following conclusions are obtained:

(1) The complex torque coefficient analysis method can effectively analyze the risk of sub-synchronous oscillation of the system, and can obtain the curve of the electrical damping coefficient change in the full frequency domain.
(2) The electrical damping obtained by the complex torque coefficient analysis method can explain the possibility of subsynchronous oscillation of the system.
(3) Different topological structures will affect the risk of sub-synchronous oscillations. The more complex the topology, the stronger the ability to resist sub-synchronous oscillations. The topological structure of the system should be reasonably designed to avoid the occurrence of sub-synchronous oscillations.

Acknowledgment. This work was supported by the National Science Foundation of China under Grant (51677057), Local University Support plan for R & D, Cultivation and Transformation of Scientific and Technological Achievements by Heilongjiang Educational Commission (TSTAU-R2018005) and Key Laboratory of Modern Power System Simulation and Control & Renewable Energy Technology, Ministry of Education (MPSS2019-05)

References

1. Cheng, S., Cao, Y., Jiang, Q.: Theory and method of subsynchronous oscillation in power system. Beijing Science Press, vol. 68, pp. 84–90 (2008)
2. Livermore, L., Ugalde-Loo, C.E., Mu, Q., et al.: Damping of subsynchronous resonance using a voltage source converter-based high-voltage direct-current link in a series-compensated great Britain transmission network. IET Gener. Transm. Distrib. **8**(3), 542–551 (2014)
3. Wang, K., Xu, Z., Du, N., et al.: Subsynchronous resonance by thyristor controlled series compensation. High Volt. Eng. **42**(1), 321–329 (2016)
4. Xu, Y.: Research on damping control of subsynchronous oscillation in HVDC transmission based on observability and controllability. Power Syst. Prot. Control **41**(9), 21–26 (2013)
5. Li, G., Zhang, S., Zhang, Z.: Analysis of the possibility of subsynchronous resonance under the high series compensation level of the 330 Kv main network in Northwest China by frequency scanning method. Proc. CSEE **12**(1), 58–61 (1992)
6. Wang, Z., Wang, L., Xie, X., et al.: A method integrating frequency scanning and complex torque coefficient to quantitatively evaluate subsynchronous resonance risk. Power Syst. Technol. **35**(7), 101–105 (2011)
7. Yang, X.: Study on network topology analysis and operation modes combination in power system, vol. 27. Huazhong University of Science and Technology, Wuhan (2007)
8. Li, H., Chen, Y., Zhao, B., et al.: Analysis and control strategies for depressing system sub-synchronous oscillation of DFIG-based wind farms. Proc. CSEE **35**(7), 1613–1620 (2015)
9. Wang, J., Guo, C., Guo, X., et al.: Realization of PSCAD/EMTDC based complex torque coeffient approach. East China Electr. Power **38**(12), 1854–1857 (2010)
10. Huang, L., Li, J., Liu, X., et al.: Realization of complex torque coefficient method in time domain based on PSCAD/EMTDC and research on the influence on electrical damping. Sensor World **24**(2), 13–18 (2018)

Summary of Fault Line Selection for Single-phase Grounding in Small Current Systems

Xianhui Zhu, Youwei Jian[✉], Nan Shi, Pin Lv, and Yue Yu

School of Electrical and Control Engineering, Heilongjiang
University of Science and Technology, Harbin 150022, China
`jianyouwei1@163.com`

Abstract. At present, fault line selection theory of small current grounding system mainly focuses on solving the reliability problem of urban power supply system. There is little research on the line selection theory of mine distribution network, but single phase ground fault in mine distribution network will cause more harm. This paper analyzes and compares the existing fault line selection theories applicable to the mine power supply system and other industrial sites, providing theoretical guidance for the study of fault line selection in special occasions such as mines.

Keywords: Fault line selection · Small current grounding system · Mine power supply system

1 Introduction

Most of the low-voltage distributed system in China adopt neutral non-grounding system, and the faults are mostly single-phase grounding fault. The increase of the unfaulted phase voltage will endanger the weak link of insulation and seriously affect the reliability of power supplies when the system has single-phase grounding fault in neutral non-grounding system [1]. Especially in mines and other special occasions, the power supply environment is more dangerous and complex, which may lead to more serious.

At present, the research of line selection theory focuses on urban power supply system, but the research on fault line selection for special occasions such as mine is relatively few. Based on the analysis and comparison of the existing fault line selection theory which is applicable to mines and other industrial sites, this paper provides theoretical guidance for the study of fault line selection in mines and other special industrial occasions.

2 Fault Signal Characteristic Analysis of Small Current Grounding System

2.1 Steady-State Characteristics of Neutral Non-grounding System

In neutral non-grounding system, we assume that the line resistance is zero, which there is no voltage drop on the load current line, and the load is treated as constant to simplify the

X. Jiang and P. Li (Eds.): GreeNets 2020, LNICST 333, pp. 24–28, 2020.
https://doi.org/10.1007/978-3-030-62483-5_3

analysis. The parameters of the distribution network system are symmetrical parameters during normal operation. The three relative ground capacitors of each line are equal to C_0, the ground capacitance of bus and back power supply is C_{0S}, and each phase the power supply is represented by E_A, E_B, E_C, respectively.

When metallic grounded fault occurs in phase A of line L2, the steady-state capacitance current distribution of the system is shown in Fig. 1. As can be seen from Fig. 1, Fig. 2 and Fig. 3, steady-state capacitive current has the following characteristics [2]:

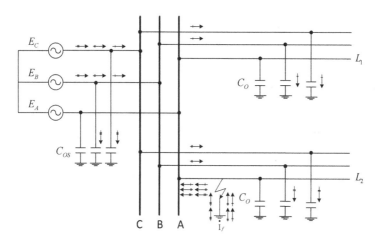

Fig. 1. Schematic diagram of single-phase ground fault

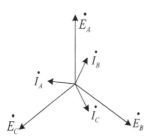

Fig. 2. Normal voltage vector relation

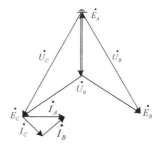

Fig. 3. Fault voltage vector relation

1. In the fault line, phase A to ground voltage drops to 0, so the capacitance current is 0. The other two phase to ground voltage rises to $\sqrt{3}$ times, and the capacitance to ground current correspondingly increases by $\sqrt{3}$ times.
2. The value of zero-sequence current in the fault line is equal to the sum of the capacitance current of the non-fault components of the whole system to the ground, and the actual direction of capacitive reactive power is line to bus.
3. The zero-sequence current of the unfaulted phase line is the grounding capacitive current of the line itself, and the zero-sequence reactive power flows from the bus to the line.

2.2 Transient Characteristics of Neutral Non-grounding System

When the neutral indirectly grounded system has ground fault, the zero-sequence current will oscillate strongly. These abundant transient signals have large amplitude and wide spectrum, which provide important criteria for identifying grounding fault. Transient characteristics of small current ground fault [3]:

1. The discharge capacitor current caused by a rapid drop in the fault phase voltage, which flows from the bus to the fault point, and the discharge is rapidly attenuated, with an oscillating frequency of up to several kilohertz.
2. The unfaulted phase voltage will rise to $\sqrt{3}$ times the phase voltage resulting in the charging capacitor current, which loops through the power supply. Due to the large inductance of the whole circuit, the charging current decays slowly and oscillates low, usually only a few hundred Hertz.

3 Fault Detection Method for Small Current Grounding System

At present, there are three kinds of fault line detection methods, namely, manual pulling method, active line selection method and passive line selection method.

3.1 Manual Wire Pulling Method

Manual wire pulling method. When an earth fault occurs, the power grid operation and maintenance personnel shall close each line one by one, and look for the fault line according to the different states after closing. This method can cause short interruption of power supply, but it is very reliable and often used in actual operation.

3.2 Active Line Selection Method

Signal injection method [4]. The method is to inject the current signal of specific frequency into the earthing circuit through the neutral point of the three-phase voltage transformer. The injected signal will be injected into the ground along the fault line through the landing site, and each line will be detected with a signal detector. When the fault occurs, the line that detects the presence of an injection signal is determined as the fault line.

The merit of this method is that it is not influenced by the arc suppression coil, does not need to install the zero sequence current transformer. It can also detect faults along the fault line with a detector to find out where the fault occurred on the overhead line. But the disadvantage is that the signal injection equipment needs to be installed. In the actual application, the fault signal is often weak and difficult to detect, resulting in misjudgment.

3.3 Passive Line Selection Method

Passive mode can be divided into fault steady-state variable line selection, fault transient variable line selection and fault transient and steady-state variable comprehensive line selection method.

Line Selection Method Based on Fault Steady State Information. Zero-sequence current amplitude method [5]. This method is based on the zero- sequence current amplitude of each line to determine the fault line. In other words, the largest amplitude is the fault line. However, the line selection may fail when the amplitude difference is not large or the bus fails. In addition, it is also influenced by unbalanced current of the current transformer, system operation mode, environmental factors and other problems. This line selection method has a small range of application and low reliability.

Zero-sequence current phase ratio method. Through fault signal characteristic analysis of neutral non-grounding system, it can be seen that the flow direction of zero-sequence current of fault line is opposite to that of healthy line. By using this characteristic, fault line can be quickly identified. However, the zero-sequence current flow of fault line is the same as that of unfaulted line when small current system is overcompensated.

Zero-sequence current group amplitude-phase method [6]. This method summarizes the above two ways of line selection. Firstly, the zero-sequence current amplitude of the line is compared, and several lines of large zero-sequence current amplitude are selected as alternative lines. Then, the phase comparison of the alternative lines is carried out. If the zero-sequence current phase of the selected line is different from that of other lines, the line is considered a fault line. If the zero-sequence current phase of each line is the same, it is judged to be bus fault. This method overcomes some disadvantages of the former two methods to some extent, but it also cannot overcome the influence of unbalanced current and transition resistance of current transformer.

Although the calculation amount of the steady-state method is small, the fault current amplitude is very small, and the current is also superimposed in the load current. The electromagnetic interference in the field is relatively large, which makes the detected fault component have a very low signal-to-noise ratio. Therefore, in the actual distribution network, the line selection method based on steady-state component is easy to cause misjudgment.

Line Selection Method Based on Fault Transient information. First half wave method [7]. There is a fixed phase relation between transient zero-sequence current and transient zero-sequence voltage at the initial stage of fault occurrence, lasting about 1/4 period. The phase polarity of the fault line is opposite, and the polarity of the healthy line is the same, so this feature can be used to identify the fault line. This method is suitable for neutral arc suppression coil grounding system and neutral indirect grounding system, and can also detect unstable grounding fault. However, the opposite polarity only holds in a very short time, and has a great influence in many aspects.

PRONY algorithm [8]. Prony algorithm can accurately analyze the fault current of grounding point. The method uses exponential fitting model for spectrum analysis. The amplitude, phase, frequency and other parameters of the transient components of fault current in the neutral non-grounding system are obviously related to the fault characteristics. The algorithm uses these characteristics to select fault line. Using prony algorithm can effectively achieve fault location, but the calculation of this algorithm is relatively large.

Line selection method of wavelet transform [9]. Wavelet transform has the reputation of "mathematical microscope" and can analyze the local characteristics of fault current signal well. Choosing the appropriate wavelet base to transform the zero sequence current, the modulus of wavelet coefficient which is much larger than the normal value can be obtained. Through the comparison of modulus, the maximum modulus is determined as the fault line.

The advantage of wavelet method is that it can be used in neutral non-grounding system and arc suppression coil grounded system. This method can also be used to deal with the situation that the fault waveform is disordered and the fault condition is complex. However, in the complex and changeable working environment of power system, all kinds of interferences are very strong and the signal-to-noise ratio is low. Wavelet analysis method is a comparative analysis of the singular points of zero-sequence current waveform, which is usually difficult to distinguish and prone to misjudgment.

4 Conclusion

This paper summarizes the method of single-phase grounding fault line selection, and introduces merit and demerit of various methods in detail, which can guide the line selection in industrial field. In the current society, whether it is urban distribution network or mine power supply reliability is becoming more and more important. With the development of fault theory research and fault signal processing, it is believed that it is not far to realize fault line selection reliably and accurately in the neutral point indirectly earth system.

References

1. Yu, Q., Chen, Z., He, Z.: Fault phase selection of mine power supply system based on frequency distribution theory. Ind. Min. Autom. **43**(08), 76–82 (2017)
2. Guo, Q., Wu, T.: Review of fault selection methods for small current grounding systems. Power Syst. Prot. Control **38**(02), 146–152 (2010)
3. Xue, Y., Feng, Z., Xu, B., et al.: Research on the selection of small current grounding line based on transient zero-sequence current comparison. Power Syst. Autom. **09**, 48–53 (2003)
4. Pan, Z., Zhang, H., Zhang, F., et al.: Analysis and improvement of earthing line selection and positioning protection for signal injection. Power Syst. Autom. **31**(04), 1–75 (2007)
5. Chen, Z., Chen, D., Yuanlong lv., et al.: Review of fault selection methods for small current grounding system. Hydraul. Electr. Mach. (10), 70–74 (2007)
6. Yao, M., An, Y., Han, H., et al.: Fault selection and location method of small current grounding system. Electric Appl. Energy Effi. Manage. Technol. (20), 12–17 (2017)
7. Gong, J., Li, Y., Wang, Y.: Realization of transient first half wave method for small current grounding system based on DSP. Electrotechnics (09), 34–36 (2006)
8. Zhang, X.: Transient line selection technology for small current ground fault based on Prony algorithm. Shandong university (2008)
9. Gu, Q., Liu, L., Yang, Y., et al.: Application of wavelet to detect fault mutation characteristics to realize small current fault line selection protection for distribution network. Chinese J. Electr. Eng. 79–83 (2001)

Design of Forum Log System Based on Big Data Analysis

Guanghua Yu$^{(\boxtimes)}$, Linan Sun, and Yongjuan Wang

Heihe University, Heihe City, Heilongjiang Province, China
Ygh2862@163.com

Abstract. Clicking on the log data saved by the website, using the big data means to analyze and mine the information stored by massive data, many crucial information of website operation can be learned. This paper adopts Hadoop distributed platform, uses HDFS data storage and Hive to analyze log big data, designs a Web log analysis system and expounds the design process of the system.

Keywords: Big data · Hadoop · The log

1 The Introduction

In the age of information and amount of data that can be accessed from web logs is getting larger and faster. However, with the traditional stand-alone analysis approach to log data, reading one T data will take a few hours, and the amount of data we need to process is much larger than this. Thus it can be seen, such a long waiting time can no longer meet the daily requirements. Based on the above problems, the big data technology can be well solved. It adopts a cluster to process massive data in parallel, compared with a single server to process data, which undoubtedly provides technical support of each forums and saves log processing costs [1].

2 Technical Introduction

2.1 Introduction to Hadoop

Hadoop is an open-source distribution framework developed by the Apache Software Foundation, which uses clusters to compete, analyze and store data. The core design of Hadoop framework is HDFS and Mapreduce. HDFS provides distributed storage for massive amounts of data, while Mapreduce provides analysis and calculations for data. At the same time, Hadoop can read the stored data quickly, while HDFS uses the method of data stream to read the data [2].

Foundation item: Philosophy and social science project of heilongjiang province (16EDC04);School-level topics(KJY202002)

X. Jiang and P. Li (Eds.): GreeNets 2020, LNICST 333, pp. 29–35, 2020.
https://doi.org/10.1007/978-3-030-62483-5_4

2.2 Introduction to Hive

Hive is an open source tool based on Hadoop for storing and processing large amounts of structured data that is presented as forms in Hive. Compared with traditional databases, Hive enjoys a larger scale of data processing and gives a support on using the collection of data such as map, struct and array. In addition, it will search for data with minimal header addressing, so it is fast in processing data over TB and PB, eliminating the energy consumption disadvantage of traditional database when searching data [3].

2.3 Introduction to Flume

Flume is a reliable, distributed log collection system that collects, aggregates and transfers data from large logs. It is a component of the Hadoop, with high ease of use and an open source tool. According to the need to modify the configuration file, system can achieve the receipt of different data sources, also can do some simple processing of the data, then the data will be transferred to the receiving place(HDFS receiving in this system).

3 The Overall Design of the System

The system is divided into three parts: data collection, data processing and data display. The overall function module is shown in Fig. 1.

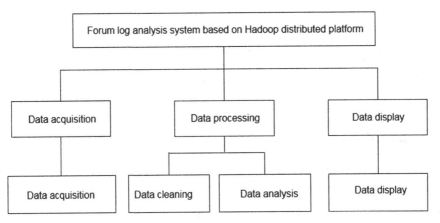

Fig. 1. Overall functional module diagram

4   Main Function Design

4.1   Data Acquisition Design

The system needs to adopt data once a day. For the convenience of classification, the system uses time as a catalogue to classify collected data and configure flume. Hadoop has the disadvantage of not being good at handling large numbers of small files. The system edit and control flume to make flume pass the data every once in a while or wait for the data to reach a certain size to pass the data once. The data acquisition process is shown in Fig. 2:

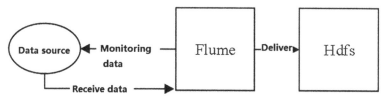

Fig. 2.   Flow chart of data acquisition module

4.2   Data Processing Design

The first step: the log data in HDFS is cleaned with the written MR algorithm, and then cleaned data stored into HDFS.

The second step: Analyze the data in hdfs using Hive's hal statements. Write hiveudf code to analyze some data that cannot be analyzed with HQL statements, and then store the analyzed data in HDFS.

The third step: sqoop is used to transform the analyzed data and transfer to the MySQL database storage. The overall flow chart of data processing is shown in Fig. 3:

4.2.1   Algorithm Design of Cleaning and Analyzing Data

This system applies MR algorithm, which has two functions of mapping and reduction (map function and reduce function). Map function is to use the form of key-value pairs to carry out preliminary processing on the transmitted data and use the key to mark each row of log data. The value is the log data [5].

The Reduce function is the further processing of the data processed by the map function and the marked data is processed by Reduction. Combine the same data into a single line and merge once to count a number. Then combine the combined data with the corresponding count times to form an array and mark the array. Finally, the array is analyzed according to the code in reduce to get the required data and the data is stored in HDFS. The flow chart of MR algorithm is shown in Fig. 4:

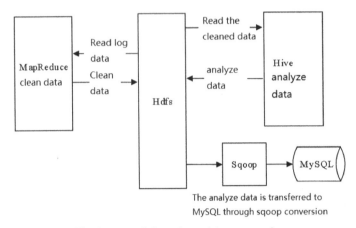

Fig. 3. Overall flow chart of data processing

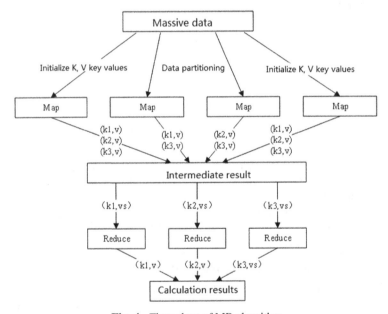

Fig. 4. Flow chart of MR algorithm

4.2.2 Detailed Design of Data Cleaning

Data cleaning is written by MapReduce to read the log files on HDFS. Then the log file is parsed into a Txt file. Clean up log data by writing map and reduce to get rid of useless fields, clutter, and clean up the data. Type the prepared MR program into a jar package, run the jar package, clean the data, and transfer the cleaned data to HDFS. The data cleaning process is shown in Fig. 5:

Fig. 5. Data cleaning flow chart

4.2.3 Detailed Design of Data Analysis

The first step in data analysis, read the cleaned data of MR in HDFS through Hive. The second step, design analysis methods based on requirements. The third step,The cleaned data is further analyzed using HQL statements and written hiveudfs. Get the data that is useful for the development of BBS, and store the analyzed data on the hdfs. The step 4, Sqoop is used to extract the data which analyzed by Hive and convert the data into a format that MySQL can recognize and store in the MySQL database. Data analysis flow chart is shown in Fig. 6:

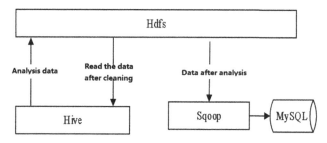

Fig. 6. Data analysis flow chart

Processing in Hive: the first step, according to the log data cleaned by MR in hdfs, establishing a Hive temporary storage framework. The second step is to analyze the log data by using HQL statements. The data that cannot be analyzed with HQL, we need to write hiveudf and put the compiled hiveudf into a jar package in hive's lib. Use the hql statements to run the self-built hiveudf jar package.

4.2.4 The Data Visualization Design

Use JSP technology and Struts2 framework to design different statistical analysis pages after obtaining the analysis results. The page provides the user with the query results. Page function module design can be based on different business needs, personalized business development. The data presentation process is shown in Fig. 7:

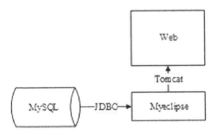

Fig. 7. Data display flow chart

5   Experimental Results

In order to verify the high efficiency of Hadoop for log analysis processing, an experimental comparison was made between a stand-alone machine and an HDFS cluster. Web logs of different file sizes were processed separately in the experiment and calculate the execution time to obtain the following data. The result is shown in Fig. 8.

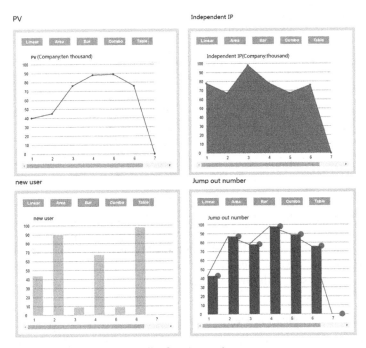

Fig. 8. The result

6   Conclusion

Data processing is the comparison of processing time between Hadoop distributed platform and stand-alone platform. As the data set grows, the processing time of the Hadoop platform becomes shorter. Since the time consumed by the system to start the MapRe-duce task is negligible in the case of large data sets, the computing efficiency is relatively high.

References

1. Yanhui, M.: Big Data Technology Foundation. Tsinghua University Press, Beijing (2016)
2. Wbite, T.: Hadoop: The Definitive Guide. O'Reilly Media, Sebastopol (2015)
3. Rutherglen, J.: Hive Programming guide.People Post Press (2013)
4. Shenzhi, S.: Research on Web Log Data Analysis System Based on Hadoop. Xidian University
5. [Apache Flume] Official document http://flume.apache.org/

SOC Estimation of Ternary Lithium Battery Based on Interpolation Method and Online Parameter Identification

Dawei Wang, Ying Yang$^{(\boxtimes)}$, Weiguang Zhao, Tianyang Yu, and Dongni Zhang

Heilongjiang University of Science and Technology, Harbin 150022, China
1050027522@qq.com

Abstract. SOC estimation is currently a function of the energy management system for new energy vehicles. Based on the SOC of batteries, the remaining available capacity of batteries can be directly determined to determine the remaining driving range of electric vehicles. Aiming at this problem, this paper use the two Resistance and Capacitance equivalent circuit model for the ternary lithium-ion battery, and then obtains the OCV-SOC curve by using spline interpolation. The improved recursive least squares (FFRLS) method with forgetting factor is used to identify parameters of the battery model. Due to the nonlinear state of the external characteristics of the battery, the linear kalman filter would lead to a large error in the estimation, which cannot meet the need for accuracy. Therefore, in this paper, EKF is improved in this paper, and spline interpolation is used to optimize the relationship between open circuit voltage and SOC in data processing, thus improving the estimation accuracy.

Keywords: SOC estimation · Spline interpolation · Recursive least squares · Extended kalman filter

1 Introduction

Nowadays, electric vehicles (EV) have gradually begun to promote the application. For the power batteries of electric vehicles, lithium batteries have higher working voltage, excellent performance, durable quality and other characteristics. Both domestic and foreign countries strongly support the domestic electric vehicle industry.

Lithium battery already used in EV are lithium iron phosphate (LiFeO4) and ternary lithium ion batteries (batteries with elements Ni, Co and Mn). LiFePO4 is a very popular lithium battery electrode material with a large discharge capacity. It is often used in new energy vehicles as a power battery [1]. But the performance of lithium iron phosphate batteries is subject to changes in temperature. Ternary lithium-ion batteries are not as safe as lithium iron phosphate batteries, but have higher energy density than lithium iron phosphate batteries. Ternary lithium-ion batteries are still subject to resource bias due to the energy density of vehicle power batteries.

© ICST Institute for Computer Sciences, Social Informatics and Telecommunications Engineering 2020
Published by Springer Nature Switzerland AG 2020. All Rights Reserved
X. Jiang and P. Li (Eds.): GreeNets 2020, LNICST 333, pp. 36–44, 2020.
https://doi.org/10.1007/978-3-030-62483-5_5

There have been basically four estimation methods for SOC. The first is to use the Coulomb method. Up to now, the utilization rate of Coulomb counting method has not been much in engineering, because it has a big limitation. The first is that the Coulomb counting method needs a sufficiently accurate initial value [8]. However, the sensor is not interference-free, so the reliability is not as high. The second is the open circuit voltage method, but this method is too high for experimental conditions, and the relationship between the open circuit voltage and the charged state may be different for different batteries. The third approach is to use machine learning or artificial neural networks. Neural network method has been applied to the estimation of SOC [2, 3]. The fourth is to use the battery equivalent model to estimate SOC. Another method is to convert the electrochemical reaction of the battery into the form of circuit for analysis by using the equivalent Thevenin model. As a kind of model convenient for SOC analysis, the Thevenin equivalent model has been widely used [4].

The state equation of two Resistance and Capacitance equivalent model of the battery is used in this paper, and then use spline interpolation to obtain the relationship curve between OCV and SOC, which can be used to estimate SOC by combining improved least square method based on forgetting factor and extended kalman filter (EKF). Experiments under UDDS conditions is used to estimate battery parameters by RLS method.

The structure of the paper can be divided into five parts. The second part is to apply the battery equivalent model to establish the equation, which lays a foundation for the following work. In the third part, the model parameters of the second part are measured based on the experiment and parameter identification algorithm. The fourth part use EKF to estimate SOC. The fifth part gives the conclusion.

2 The Establishment of Battery Model

For lithium-ion batteries, it is impractical to model them directly, because charging or discharging a battery is not a linear time-invariant model, and the equations used to describe internal chemical reactions cannot be used directly in engineering [5]. The two Resistance and Capacitance equivalent circuit model takes into account the characteristics of the inside of the battery and can accurately describe its dynamic characteristics.

Figure of two Resistance and Capacitance model is shown in Fig. 1. C_b is an ideal voltage source, and the voltage across it is the battery's open circuit voltage. I_t represents the terminal current of battery; The two sets of RC represent the polarization characteristics of the cell in the Thevenin model.

According to the battery equivalent model, the mathematical expression can be obtained as follows,

$$\begin{cases} U_o = U_{oc} + U_1 + U_2 + I_t R_0 \\ I_t = U_1/R_1 + C_1 dU_1/dt = U_2/R_2 + C_2 dU_2/dt \end{cases} \tag{1}$$

where U_t is port voltage, U_{oc} is open circuit voltage, and U_1, U_2 is voltage of the two RC parts respectively.

According to the time integration method, the expression of the battery SOC is as follows,

$$SOC = SOC_0 - \frac{\eta}{Q_c} \int_0^t i(\tau) d\tau \tag{2}$$

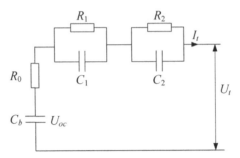

Fig. 1. Cell model equivalent circuit diagram

where SOC_0 is the initial value of SOC, and Q_c is battery capacity. Parameter η is the Coulomb efficiency. The equations in the discrete state of two RC model are,

$$\begin{bmatrix} SOC_{k+1} \\ U_{1k+1} \\ U_{2k+1} \end{bmatrix} = A \begin{bmatrix} SOC_k \\ U_{1k} \\ U_{2k} \end{bmatrix} + BI_{t,k} + \begin{bmatrix} w_{1,k} \\ w_{1,k} \\ w_{1,k} \end{bmatrix} \tag{3}$$

$$U_{t,k} = U_{oc}(SOC_k) - U_{1,k} - U_{2,k} - R_0 I_{t,k} + v_k \tag{4}$$

where $A = \begin{bmatrix} 1 & 0 & 0 \\ 0 & e^{\frac{-T}{\tau 1}} & 0 \\ 0 & 0 & e^{\frac{-T}{\tau 2}} \end{bmatrix}, B = \begin{bmatrix} -\eta T/Q_c \\ R_1(1 - e^{\frac{-T}{\tau 1}}) \\ R_2(1 - e^{\frac{-T}{\tau 2}}) \end{bmatrix}$. The product of each set of resistance

and capacitance values corresponding to the time constant τ. System noise is w, and observation noise is v.

3   Identification of Battery Model Parameters

The capacity of lithium battery is 32 Ah. The functional relationship between the open circuit voltage (OCV) and SOC can be displayed through a polynomial fitting method, but it still has its limitations for the polynomial fitting, so the fitting method is changed to use spline interpolation method.

In this paper, piecewise spline interpolation is adopted to fit the experimental data. The advantage of using spline interpolation to establish OCV model is to make the data smoother, and the whole OCV curve is continuous, so that the accuracy is higher.

According to the sampled data points, two different methods are used for data fitting. It can be seen that the degree of fitting of the 6th polynomial is significantly less than that of the cubic Spline interpolation. Therefore, Spline interpolation is used to fit the OCV-SOC curve. The interpolation results are shown in Fig. 2.

The experiment uses HPPC working conditions to identify the parameters. The test data of HPPC is brought into the FFRLS algorithm for identification, and the battery parameters can be obtained. Finally, the performance of SOC algorithm combining spline interpolation and EKF is tested under UDDS condition. Figure 3 shows the terminal voltage variation of the battery under typical UDDS operating conditions.

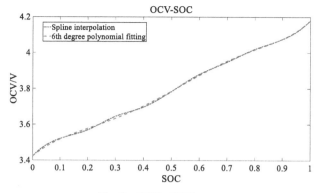

Fig. 2. OCV - SOC curve

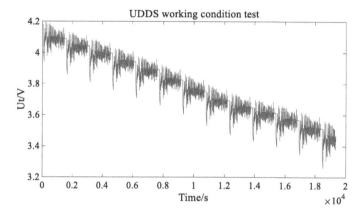

Fig. 3. Battery terminal voltage under UDDS condition

The characteristic of least square method is that it can be widely used for system identification [7]. The RLS method is an improvement on the LS method. In order to solve the problem that the number of recursive steps of least square method is too large to be corrected normally, the forgetting factor is added to eliminate this effect. The process of RLS is as follows:

If a system expression in discrete form can be written:

$$\theta_k = \sum_{i=1}^{n} m_i \theta_{k-i} + \sum_{i=1}^{n} p_i u_{k-i} + v_k \tag{5}$$

The parameter to be estimated is m_i, p_i. F is the sum of the squares of the residuals:

$$F(\psi) = \sum_{i=1}^{N} (\theta - \Phi \psi)^2 \tag{6}$$

The purpose of recursive least squares is to minimize $F(\psi)$. The equation after combining the recursive least square method with the battery model is as follows:

$$U_t = U_{oc} + \left(\frac{R_1}{\tau_1(x_k - x_{k-1})/T + 1} + \frac{R_2}{\tau_2(x_k - x_{k-1})/T + 1}\right)I_t + R_0 I_t \qquad (7)$$

Equation (8) can be simplified into the following form:

$$m_1 U_t v^2 + m_2 U_t v + U_t = m_1 U_{oc} v^2 + m_2 U_{oc} v + U_{oc} + m_1 R_0 I v^2 + m_3 I v^2 + m_4 I \qquad (8)$$

where $m_1 = \tau_1 \tau_2$, $m_2 = \tau_1 + \tau_2$, $m_3 = R_0 + R_1 + R_2$, $m_4 = R_0 m_1 + R_1 \tau_1 + R_2 \tau_2$, $v = (x_k\text{-}x_{k-1})/T$.

Finally, the equation can be converted to the following form:

$$\begin{cases} U_{t,k} - U_{oc,k} = a_1(U_{oc,k-1} - U_{t,k-1}) + a_2(U_{oc,k-2} - U_{t,k-2}) + a_3 I_k + a_4 I_{k-1} + a_5 I_{k-2} \\ a_1 = \dfrac{-(2m_1 + m_2 T)}{T^2 + m_2 T + m_1} \\ a_2 = \dfrac{m_1}{T^2 + m_2 T + m_1} \\ a_3 = \dfrac{m_1 R_0 + m_3 T^2 + m_4 T}{T^2 + m_2 T + m_1} \\ a_4 = \dfrac{-(2m_1 R_0 + m_4 T)}{T^2 + m_2 T + m_1} \\ a_5 = \dfrac{m_1 R_0}{T^2 + m_2 T + m_1} \end{cases} \qquad (9)$$

After a_1–a_5 are identified, the parameters of two RC model can be obtained according to the parameter relationship. The recognition image is shown below. The identification results of R_0 is shown in Fig. 4, and other parameter identification results are shown in Fig. 5, Fig. 6.

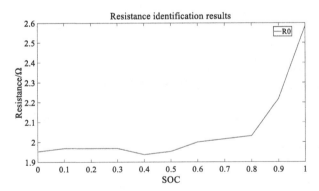

Fig. 4. R_0 identification

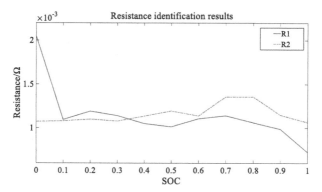

Fig. 5. $R_{1,2}$ identification

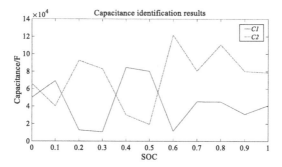

Fig. 6. $C_{1,2}$ identification

4 Battery SOC Estimated by EKF

Due to the immeasurability of SOC, in order to have a relatively accurate monitoring of battery state, a specific algorithm is needed to calculate the accurate SOC. Kalman filter algorithm is widely used in system control. The filter of the system is realized by using the principle of recursion [6]. The basic principle of kalman filter is as follows,

$$\begin{cases} x_{k+1} = Ax_k + Bu_k + w_k \\ z_k = Mx_k + Nu_k + v_k \end{cases} \tag{10}$$

where A, B, M, N are the coefficient matrixs, and w and v are the process noise and measurement noise.

State prediction, covariance prediction and K gain are shown below:

$$\begin{cases} \hat{x}_{k+1} = A_k\hat{x}_k + B_ku_k \\ P_{pre,k+1} = A_kP_kA_k^T + Q \\ K_{kal} = P_{pre,k+1}C_{k+1}^T \left(C_{k+1}P_{pre,k+1}C_{k+1}^T + R\right)^{-1} \end{cases} \tag{11}$$

At the end of these steps, the latest state and the estimation of the covariance of the states can be calculated with the following equation:

$$\begin{cases} \hat{x}'_{k+1} = \hat{x}_{k+1} + L_{k+1}[z_{k+1} - M_{k+1}\hat{x}_{k+1} - N_{k+1}u_{k+1}] \\ P'_{pre,k+1} = (I - L_{k+1}M_{k+1})P_{pre,k+1} \end{cases} \tag{12}$$

At first, the initial value is set for the whole nonlinear system. After the initial value is set, the state of the system is predicted and its covariance is estimated for the next prediction. After iteration, the estimated result can be obtained as follows,

$$x_{k+1} \approx A'_k x_k + f(u_k, x_k) - A'_k x_k + w_k \tag{13}$$

$$z_{k+1} \approx M'_k x_k + h(u_k, x_k) - N'_k x_k + v_k \tag{14}$$

where f and h are nonlinear functions respectively, and the partial derivative of the state variable x_k is taken to carry out Taylor expansion. Their first partial derivatives are respectively used as the new coefficient matrix:

$$\begin{cases} A'_k = \dfrac{\partial f}{\partial x_k}\Big|_{x_k = \hat{x}_k}(x_k - \hat{x}_k) \\ M'_k = \dfrac{\partial h}{\partial x_k}\Big|_{x_k = \hat{x}_k}(x_k - \hat{x}_k) \end{cases} \tag{15}$$

The iteration of status updates is complete. The images of SOC accuracy value, estimated value of the battery tested under UDDS condition and the images of error curve are shown in Fig. 7, Fig. 8.

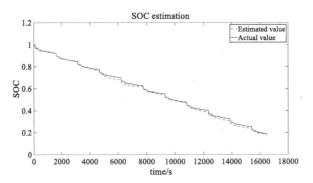

Fig. 7. SOC estimation curve

It can be seen that the estimated value at the beginning of the image still maintains a good tracking effect, but the error gradually increases at the middle and the end of the image, and the error at the end of the image decreases. After analysis, it may be related to OCV There is a problem with the fitting of the OCV-SOC curve relationship. In the interpolation curve of OCV-SOC, the curve inputted by the sample point and then spline interpolation is not accurate enough. In the subsequent work, it is necessary to use the OCV-SOC curve. Better methods will improve its accuracy in order to expect more accurate estimation results.

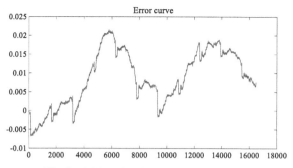

Fig. 8. SOC estimation relative error curve

5 Conclusion

For the ternary lithium-ion battery of this experiment, an equivalent two RC model is established for its characteristics, and its resistance and capacitance were identified using an improved RLS regression method with a forgetting factor for its equivalent circuit. Compared with offline parameter identification, the accuracy of this method is improved, and online parameter identification has the advantages of simple operation and less program volume. For the non-linear battery model, EKF is used to estimate the SOC, which improves the shortcomings of the extended Kalman filter that uses offline parameter identification for the estimation accuracy of the non-linear system.

Acknowledgment. This work was financially supported by the Heilongjiang Provincial Colleges and Universities Basic Scientific Research Business Expense Project(Hkdqg201908).

References

1. Lian, B., Adam, S., Li, X., Yu, D., Wang, C., Dunn, R.W.: Optimizing LiFePO4 battery energy storage systems for frequency response in the UK system. IEEE Trans. Sustainable Energy, **8**(1), 385–394 (2017)
2. Chemali, E., Kollmeyer, P.J., Preindl, M., Ahmed, R., Emadi, A.: Long short-term memory-networks for accurate state of charge estimation of li-ion batteries. IEEE Trans. Ind. Electron. **65**(8), 6730–6739 (2018)
3. Xiong, R., Cao, J., Yu, Q., He, H., Sun, F.: Critical review on the battery state of charge estimation methods for electric vehicles. IEEE Access, **6**, 1832–1843 (2017)
4. Meng, J., et al.: An overview and comparison of online implementable SOC estimation methods for Lithium-Ion battery. IEEE Trans. Ind. Appl. **54**(2), 1583–1591 (2018)
5. Wang, W., Wang, X., Xiang, C., Wei, C., Zhao, Y.: Unscented kalman filter-based battery SOC estimation and peak power prediction method for power distribution of hybrid electric vehicles. IEEE Access, **6**, 35957–35965 (2018)
6. El Din, M.S., Hussein, A.A., Abdel-Hafez, M.F.: Improved battery SOC estimation accuracy using a modified UKF with an adaptive cell model under real EV operating conditions. IEEE Trans. Transport. Electrification, **4**(2), 408–417 (2018)

7. Rahimi-Eichi, H., Baronti, F., Chow, M.-Y.: Online adaptive parameter identification and state-of-charge coestimation for lithium-polymer battery cells. IEEE Trans. Ind. Electron. **64**(4), 2053–2061 (2014)
8. Meng, J., Stroe, D.J., Ricco, M., Luo, G., Teodorescu, R.: A simplified model based state-of-charge estimation approach for lithium-ion battery with dynamic linear model. IEEE Trans. Ind. Electron. **66**(10), 7717–7727 (2016)

Parameter Identification of Six-Order Synchronous Motor Model Based on Grey Box Modeling

Xianzhong Xu[✉], Xunwen Su, Dongni Zhang, Pengyu An, and Jian Sun

Heilongjiang University of Science and Technology, Harbin 150022, China
36061636@qq.com

Abstract. As the "heart" of power system, synchronous generator's accurate model parameters are the basis of power system simulation, operation analysis and fault diagnosis. These parameters also have a very important impact on the operation analysis of power grid. This paper introduces the mathematical model of the sixth-order synchronous generator and establishes the incremental model for its identification. The methods of grey box modeling and nonlinear least square are used to identify the parameters of the sixth-order synchronous generator. When a single - phase short - circuit fault occurs in the power system, the response data of the generator is simulated in the PSASP software. When a program for synchronous machine parameter identification is written, the result will verify the validity of this approach.

Keywords: Synchronous machine · Parameter identification · Least square method

1 Introduction

Synchronous generator is the core of power system. The establishment of accurate synchronous generator model is crucial to accurate calculation and the analysis of power system dynamic characteristics. At present, most power system analysis and calculation can only be based on data provided by the manufacturer or user manual. However, as the manual data can't consider the influence of actual working conditions, the results are often inconsistent with actual working conditions, which seriously affects the accuracy and reliability of system analysis and calculation. Therefore, it is of great practical significance to study the modeling of real-time synchronous generator set [1, 2].

Through the GPS with the Phasor Measurement unit and the wide-area Measurement System (Wide Area Measurement System of WAMS), wide-area generator power Angle and bus voltage can be monitored in real time. Phasor's observation of the real running state of the power grid provides a key to the safe and stable operation of power grid in the measurements. It has been widely used in the monitoring of normal and abnormal state of power system, auxiliary decision-making of accident handling and dynamic process

X. Jiang and P. Li (Eds.): GreeNets 2020, LNICST 333, pp. 45–52, 2020.
https://doi.org/10.1007/978-3-030-62483-5_6

control. Using the synchronous phasor measurement information provided by the wide-area measurement system can improve the accuracy and reliability of the power system simulation model, and the reliability and economic benefit of its operation, which has far-reaching significance for the power system to adapt to the development [3, 4].

Therefore, this paper carries on the Matrix modeling to the sixth-order Synchronous Motor Mathematical Model. Based on the measurement information of synchronous phasor, grey box modeling and least square method are used to identify the parameters of synchronous motor. Finally, the validity of parameter identification is proved.

2 Mathematical Modeling of Six-Order Synchronous Motor

2.1 Mathematical Model of Six-Order Synchronous Motor

According to the choice of different state vectors, different identification models (i.e. state equations) can be constructed. The sixth order model of synchronous generator is selected in this paper, which is as follows.

Generator electrical model of d axis:

$$
\begin{cases}
T_{d0}' \cdot \dfrac{dE_q'}{dt} = E_f - E_q' - \dfrac{x_d - x_d'}{x_d' - x_d''}(E_q' - E_q'') \\
T_{d0}'' \cdot \dfrac{dE_q''}{dt} = E_q' - E_q'' - (x_d' - x_d'')i_d + T_{d0}'' \cdot \dfrac{dE_q'}{dt} \\
u_q = E_q'' - x_d'' \cdot i_d
\end{cases}
\tag{1}
$$

Generator electrical model of q axis:

$$
\begin{cases}
T_{q0}' \cdot \dfrac{dE_d'}{dt} = -E_d' - \dfrac{x_q - x_q'}{x_q' - x_q''}(E_d' - E_d'') \\
T_{q0}'' \cdot \dfrac{dE_d''}{dt} = E_d' - E_d'' + (x_q' - x_q'')i_q + T_{q0}'' \cdot \dfrac{dE_d'}{dt} \\
u_d = E_d'' + x_q'' \cdot i_q
\end{cases}
\tag{2}
$$

Rotor motion equation:

$$
\begin{cases}
\dfrac{d\delta}{dt} = \omega - 1 \\
M \cdot \dfrac{d\omega}{dt} = T_m - T_e - D(\omega - 1)
\end{cases}
\tag{3}
$$

Where E_d'', E_q'', E_d', E_q' are subtransient and transient potient of d_q axis; x_d', x_q', x_d'', x_q'' are substransient, transient and synchronous reactance of d_q axis, T_{d0}'', T_{q0}', T_{d0}', T_{q0}'' are subtransient and transient open circuit time constant of d_q axis, E_f is exciting voltage, δ is generator power angle, ω is nominal angular, M is inertia time constant, D is damping coefficient, T_m, T_e is mechanical and electromagnetic torque.

2.2 Incremental State Model of Six-Order Synchronous

The matrix form of six-order synchronous motor's state incremental equation can be obtained by converting the mathematical model of the synchronous motor into an incremental equation and then expressing it in matrix form [5, 6].

From formula (1), the matrix expression of the d-axis increment equation can be obtained as follows:

$$
\begin{cases}
\begin{pmatrix} \Delta \dot{E_q'} \\ \Delta \dot{E_q''} \end{pmatrix} = \begin{pmatrix} -\dfrac{1}{T_{d0}'} \dfrac{x_d-x_d''}{x_q'-x_d''} & \dfrac{1}{T_{d0}'} \dfrac{x_d-x_d'}{x_d'-x_d''} \\ -\dfrac{1}{T_{d0}'} \dfrac{x_d-x_d'}{x_d'-x_d''} + \dfrac{1}{T_{d0}''} \dfrac{1}{T_{d0}'} \dfrac{x_d-x_d'}{x_d'-x_d''} & -\dfrac{1}{T_{d0}''} \dfrac{x_d'}{x_d''} \end{pmatrix} \cdot \begin{pmatrix} \Delta E_q' \\ \Delta E_q'' \end{pmatrix} + \begin{pmatrix} 0 & \dfrac{1}{T_{d0}'} \\ \dfrac{x_d'-x_d''}{T_{d0}''x_d''} & \dfrac{1}{T_{d0}'} \end{pmatrix} \cdot \begin{pmatrix} \Delta u_q \\ \Delta E_{fd} \end{pmatrix} \\
\Delta i_d = \begin{pmatrix} 0 & \dfrac{1}{x_d''} \end{pmatrix} \begin{pmatrix} \Delta E_q' \\ \Delta E_q'' \end{pmatrix} - \begin{pmatrix} -\dfrac{1}{x_d''} & 0 \end{pmatrix} \cdot \begin{pmatrix} \Delta u_q \\ \Delta E_{fd} \end{pmatrix}
\end{cases}
$$

(4)

In the identification process, the input vector is $[\Delta u_q, \Delta E_{fd}]$, the state variable is $\left[\Delta E_q' \ \Delta E_q'' \right]$, Output is Δi_d, The parameter to be identified is $\alpha = [x_d, x_d', x_d'', T_{d0}', T_{d0}'']^T$.

From formula (2), the matrix expression of the q axis increment equation can be obtained as follows:

$$
\begin{cases}
\begin{pmatrix} \Delta \dot{E_q'} \\ \Delta \dot{E_d''} \end{pmatrix} = \begin{pmatrix} -\dfrac{1}{T_{q0}'} \dfrac{x_q-x_q''}{x_q-x_q''} & \dfrac{1}{T_{q0}'} \dfrac{x_q-x_q'}{x_q-x_q''} \\ \dfrac{1}{T_{q0}''} - \dfrac{1}{T_{q0}'} \dfrac{x_q-x_q'}{x_q-x_q''} & \dfrac{1}{T_{q0}''} \dfrac{x_q-x_q'}{x_q-x_q''} - \dfrac{1}{T_{q0}''} \dfrac{x_q'}{x_q} \end{pmatrix} \cdot \begin{pmatrix} \Delta E_q' \\ \Delta E_d'' \end{pmatrix} + \begin{pmatrix} 0 \\ -\dfrac{1}{T_{q0}''} \dfrac{x_q-x_q''}{x_q} \end{pmatrix} \cdot (\Delta u_d) \\
\Delta i_q = \begin{pmatrix} 0 & -\dfrac{1}{x_q''} \end{pmatrix} \cdot \begin{pmatrix} \Delta E_q' \\ \Delta E_d'' \end{pmatrix} + \dfrac{1}{x_q''} \cdot \Delta u_d
\end{cases}
$$

(5)

In the identification process, the input vector is Δu_q, State the amount of $\left[\Delta E_d' \ \Delta E_d'' \right]$, Output is Δi_q. The quantity to be recognized is $\alpha = \left[x_q, x_q', x_q'', T_{q0}', T_{q0}'' \right]^T$.

From formula (3), the matrix expression of generator rotor increment equation can be obtained as follows:

$$
\begin{cases}
\begin{pmatrix} \Delta \dot{\omega} \\ \Delta \dot{\delta} \end{pmatrix} = \begin{pmatrix} -\dfrac{D}{M} & 0 \\ 1 & 0 \end{pmatrix} \cdot \begin{pmatrix} \Delta \omega \\ \Delta \delta \end{pmatrix} + \begin{pmatrix} -\dfrac{1}{M} \\ 0 \end{pmatrix} \cdot \Delta T_e \\
\Delta \delta = \Delta \delta
\end{cases}
$$

(6)

In the identification process, the input vector is ΔT_e, The output is $\Delta \delta$, The quantity to be recognized is $\alpha = [M, D]^T$.

3 The Principle and Process of Parameter Identification of Synchronous Motor

3.1 The Principle of Parameter Identification

Parameter identification problem is to determine the mathematical model of a system or a process by observing its input-output relationship. Parameter identification based on grey box modeling is selected by the rules in a class of models which are best suited to the data [7, 8].

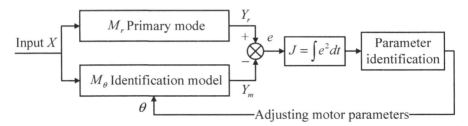

Fig. 1. The process and principle of synchronous motor parameter identification based on grey box modeling

The basic process of synchronous generator identification is shown in Fig. 1 which uses the input and output data provided by the dynamic process of the test system. The structure and parameters of the adjustment model of the synchronous motor mathematical model are constantly adjusted. The results of the model are as close as possible. The X is the input vector, Y_r is the output vector of the prototype system, Y_m is the output vector of the model, and e is the error vector, θ is the model parameter vector.

In order to evaluate the model of parameter identification, an equivalent criterion J is defined to measure error e. Under the action of an input signal X, the error between the actual system output and the model output is e. After the calculation identification criteria, the model parameters will be corrected and repeated until the error e is small enough.

3.2 The Process of Parameter Identification

1. Let the differential equation of the continuous system be as shown in Eq. (7).

$$\dot{X}(\alpha) = A(\alpha)X(\alpha) + B(\alpha)U$$
$$Y(\alpha) = C(\alpha)X(\alpha) + D(\alpha)U \tag{7}$$

Where $X(\alpha)$ is a state vector, $Y(\alpha)$ is an output observation vector, U is a known input vector, $A(\alpha), B(\alpha), C(\alpha), D(\alpha)$ are matrix contains parameters that need to be identified.

2. Give the initial value of each state vector and quantity to be identified.
3. Calculate the dynamic process curve Y_m of the output vector according to the incremental model of parameter identification by fourth order runge-kutta method.
4. The value of objective function $J(\Delta\alpha)$ is calculated from the measured data of simulation and the data of identification model. The objective function is shown in Eq. (8).

$$J(\Delta\alpha)_{\alpha_0} = \int_{T_0}^{T} [Y_r - Y_M(\alpha_0) - (\frac{\partial Y_M}{\partial \alpha^T})_{\alpha_0} \Delta\alpha]^T W[Y_r - Y_M(\alpha_0) - (\frac{\partial Y_M}{\partial \alpha^T})_{\alpha_0} \Delta\alpha]dt \tag{8}$$

5. In order to use the least square method, the partial derivatives of the output vector Y_m are obtained for the identification parameters. The process of finding the partial derivative is shown in Eq. (9).

$$(\frac{\partial Y_M}{\partial \alpha^T})_{\alpha_0} = C(\alpha)(\frac{\partial X_M}{\partial \alpha^T})_{\alpha_0} + (\frac{\partial C(\alpha)}{\partial \alpha^T})_{\alpha_0} X_M(\alpha_0) + (\frac{\partial D(\alpha)}{\partial \alpha^T})_{\alpha_0} U \qquad (9)$$

Where $(\frac{\partial C(\alpha)}{\partial \alpha^T})$, $(\frac{\partial D(\alpha)}{\partial \alpha^T})$ can be obtained directly, α_0 is the initial value of quantity to be identified. $\frac{\partial X_M}{\partial \alpha^T}$ is expected to solve the following differential Eqs. (10).

$$\Delta \alpha = [\int_{T_0}^{T} (\frac{\partial Y_M}{\partial \alpha^T})_{\alpha_0}^T W (\frac{\partial Y_M}{\partial \alpha^T})_{\alpha_0} dt]^{-1} \cdot \int_{T_0}^{T} (\frac{\partial Y_M}{\partial \alpha^T})_{\alpha_0}^T W (Y_r - Y_M(\alpha_0)) dt \qquad (10)$$

6. Calculate $J(\hat{\alpha})$ by Eqs. (11).

$$\hat{\alpha} = \alpha_n + \Delta \alpha \qquad (11)$$

7. If $J(\hat{\alpha}) < J(\alpha_n)$, $\alpha_{n+1} = \hat{\alpha}$, $k = 1$, turn to step 9. Otherwise, continue.
8. Take $k = k + 1$ and move on to step 5.
9. If $\max_{1 \le j \le m} (\frac{\Delta \alpha_n^j}{\alpha_n^j}) < \varepsilon$, Where $\Delta \alpha_n^j$ is the jth component of $\Delta \alpha_n$, Take α_{n+1} as the identification result, otherwise $n = n + 1$, go back to step 2.

4 Simulation Examples and Identification Results

In this paper, the integrated power system simulation program (PSASP) is applied to carry out dynamic digital simulation test with EPRI-7 node system as an example. The wiring diagram of the system is shown in Fig. 2. The process and principle of the synchronous motor parameter identification are based on grey box modeling. The method described above is applied to identify the parameters of generator G1.

The synchronous generator uses a detailed six - order model. The electrical damping of the damping winding has been taken into account in detail in this model. Only when the mechanical damping is very small, can D can be zero.

The disturbance data is based on the single-phase short-circuit grounding fault data on the line. At 0.02 s, the fault occurs on the line between bus b4-500 and bus b3-500. At 0.1 s, the fault is eliminated and the data interval is 0.01 s (Fig. 2).

The simulation results show that when using the line fault by the measurement of the generator voltage, current and Angle information, time-domain nonlinear least squares method can identify respectively d shaft generator and q axis and the rotor equations of motion parameters. Test results show that the algorithm has strong robustness, and the method is used to identify the effectiveness of the synchronous generator parameters (Table 1 and Figs. 3, 4).

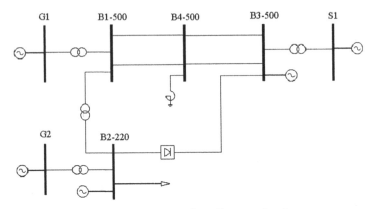

Fig. 2. EPRI (China) Seven lines diagram of node system

Table 1. Parameter identification results

Name of parameter	Set Value	Given initial value	Results of identification	Given initial value	Results of identification
x_d (pu)	0.162	0.150	0.1714	0.3564	0.1711
x'_d (pu)	0.0199	0.016	0.0197	0.044	0.0197
x''_d (pu)	0.0154	0.013	0.0154	0.0339	0.0154
x_q (pu)	0.162	0.14	0.1667	0.324	0.1627
x'_q (pu)	0.0398	0.032	0.0357	0.08	0.0385
x''_q (pu)	0.0154	0.013	0.0153	0.0308	0.0153
T'_{d0} (s)	8.62	6.79	8.5436	18.96	8.5436
T''_{d0} (s)	0.05	0.05	0.0492	0.11	0.0492
T'_{q0} (s)	2.2	1.77	2.1482	4.4	2.1314
T''_{q0} (s)	0.07	0.05	0.0727	0.14	0.00726
M	8	5	7.9757	24	7.9757

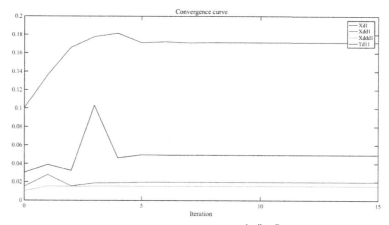

Figure 3 The convergence curve of $x_d x'_d x''_d \; T''_{d0}$ parameters

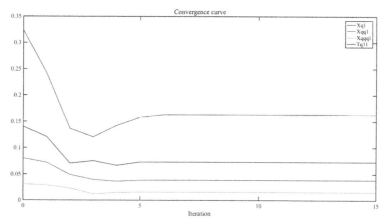

Fig. 4 The convergence curve of $x_q x'_q x''_q \; T''_{q0}$ parameters

Acknowledgment. This work is supported by the National Science Foundation of China under Grant (51677057), Local University Support plan for R & D, Cultivation and Transformation of Scientific and Technological Achievements by Heilongjiang Educational Commission (TSTAU-R2018005) and Key Laboratory of Modern Power System Simulation and Control & Renewable Energy Technology, Ministry of Education (MPSS2019-05).

References

1. Mi, Z., Chen, Z., Nan, Z., et al.: Dynamic parameter identification of synchronous machines **18**(2), 100–105 (1998)
2. Zhao, Z., Zheng, F., et al.: Online parameter identification of big synchronous machines. J. Tsinghua Univ. (Sci& Tech) **38**(9), 10–13 (1998)

3. Xu, S., Xie, X., Xin, Y.: Present application situation and development tendency of synchronous phasor measurement technology based wide area measurement system. Grid Technol. **29**(2), 44–49 (2005)

4. Ju, P., Zheng, S., et al.: Steady state voltage stability analysis for Guangxi power grid. Electric Power Automation Equipment (Sci& Tech **24**(7), 37–40 (2004)

5. Ni, Y., Chen, S., Zhang, B.: Beijing: Theory and Analysis of Dynamic Power System. Tsinghua University Press (2002)

6. Ju, P.: Nonlinear Identification of Power Systems. Hohai University Press, Nanjing (1999)

7. Lee, C.C., Tan, O.T.: A weighted least squares parameter estimator for synchronous machine. IEEE Trans **1**(1), 96–111 (1997)

8. Kyriakides, E., Heydt, G.T.: An observer for the estimation of synchronous generator damper currents for use in parameter identification. IEEE Trans. Energy Convers. **18**(1), 175–177 (2003)

Ultrasonic Power Supply of Oil-Water Separation System

AnHua Wang[✉], DongDong Wan, and HongKai Ding

Electronic and Information Engineering Institute, Heilongjiang
University of Science and Technology, Harbin 150022, China
3036361_cn@sina.com

Abstract. With the continuous development of oil field, the recovery and recovery of crude oil are gradually reduced, and the water content ratio of oil field production is higher and higher. The ultrasonic oil-water separation technology can improve this situation. The performance of ultrasonic power supply directly affects the reliability and economic benefit of oil-water separation system, which is an important part of the system. This paper introduces an ultrasonic power supply with high efficiency and strong stability. This power supply mainly adopts inverter technology to realize the continuous regulation of output voltage and working frequency. Matlab simulation experiment shows that the output voltage of H-bridge is positive and negative pulse waveform, and the output current is sine wave; reasonable setting of parameters can ensure that the output voltage amplitude is 0 V–00 V, and the frequency is 15 kHz–35 kHz, which meets the actual production demand. The results show that the ultrasonic power supply designed in this paper has good output characteristics and can provide energy to the transducer efficiently and reliably, which is of great significance to the practical application of the ultrasonic oil-water separation system.

Keywords: Oil water separation · Ultrasonic power · Inverter technology · Positive and negative pulse

1 Introduction

The crude oil extracted from oil wells contains not only a small amount of solid impurities such as silt and rust, but also a large amount of water. These moisture will increase the fuel consumption during heating and smelting. If working in this condition for a long time, it will cause the corrosion of equipment and pipelines and even the blockage of pipelines. This problem can be effectively improved by using ultrasonic technology. Because the common power supply can not provide the high frequency alternating current matching with the ultrasonic transducer, the research on the high-efficiency and reliable ultrasonic special power supply [1–4] has become a key problem.

At present, some achievements have been made in the field of ultrasonic power supply at home and abroad. It has gone through two stages of electronic tube generator and transistor analog generator [5, 6], but the former has been eliminated due to its too many

X. Jiang and P. Li (Eds.): GreeNets 2020, LNICST 333, pp. 53–65, 2020.
https://doi.org/10.1007/978-3-030-62483-5_7

shortcomings, and the latter is suitable for low-power occasions (below 200 W), and its cost is low and can be used in a certain range [7]. Transistor switch generator [8] is the current mainstream development direction, which can be used in high-power situations (more than 200 W) and can be easily combined with micro electric processor [9]. The switch generator uses the duty cycle of control signal to control the output voltage, which has low power consumption and high efficiency. Thus, the heat dissipation requirements are reduced so that the volume and weight per unit power are smaller [10].

In this paper, the ultrasonic power supply for oil-water separation is taken as the research object. Combined with the current research trend, the DSP with faster CPU and higher integration is used for control, and the new inverter technology is used to realize the continuous change of output voltage and frequency, improve work efficiency and reduce cost.

2 Overall Plan

The structure of the ultrasonic power system is shown in Fig. 1. The input voltage is 220 V/50 Hz AC. the DC voltage is output through the diode uncontrolled rectifier circuit. The DC voltage is chopped and reduced by the forward converter to obtain a stable DC voltage to the H-bridge transmitter. The H-bridge transmitter outputs voltage square wave and current sine wave. The frequency and amplitude of the output voltage are sent to the DSP by the sampling circuit, and then to the driver circuit after the DSP processing. The driver circuit drives the switch tube according to the DSP signal.

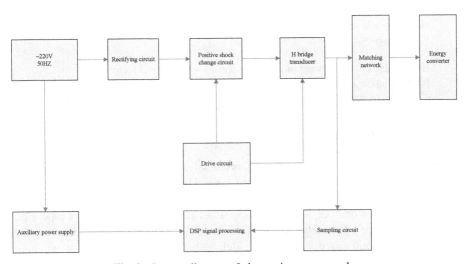

Fig. 1. System diagram of ultrasonic power supply.

3 Hardware Circuit Design

3.1 Main Circuit

The main circuit is as shown in Fig. 2. The input is 220 V 50 Hz AC. the DC voltage is output through a single-phase uncontrolled rectifier circuit. The chopper step-down circuit is a single switch forward circuit, and the forward excitation is output to the single-phase H-bridge transmitter, so as to realize positive and negative pulse voltage and sinusoidal current output.

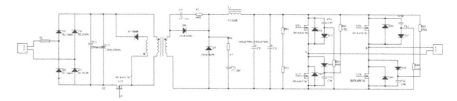

Fig. 2. Schematic diagram of main circuit.

The driving chip of H transmitter is IR2110. IR2110 is highly integrated, with external protection and blocking port, which can be used to drive power tube of - 4–500 V bus voltage system, and can drive two switch tubes of the same bridge arm. And IR2110 can drive power switches with frequency up to 500 kHz [11–14] to meet the system requirements. The driving circuit is mainly composed of bootstrap diode, bootstrap capacitor and other peripheral devices, as shown in Fig. 3.

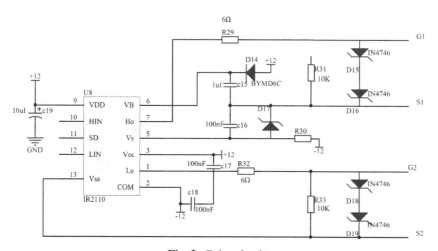

Fig. 3. Drive circuit.

3.2 Auxiliary Power Circuit

Because the digital chip in the control circuit needs positive and negative 12 V voltage and positive 5 V voltage, it needs to design an efficient auxiliary power supply for the control circuit. Flyback converter is selected as the main circuit of the auxiliary power supply, which has the advantages of simple structure, high efficiency and small loss, and is suitable for outputting multi-channel DC [15]. As shown in Fig. 4, the schematic diagram of auxiliary power supply is 220 AC, 36 V AC is obtained through power frequency transformer, and 43 V DC is obtained after diode rectification and capacitance filtering. Finally, the flyback converter is used to obtain positive and negative 12 V voltage, and the switching frequency is 82 kHz.

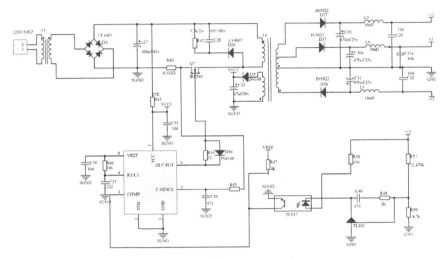

Fig. 4. Auxiliary power supply.

3.3 Detection Circuit

Figure 5 shows the output current detection circuit. The detected current should be fed back to the inverter control circuit for frequency tracking, so it must be selected from the resonance channel. So I choose the current of the channel as the frequency feedback signal. Hall sensor is used to detect the current of the channel. Because the output current of the ultrasonic power supply is sine wave, the detected current should be processed into DC signal and sent to the A / D port of DSP.

The output voltage detection circuit is shown in Fig. 6. By setting the appropriate magnification, adjust the linear proportion relationship between the sampling voltage input and output. Since the working voltage range of DSP is 0 to 3.3 V, the amplitude of the voltage signal entering DSP should be controlled between 0 to 3.3 V. So the limiting circuit is designed to prevent the high amplitude of the voltage signal entering the DSP from burning the chip.

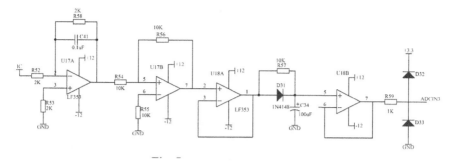

Fig. 5. Current detection circuit.

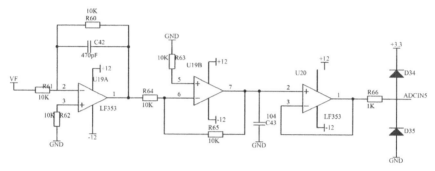

Fig. 6. Voltage detection circuit.

3.4 Protection Circuit

Figure 7 shows the over-current and over-voltage protection circuit. The 5-pin lm319 chip is led out by the current sampling circuit in Fig. 5. If the input voltage of 5-pin is higher than the 4-pin positive phase input voltage of the over-current comparator, the 12 pin of the comparator will output a low level, and the LED D39 is on, indicating the line over-current. Since the 1 pin of the optocoupler pc787 is high level, the LED in the optocoupler will turn on the light, then the 4 and 3 pins of the optocoupler will turn on. The 3-pin optocoupler is connected to the power protection interrupt pin pdpinta of DSP, and the potential of 3-pin changes to low level, which will produce a falling edge, resulting in the effective interruption. DSP immediately blocks PWM drive signal to realize protection. The working principle of over-voltage protection is similar to that of over-current protection.

3.5 Buffer Circuit

Because the switch tube flows a lot of current when it is turned on and bears a lot of impulse voltage when it is turned off, it is necessary to design a buffer circuit to restrain the impulse of current and voltage. The buffer circuit can also improve the reliability of the circuit, reduce the loss of the switch and suppress the electromagnetic interference. The buffer circuit is shown in Fig. 8.

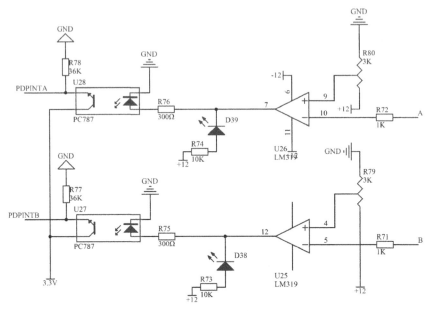

Fig. 7. Protection circuit of over voltage and over current.

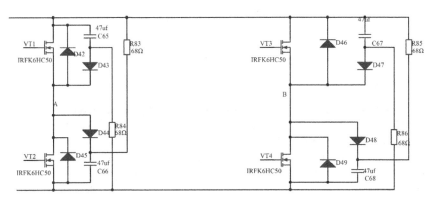

Fig. 8. Buffer circuit.

4 Software Design and Simulation Analysis

4.1 Flow Chart Design of Current and Voltage Sampling Program

TMS320LF2407A chip has integrated a/D A/D conversion module, so the collected voltage signal is directly sent to a/D pin adcin00 of DSP. The timer in the event manager EVA is used to generate 1 cycle ADC conversion trigger signal and read out the conversion value in the ADC interrupt program. In order to improve the accuracy of the sampling value, the average value of the sampling value is taken by setting the sampling times [17, 18]. Figure 9 shows the current and voltage sampling flow chart.

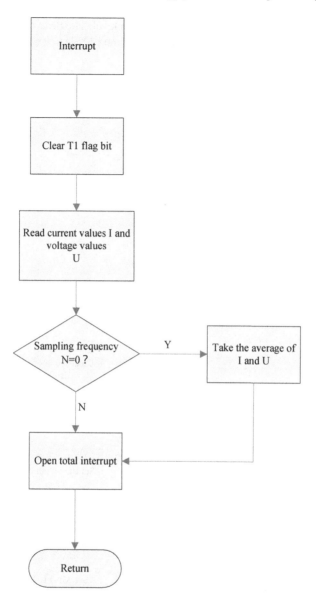

Fig. 9. Flow chart of current detection.

4.2 Flow Chart Design of External Pin Interrupt Program

The external pin interrupt flow is shown in Fig. 10. The external interrupt XINT1 is used to detect whether the waveform output by the optocoupler isolator pc787 is the rising edge or the falling edge. The external interrupt XINT1 pin status is monitored by xint1cr in the DSP chip [19]. When rising edge is detected, iopa6 pin outputs low level, and when rising edge is detected, iopa6 pin outputs high level. The dead time is equal to the delay time plus the program instruction execution time [20, 21].

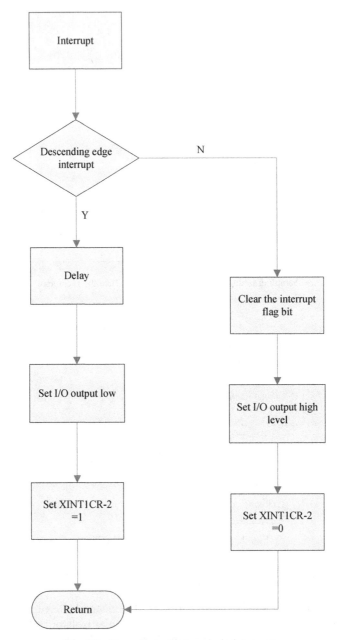

Fig. 10. Flow chart of external pin interrupt.

4.3 Simulation Analysis

The simulation experiment is carried out on the Simulink platform, which is an integrated environment of dynamic system modeling, simulation and comprehensive analysis provided by MATLAB [22, 23]. As shown in Fig. 11, it is the output voltage waveform in the single switch forward open-loop state. The output voltage changes with the input voltage, and the output voltage drops seriously under load. Under the open-loop control, the output voltage tends to be stable after 0.4 s. Since the open-loop control has large defects, the closed-loop control is adopted, and the stable time waveform is shown in Fig. 12. The output stabilization time is 0.0037 s. Compared with the open-loop control, the proportional feedback control greatly improves the control speed. In a certain input range, the output voltage is stable and the load adjustment rate is small.

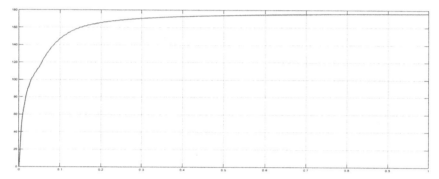

Fig. 11. Output voltage wave of open loop.

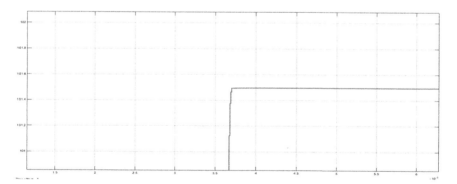

Fig. 12. Output voltage wave of proportional control.

Although the proportional control is better than the open-loop control, but the output has a large error [24, 25], which can not meet the actual requirements, so the proportional integral control is chosen to reduce the steady-state error of the system. The output voltage is as shown in Fig. 13. The steady-state error is 0.3, the stable time is 0.0106 s, and the overshoot is 0.01%.

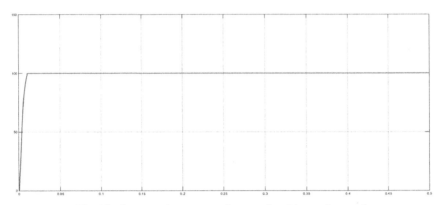

Fig. 13. Output voltage wave of proportional integral control.

The four switches of the H-bridge unit adopt phase-shifting control, and there is a certain dead zone between the pulses to prevent the straight through of the same bridge arm. The driving waveform is shown in Fig. 14.

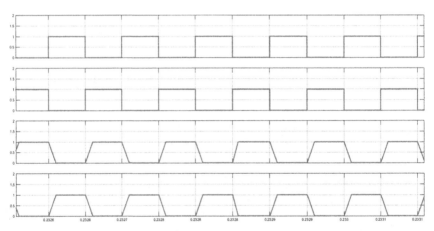

Fig. 14. Wave of drive signal.

The output voltage of H bridge is square wave, the frequency is 20 kHz, and the amplitude is plus or minus 100 square wave. The output voltage meets the actual demand, and its waveform is shown in Fig. 15.

The output matching network is LC resonance [26]. The resonance frequency can be obtained by setting the parameters of inductance and capacitance reasonably. The resonance frequency is set in the controllable range of H-bridge. The output current waveform after resonance matching is shown in Fig. 16, which is similar to sine wave. In the experiment, the capacitance is set as 500 pf, the inductance is set as 1000 μ h, and the resonance frequency is 22.5 kHz.

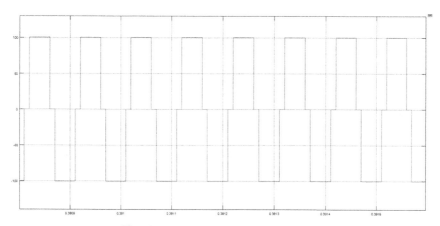

Fig. 15. Output voltage wave of H-bridge.

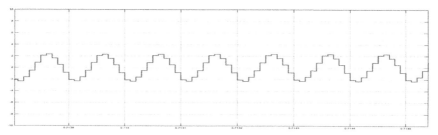

Fig. 16. Output current wave.

5 Conclusion

In this paper, the ultrasonic power supply for oil-water separation device is studied. The main structure of the system is composed of single-phase uncontrolled rectifier circuit, single switch forward circuit and single-phase H-bridge transmitter, and DSP is used as the control chip. By using the typical PID control algorithm, the output voltage and frequency are changed continuously, and the power density of the power supply is greatly improved. Through the MATLAB simulation experiment, it is found that the output voltage and output current waveform are consistent with the expectation after the resonance matching; in the appropriate frequency and amplitude range, it can meet the actual work needs. It is proved that the scheme of ultrasonic power supply proposed in this paper is feasible and has certain reference value for practical application.

Acknowledgment. This work was supported by Harbin Science and Technology Innovation Talents Special Project (NO. 2017RAQXJ031); Heilongjiang Fundamental Research Foundation for the Local Universities in 2018 (2018KYYWF1189); 2017 National Nature Fund, (NO. 51674109); Key project Task of Public Safety Risk Control and Emergency Technical Equipment of National Key R&D Program (NO. 2017YFC0805208).

References

1. Ji, H.J., Zhou, J.B., Meng, L.: Ultra-low temperature sintering of Cu@Ag core-shell nanoparticle paste by ultrasonic in air for high-temperature power device packaging. Ultrason. Sonochem. **41**, 375 (2017)
2. Li, H., Cao, B., Liu, J.: Modeling of high-power ultrasonic welding of Cu/Al joint. Int. J. Adv. Manuf. Technol. **97**(1) (2018)
3. Ye, H., Li, Y.G., Yuan, Z.G.: Ultrasonic-assisted wet chemical etching of fused silica for high-power laser systems. Int. J. Appl. Glass Sci. **9**(2) (2018)
4. Qu, B.D., Ma, J.X., Du, S.: Ultrasonic power based on SG3525 FM control. Power Technol. **38**(7), 1358–1360 (2014)
5. Li, C.L.: Principles and common faults of CSF-7 ultrasonic generator. Textile Equipment **24**(2), 25–26
6. Huang L.: Technical modification of ultrasonic generator for electron tube **2**(8), 35–37 (2004)
7. Xu, T.: Development of ultrasonic generator power technology. Clean. Technol. (4), 10–15
8. Pei, X., Cwikowski, O., Smith, A.C., et al.: Design and experimental tests of a superconducting hybrid DC circuit breaker. IEEE Trans. Appl. Superconductivity **PP**(99), 1 (2018)
9. Jing, N., Meng, X.F., Li, N.: Quartz crystal sensor using direct digital synthesis for dew point measurement. Measurement **117**, 73–79 (2018)
10. Yu, W.J., Wang, R., Zhang, Z.X., et al.: Relationship between total dose effect of partially exhausted SOI MOSFET and bias state (English). Chin. Phys. C **31**(9), 819–822 (2007)
11. Xu, J.: Research on wheel motor driven pure electric vehicle control system. Hangzhou University of Electronic Science and Technology (2014)
12. Huang, Q.B.: Research on brake control strategy and controller development of four-wheel hub-motor driven electric vehicle. Hangzhou University of Electronic Science and Technology (2015)
13. Liu, Q.X.: Research on inverter control and power modulation system of series resonant high frequency power supply. Xi'an University of Technology (2005)
14. Pei, J.L.: Research on digital controlled ultrasonic power supply. Jiangnan University (2008)
15. Lu, Y., Zheng, C.B., Deng, Y.C., Li, X.Q.: Multi-channel isolated output switching power supply with backup battery. Appl. Electr. Technol. **38**(08), 69–72 (2012)
16. Deng, M., Yan, J., Liao, Y.: Design of constant-current electronic load based on UC3843PWM control. Electric switch **49**(03), 20–22+25 (2011)
17. Li, M.Y., Zhang, S.R.: Optimal dead zone frequency following system for series resonant inverter. Power Electr. Technol. **03**, 45–47 (2004)
18. Chang, S., Li, M.: DSP based series resonance induction heating system with optimal dead zone. Electric Drive Autom. (06), 23–27 (2004)
19. Chen, L.: TMS320C5402DSP interrupt resource and its application. J. Xiamen Inst. Technol. **12**(2), 61–66 (2004)
20. Zhang, S.R.: Research on high frequency induction heating power control system based on DSP. Xi'an University of Technology (2004)
21. Zheng, H., Qin, X.D., Tao, H.J.: Design of DC/DC converter digital control system for on-board charging power supply. Autom. Technol. (7) (2019)
22. Liaw, C.M, Chen, T.H, Lin, W.L.: Dynamic modelling and control of a step up/down switching-mode rectifier. **146**(3), 317-0
23. Dong, Y.H., Zhang, C., Xu, M.Z., et al.: Step-down dc switching regulator power supply. Commun. World **1**, 270–271 (2017)
24. Li, H., Wang, S., Li, D., et al.: Intelligent ground control at longwall working face. Meitan Xuebao/J. Chin. Coal Soc. **44**(1), 127–140 (2019)

25. Cheng, J., Ju, H., Park, X.D.: Static output feedback control of switched systems with quantization: A nonhomogeneous sojourn probability approach. Int. J. Robust Nonlinear Control (2019)
26. Wang, Y.Y.: The role of LC parallel resonance circuit in communication electronic circuit. Digital World **5**, 136–137 (2017)

Design and Analysis of a New Logistic Chaotic Digital Generation Circuit

Juan Wang[✉], Liu Wenbin, Han Tongzhuang, and Zhou Xin

Electronic and Information Engineering Institute, Heilongjiang University of Science and Technology, Harbin 150027, People's Republic of China
76115347@qq.com

Abstract. Chaotic signals have the characteristics of noise-like, difficult to predict, and very sensitive to the initial state. They have broad application prospects in the fields of communication and cryptography. Aiming at the shortcomings of the traditional one-dimensional discrete logistic chaotic map, the full mapping space is small, and the complexity is not high. In this paper, a new type of logistic digital chaos generation circuit with better performance is modeled and simulated. By testing and analyzing the initial value sensitivity, autocorrelation and other characteristics of the generated sequence, the influence and rules of circuit parameter setting on the randomness of digital chaotic sequences are obtained, so as to provide the necessary theoretical basis for the application of digital chaotic sequences.

Keywords: New logistic chaos · Digital circuit · NIST test

1 Introduction

Chaos is a common motion form of nonlinear dynamics, and is a seemingly random behavior generated by deterministic nonlinear systems. This random behavior is different from the general random phenomenon and is determined only by the randomness of the system [1]. Owing to the unpredictability, anti-interception, high randomness, high complexity and easy implementation of chaotic signals, it has an excellent application prospect in secure communication [2]. Although the traditional logistic mapping is simple in form, it has the problems of small chaotic parameter range, small full mapping space, uneven iterative distribution, and low complexity, etc., poor security greatly limits the application of chaotic circuits in practice [3–6]. A new type of logistic mapping can be proposed in Literature [7], which has the advantages of high complexity, difficult prediction, strong sensitivity, full mapping parameter range, uniform iterative distribution, and strong randomness, and can better meet the chaotic secure communication field application requirements.

Due to Field Programmable Gate Array (FPGA) has the advantages of short development cycle, low input cost, reprogrammable, erasable, etc. [8], Many researchers have tried to use digital circuit technology to implement chaotic systems and applications

X. Jiang and P. Li (Eds.): GreeNets 2020, LNICST 333, pp. 66–74, 2020.
https://doi.org/10.1007/978-3-030-62483-5_8

in FPGA circuit boards. In this paper, the new logistic digital chaotic circuit in Matlab/Simulink environment can be achieved by modeling and simulating via using the DSP-Builder toolbox. Via reasonable circuit design and parameter setting, simulation analysis and NIST test of the new logistic chaotic circuit can be experimented, and the potential security risks are reduced and eliminated by studying the influence of the circuit parameter settings on the randomness of the chaotic sequence and its regularity. Thus, the digital chaotic signals generated by the system model can better meet the requirements in the application of chaotic secure communication systems.

2   New Logistic Chaotic Map

Discrete chaotic mapping refers to a system described by a difference equation, capable of generating discrete time domain and continuous amplitude chaotic signals. Owing to only through mapping functions, generation rules, and initial conditions, chaotic sequences can be generated. Thus, it has the advantages of simple, rapid, and easy control, and is widely used in the field of secure communications.

The traditional logistic chaotic map has become a widely studied and applied chaotic map because of its simple mathematical model. Although it has good pseudo-randomness and correlation and simple circuit implementation, there are still many problems, such as infinitely many fixed point attractors, small mapping space, small chaotic parameter range, uneven iterative distribution, and low complexity. Thereby reducing system security and anti-interference [9]. The new logistic chaotic mapping equation is:

$$x_{n+1} = \mu x_n (1 - x_n^2) \quad n = 1, 2, 3 \ldots \tag{1}$$

In Eq. (1), initial value $x_0 \in (0, 1)$, x_n is the value of the n iteration, x_{n+1} is the value of the $n + 1$ iteration, $x_n \in (-1, 1)$, system parameters $\mu \in (0, 4]$, when $\mu \in [2.38, 4]$, the new logistic map is in a chaotic state.

3   Circuit Design of New Logistic Chaotic Map

As shown in Fig. 1, the new logistic chaotic circuit is built using the DSP Builder component library in the Matlab/Simulink environment. When the input pulse of the data selector is high, Constant $= 0.15$ is used as the initial value of the chaotic circuit. When the input pulse of the data selector is low, the new logistic chaotic iteration is started. The delayer ensures stable data output. The multiplier completes the data multiplication, the adder completes the data addition, the amplifier completes the data amplification and completes an iterative output, and finally realizes the interval quantization output through the barrel shift register and the bit decimator. The interval quantization function is defined as follows:

$$T[\mathrm{x}(n)] = \begin{cases} 0, x(n) \in \bigcup\limits_{k=0}^{2^m-1} I_{2k}^m \\ 1, x(n) \in \bigcup\limits_{k=0}^{2^m-1} I_{2k-1}^m \end{cases} \tag{2}$$

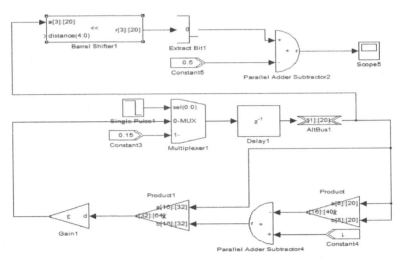

Fig. 1. Digital generation circuit of new logistic chaotic map

In Eq. (2), $m > 0$ and any positive integer. $I_0^m, I_1^m, I_2^m, \ldots$ is 0, 1 interval 2^m consecutive equal molecular interval. If the conversion value falls in the odd interval, it is quantized to 1, and if the conversion value falls in the even interval, it is quantized to 0.

In the case of ensuring that the bit width is sufficient and data will not overflow, minimizing the width of the bit width as much as possible can save resources inside the FPGA chip. The Altbus module (a:b) can map floating-point signals to fixed-point data. a represents the number of binary digits before the decimal point, and b represents the number of binary digits after the decimal point. By changing the settings of the Altbus module, the optimal parameter settings of the system are obtained, and achieve circuit design on target hardware.

4   Characteristics Analysis of New Logistic Chaotic Map

4.1   Numerical Distribution of New Logistic Chaotic Map

The numerical distribution of chaotic sequences is the criterion for judging the system's pseudo-randomness and the ability to resist statistical analysis attacks. The more uniform the distribution, the stronger its performance [10]. The new logistic numerical distribution is simulated. As shown in Fig. 2, the simulation time is set to 100, the gain is set to 3, and the initial value is set to 0.15. Through simulation analysis, it can be seen that when the Altbus module parameter $b < 20$, the numerical distribution of the new logistic chaotic sequence is not ideal; when the Altbus module parameter $b \geq 20$, the new logistic chaotic sequence has a relatively ideal numerical distribution. Therefore, it is concluded that the larger the b value of the Altbus module parameter is set, the better the numerical distribution of the new logistic. In order to balance system performance and hardware resources, the b value should generally be set to 20.

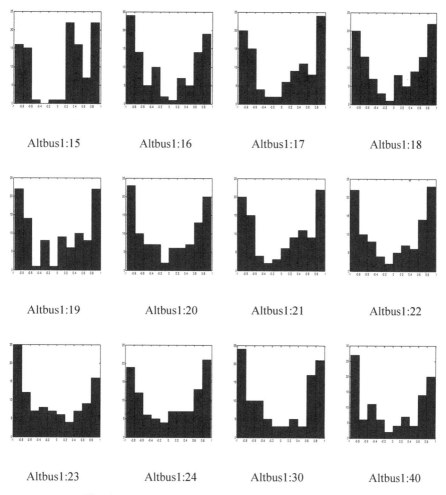

Fig. 2. Numerical distribution of new logistic chaotic map

4.2 Initial Value Sensitivity of New Logistic Chaotic Map

The initial value sensitivity of chaos is an objective reflection of the change of the system trajectory when the system undergoes small changes in the initial conditions [11]. The strength of the randomness of the system depends on the initial value sensitivity. Simulation experiments are performed on the initial sensitivity of the new logistic chaotic model. As shown in Fig. 3, the simulation time is set to 100, the gain is set to 3, and the initial values are set to 0.15 and 0.150001. Fig(a) shows the iteration values, and Fig(b) shows the difference between the two. Through simulation analysis, it can be seen that when the parameter b of the Altbus module is less than 20, the initial value sensitivity of the new logistic chaotic sequence is not ideal; when the parameter b of the Altbus module is greater than or equal to 20, the initial chaotic value differs only by 0.000001 and its iterative output will be completely different. Therefore, it is concluded that the

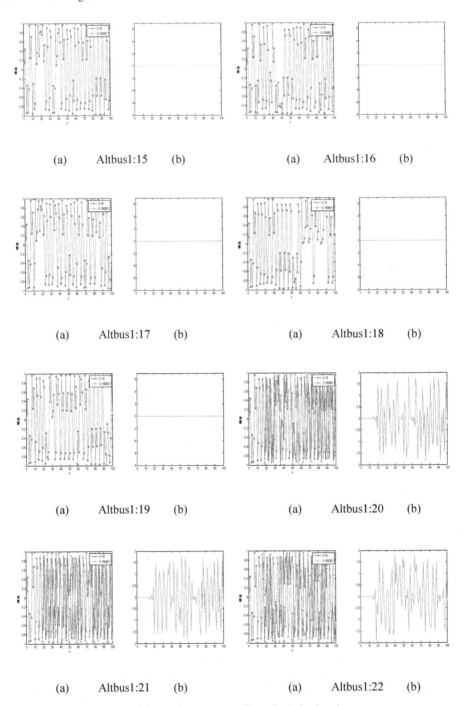

(a)   Altbus1:15   (b)   (a)   Altbus1:16   (b)

(a)   Altbus1:17   (b)   (a)   Altbus1:18   (b)

(a)   Altbus1:19   (b)   (a)   Altbus1:20   (b)

(a)   Altbus1:21   (b)   (a)   Altbus1:22   (b)

Fig. 3. Initial value iteration of new logistic chaotic map

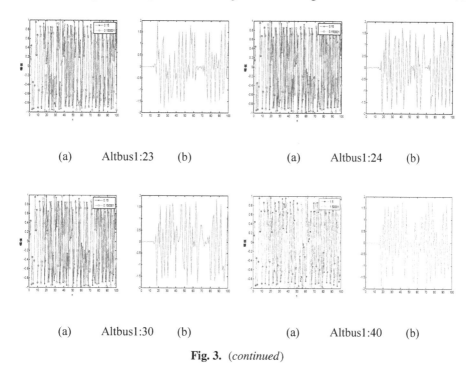

(a)   Altbus1:23   (b)   (a)   Altbus1:24   (b)

(a)   Altbus1:30   (b)   (a)   Altbus1:40   (b)

Fig. 3. (*continued*)

larger the b value of the Altbus module parameter, the better the initial value sensitivity of the new logistic. In order to balance system performance and hardware resources, the b value should generally be set to 20.

4.3 Autocorrelation of New Logistic Chaotic Map

Autocorrelation is a measure of the time-delay sequence similarity of the random sequence itself, and is an important index for measuring the security of chaotic systems [12]. In order to verify the influence of autocorrelation on the random characteristics of chaotic systems, a simulation experiment is performed on the autocorrelation of the new logistic chaotic model. As shown in Fig. 4, the simulation time is set to 1000, the gain is set to 3, and the initial value is set to 0.15. The simulation analysis shows that when the Altbus module parameter $b < 20$, the autocorrelation of the new logistic chaotic sequence is not ideal; when the Altbus module parameter $b \geq 20$, the new logistic chaotic sequence has ideal autocorrelation. Therefore, it is concluded that the larger the b value of the Altbus module parameter, the better the autocorrelation of the new logistic. In order to balance system performance and hardware resources, the b value should generally be set to 20.

4.4 NIST Test of New Logistic Chaotic Map

In order to further verify that the performance of the new logistic chaotic sequence is affected by the Altbus module parameter settings, according to the 16 test standards

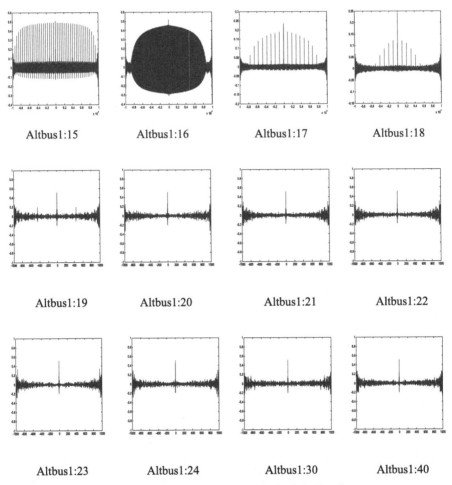

Altbus1:15   Altbus1:16   Altbus1:17   Altbus1:18

Altbus1:19   Altbus1:20   Altbus1:21   Altbus1:22

Altbus1:23   Altbus1:24   Altbus1:30   Altbus1:40

Fig. 4. Autocorrelation analysis of new logistic chaotic map

prepared by the National Institute of Standards and Technology (NIST), this paper conducts four tests: single-bit frequency, block frequency, runs, and longest run [13]. Among them, the p value * indicates that it does not meet the test requirements. If the p value is a number, it indicates that the test is passed. A larger p value indicates a better test result [14]. As can be seen from Table 1, when the b value of the Altbus module is set to 15–19, the test result of the new logistic chaotic sequence is poor, and the test result is better when the b value of the Altbus module is set to 20–25. Therefore, it is concluded that the larger the b value of the Altbus module parameter is set, the better the test result of the new logistic. In order to balance system performance and hardware resources, the b value should be set to 20 in combination with the above experiments.

Table 1. NIST test of new logistic chaotic map

Altbus a:b	Single-bit frequency		Block frequency		Runs		Longest run	
	p value	Proportion	p value	Proportion	p value	Proportion	p value	Proportion
1:15	*	100/100	*	100/100	*	100/100	*	99/100
1:16	*	100/100	*	100/100	*	98/100	*	100/100
1:17	*	98/100	*	97/100	*	98/100	*	98/100
1:18	*	99/100	*	100/100	0.304126	100/100	*	99/100
1:19	*	100/100	*	98/100	0.401996	100/100	*	100/100
1:20	0.381557	100/100	0.897762	100/100	0.237811	100/100	0.350485	100/100
1:21	0.319084	97/100	0.534146	97/100	0.162606	97/100	0.201998	97/100
1:22	0.619747	98/100	0.595549	98/100	0.334408	98/100	0.384125	98/100
1:23	0.712603	100/100	0.616305	100/100	0.498671	100/100	0.667396	100/100
1:24	0.892042	98/100	0.743126	98/100	0.868922	98/100	0.732411	98/100
1:25	0.912033	99/100	0.754439	99/100	0.877469	99/100	0.886579	99/100

5 Conclusion

Traditional logistic chaos, which has its unique characteristics and simple calculation rules, has very significant advantages in practical applications. The design and implementation of chaotic circuits is the key to the research of chaotic applications. This paper utilizes DSP-Builder toolbox to realize the modeling and simulation of the new logistic digital simulation circuit in matlab/simulink environment. It overcomes the difficulty and instability of the design of the analog chaotic circuit, and focus on the influence of chaotic circuit characteristics on the Altbus parameters setting of the conversion module, thereby reducing and eliminating potential safety hazards. The conclusions in this paper provide a certain theoretical basis for the generation of digital chaotic circuits, and have excellent application prospects in chaotic secure communication systems. The next step is to study the frequency domain characteristics of the new logistic digital circuit by changing the parameter settings, and obtain more accurate and comprehensive new logistic chaotic circuit characteristics.

Funding. This work was supported by Heilongjiang Fundamental Research Foundation for the Local Universities (2018KYYWF1189) and Science and Technology Innovation Foundation of Harbin (2017RAQXJ082).

References

1. Qin, J.C.: Realization of chaotic system and its control digital circuit. Masterundefineds degree thesis of Yunnan University (2014)
2. Wang, J., Ding, Q.: Dynamic rounds chaotic block cipher based on keyword abstract extraction. Entropy **20**(9), 693 (2018)
3. Li, Z.B., Li, F.Q., Zhu, X.M.: Chaotic secure communication simulation of deformed Chua's circuit. J. West Anhui Univ. **05**, 61–64 (2014)

4. Wu, Y.: Improvement of encryption scheme of classic logistic map image and exploration of double chaos encryption technology of digital image. Xinjiang University of Finance and Economics (2016)
5. Xu, D., Cui, X.X., Tian, W., et al.: Research on chaotic random number generator based on Logistic mapping. Microelectr. Comput. **33**(2), 1–6 (2016)
6. Jia, Y.J., Wang, Y.Y.: Design and implementation of improved logistic chaotic sequence generator. Mach. Electr. **37**(04), 24–29 (2019)
7. Wang, J., Ding, Q.: Excellent Performances of the Third-Level disturbed chaos in the cryptography algorithm and the spread spectrum communication. J. Inf. Hiding Multimed. Signal Process **7**, 826–835 (2016)
8. Wang, J., Wang, Y., Yin, C., et al.: Investigation on the simulation of one-dimensional discrete chaotic digital generation circuit. In: 2015 Third International Conference on Robot, Vision and Signal Processing (RVSP), pp. 195–199. IEEE (2015)
9. Wang, J., Yang, T., Li, Y., et al.: Design of integer chaotic key generator for wireless sensor network. Int. J. Future Gen. Commun. Network. **9**(11), 327–336 (2016)
10. Chen, Z.G., Liang, D.Q., Deng, X., et al.: Performance analysis and improvement of Logistic chaotic mapping. J. Electr. Inf. Technol. **38**(6), 1547–1551 (2016)
11. Dang, X.Y., Li, H.T., Yuan, Z., et al.: Implementation of chaotic mapping based on digital-analog mixing. Acta Phys. Sin **64**(16), 160501–160501 (2015)
12. Fan, C.L., Ding, Q.: Improved algorithm based on Logistic chaotic sequence and its performance analysis. Electr. Device **38**(4), 759–763 (2015)
13. Cai, D., Ji, X.Y., Shi, H., et al.: Improved method and performance analysis of piecewise Logistic chaotic mapping. J. Nanjing Univ. (Nat. Sci.) **52**(5), 809–815 (2016)
14. Qi, Y.B., Sun, K.H., Wang, H.H., et al.: Design and performance analysis of a hyperchaotic pseudorandom sequence generator. Comput. Eng. Appl. 135–139 (2017)

The Device of Graphite Washed off Acid-Base Automatically Based on MCU

Huang Yaoqun[✉], Zhou Bo, Wu Yan, and Zhang Haijiao

School of Electronics and Information Engineering, Heilongjiang University of Science and Technology, Harbin 150022, China
huangyaoqun@126.com

Abstract. Graphite water elution acid and alkali is the indispensible key link of producing high purity graphite, expanded graphite and other graphite deep processing. This design adopts the AT89S52 single chip microcomputer as the control core, supplemented by pH sensors, temperature sensors, power amplification circuit, display circuit, buttons, and control circuit, controlling the water pump, motor, scraper centrifuge, air knife and mechanical device to achieve the automation of graphite automatic washing to take off the acid and alkali, solving the problems such as using artificial detection graphite liquid pH to test whether it can achieve specific standard, causing inaccurate measurement, inefficiency and waste of human resources and other issue. Through the application of the graphite automatic water elution acid-base device, it can significantly improve graphite washing steps in the process of deep processing of efficiency and effectiveness, shortening the process time, reducing the labor costs and electricity consumption. The results show than compared with the conventional similar equipment this design can reduce the electricity consumption nearly 35% and improve work efficiency about 2.5 times, have a high practical application value of engineering.

Keywords: Graphite washed · Automation · pH sensor

1 Overview

At present, washing the centrifugal equipment at home and abroad can only be washed after the material by means of centrifugal dry, and do not have the function of the pH detection for the washing material. Each wash must be manually sampled on the lotion. PH meter is used to measure the discharge of the pH of the water. And then it is judged whether to mark, if not up to standard, it requires manpower and the material to water, centrifugal, repeat the above steps, which is time-consuming. If washing time is not long enough, it will make drainage pH value not standard, which needs to adjust the time of centrifugal, a new water centrifugal. If washing time is too long, it is a waste of time, and the efficiency is low. The equipment can be concluded a kind of material washing process after many washing experiments, which both increased the burden of human, and operation is extremely complicated [1, 2].

© ICST Institute for Computer Sciences, Social Informatics and Telecommunications Engineering 2020
Published by Springer Nature Switzerland AG 2020. All Rights Reserved
X. Jiang and P. Li (Eds.): GreeNets 2020, LNICST 333, pp. 75–83, 2020.
https://doi.org/10.1007/978-3-030-62483-5_9

This design mainly adopts intelligent real-time monitoring measures, which will make washing machine and washing centrifuge become one, achieving automatic control of the equipment by adding a pH sensor and temperature sensor. When washing, electrical machine inside the washing machine can turn positive and negative two directions in turn, mixing fully to material, which will make part of the glue material conduct a thorough break by washing machine's first wash, preventing incomplete washing of materials' getting together in the process of washing materials. After washing, materials through a water pump will be broken and lotion with import water washing in the centrifuge, the centrifugal dehydration, and set a pH sensor and a temperature sensor in the outlet of the centrifuge, collecting the information of lotion's PH in time when washing centrifuge dehydration. Then to judge whether the PH of lotion is standard through the single chip microcomputer. If it is not up to standard, washing centrifuge will continue to run until the PH of lotion is up to standard, it will stop working automatically and then the single-chip microcomputer control water centrifuge spatula to scrape down in the material [3], and is accompanied by the end of the prompt, suggesting it is the end of the washing. It not only can solve common equipment which can not be real-time monitoring of pH value and achieve full automatic difficulty, and also can lighten the burden of human, save water resources.

2   System Design Scheme and the Key Circuit

This design system regards AT89S52 single chip microcomputer as the control core of the system. Circuit is controlled by button, PH value detection and power amplifier circuit, A/D conversion circuit, temperature compensation circuit and the water level sensor circuit, etc. and pH sensors, temperature sensors and water pumps, pneumatic scraper centrifuge [4], the mechanism structure diagram of system is as shown in Fig. 1.

When system is working, it puts graphite raw material into the washing machine, doing washing mixing. Blades in the washing machine have positive and negative rotation alternately, which can break materials thoroughly, preventing materials glue together because of going bad [5, 6]. After washing mixing water pump will make graphite material in the washing machine and lotion smoke into the centrifuge, and open water and centrifugal centrifuge. Placing a pH sensors at the outlet of the centrifuge, lotion for real-time measurement of discharge outlet, measuring the size of the PH value, transmitting numerical value to single chip microcomputer to judge whether graphite materials washed completely through single chip microcomputer. If the pH value is not up to standard, centrifuge will keep water for material and the status of centrifugal, until single chip microcomputer judges that the pH sensors detect the liquid pH value standard, centrifuge begins to stop work. Then open air knife, sweep the materials on the walls of the centrifuge, complete the automated processing of graphite automatic washing to take off the acid and alkali. This design also chooses LCD to display the current working state, we can intuitively understand the working information of system. At the same time, it can also show washing or centrifuge time. And also it can be adjusted by button scanning circuit of washing or centrifugal time and for other operating systems. When washing is over and through the test, the buzzer will beep, warning that staff's washing ends, reducing the human, improving efficiency.

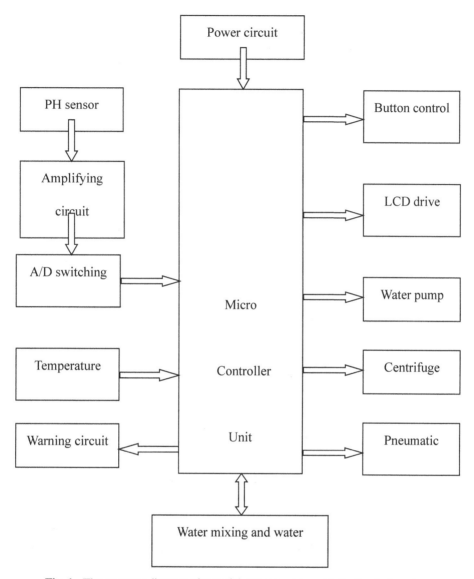

Fig. 1. The structure diagram of material remove waste acid alkaline bath system

This design also chooses LCD to display the current working state, we can intuitively understand the working information of system. At the same time, it can also show washing or centrifuge time. And also it can be adjusted by button scanning circuit of washing or centrifugal time and for other operating systems. When washing is over and through the test, the buzzer will beep, warning that staff's washing ends, reducing the human, improving the process efficiency.

2.1 Button Control Circuit

Button circuit is mainly used for washing time of systems and the setting of graphite lotion PH reference. This design adopts the independent key, key input with low level effectively, when the switch is not pressed, the input for the high level, after closing, input is low level. Connection circuit is shown in Fig. 2.

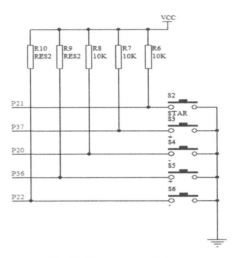

Fig. 2. Button control circuit

2.2 PH Value Detection and Power Amplifier Circuit

PH sensors are used to detect the centrifuge after washing lotion of pH value, to determine the graphite raw water elution the effect of acid and alkali, but as a result of general pH sensor output voltage is in the Howe v level, if it is directly transmitted to MCU and it cannot identify [7], so it is necessary to carry out amplification on its output voltage driving. PH sensor resistance generally in more than 10m euro, ordinary amplifying circuit resistance can never reach. So you need to select input fault of power amplifier, so that you can need not consider to carry out amplification on its pH sensor resistance. This design chooses OP07 power amplifier to achieve as shown in Fig. 3, which has characteristic of low disorders, high open-loop gain especially suitable for high gain measurement equipment and the weak signal amplification. At the same time, it adopted to eliminate the zero wave potentiometer, which can effectively eliminate the zero drift.

2.3 A/D Conversion Circuit Design

PH value detection signal which is zoomed is a an analog voltage signal, needing to convert to digital signals, in order to conducting detection and management. The design adopts TLC1549 as the core of the A/D conversion circuit. It is 10 AD converter, internal converter with high speed (10 μs conversion time), high precision (10 resolution,

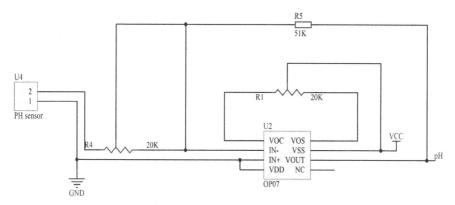

Fig. 3. PH value detection and power amplifier circuit

maximum plus or minus 1 LSB not adjust error) and the characteristics of low noise. TLC1543 chip of the three input and an output terminal with 51 series microcontroller I/O ports can be connected directly.

2.4 Temperature Compensation Circuit Design

This system adopts the pH sensor as washing graphite material detection device, because the change of medium temperature has the influence on the electrode pH measurement, which will lead to the measurement deviation, affect the washing effect. So you need to use temperature compensation correction ways for pH sensor calibration to measure precise pH value. Considering the design needs to measure the temperature of the acid and alkali liquid, so using strong acid and alkali resistance and with industrial water-proof outer packing of the digital thermometer DS18B20, the thermometer provides nine (binary) temperature readings, measuring range from −55 °C to +125 °C, the incremental value of 0.5 °C, is available in ls (typical) within the temperature transform into digital. At the same time, adopting the mode of external power supply, speeding up the measuring [8].

2.5 Water Level Sensor Circuit Design

Considering the design needs to measure the liquid level of the acid and alkali liquid, so we should choose the sensors which are strong acid and alkali resistance, high temperature resistant, anti-aging, tearing resistance, difficult to produce scale, no water seepage. Therefore, this design uses industrial water level sensor. It adopts special composite material and design pulse of alternating voltage internally, and use high temperature resistant material encapsulation, make the sensing component with super corrosion resistance, anti scaling function, stable work. The core parts of sensor adopt foreign latest win min element has high precision, to ensure the measurement precision in bad environment. Water level sensor is mainly composed of four resistors in series, from the bottom to up, in turn, is 30k, 10k, 10k, 10k, and the corresponding electrode from bottom to top respectively are: the public, 75% electrode, 60% electrode, 50% electrode, 30% electrode. No water in the tank, the water resistance is 60k, with the increase of water level, the resistance will decrease in turn. When measuring water level, the sensor is inserted into the water, the deeper the resistance is smaller.

3 The Design of System Software

The design of System software in the Windows 7 environment adopts Keil Vision4 software and makes use of C language to write in order to realize the single chip microcomputer control functions. Software control programs include the master control program, the pH reading, button scanning, liquid crystal display subroutine. Master control program controls the overall program. Through their procedure call, the function of each part of the system is realized.

3.1 Main Program Design

After the system is powered on, it begins to initialize. After manual work puts graphite material into washing machine, washing time and PH value of standard setting can be set. After it begins, single-chip microcomputer controls washing machine to wash, which includes the forward and reverse of washing machine. When the washing time ends, it shows washing time's over through the liquid crystal display, accompanied by voice prompt, and then water pump starts to work, which make a mixture of graphite materials and water in the washing machine pump into the centrifuge. At the same time, the centrifugal is opened to do centrifugal dehydration of graphite and to go on washing from the centrifuge internal to the graphite surface water. When washing. PH sensor tests the water solution PH value through real-time monitoring, to determine whether the washing is over or not. If it is not up to standard, washing should be continued. If the PH value of testing standards is up to standard, washing will be over. And then return to the main program and continues to scan button, preparing for the next operation. The system main program flow chart is shown in Fig. 4.

3.2 PH Value Setting Subroutine

When operating procedures are for setting pH, through scanning the buttons, to determine whether the corresponding key is pressed or not, if pressed, then bring up the corresponding subroutine to control system and through the LCD12864 to display. If no press, then continue to scan. S2 for the subsequent operation keys, the S3 key for pH value, minimum value minus 0.1; S4 for pH value, minimum value will be increased by 0.1; S5 for pH value maximum reduced 0.1; A maximum S6 for pH value will be increased by 0.1. PH value set subroutine flow chart shown in Fig. 5.

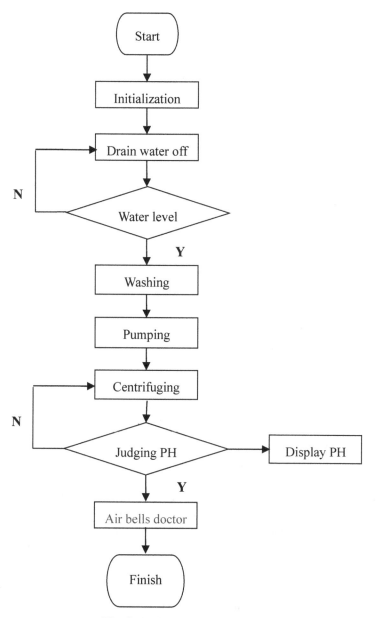

Fig. 4. Main program flow chart

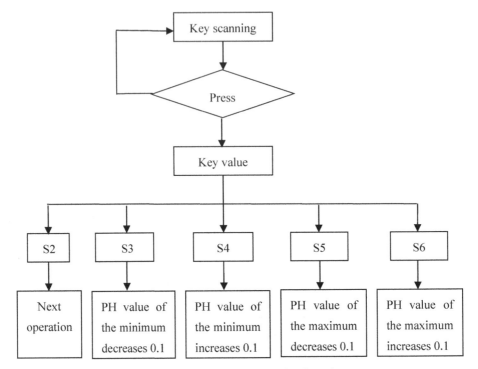

Fig. 5. PH value setting subroutine flow chart

4 Conclusion

The design puts AT89S52 single chip microcomputer as control core, combing with high precision dynamic real-time monitor of the pH value, realizing the full automatic control for the graphite water elution acid-base device. The system hardware structure is compact and simple, which is easy to control, shortening the time of preparation process of high purity graphite and expanded graphite, significantly improving the effect of graphite washed link, improving the production efficiency, solving the traditional process using artificial detection caused by inaccurate measurement, such problems as low efficiency and waste of human resources, reducing the labor costs and electricity consumption. In order to reach the purpose of rapid graphite washing in the industry, compared with the conventional similar centrifuge, the motor power have been increased from 30 kw to 55 kw, the rotate speed have been increased from 970 RPM to 2400 RPM, and realized the PH value automatic accurate measurement and automatic control. The experiment result shows that compared with the traditional similar equipment, the power consumption of this equipment can be reduced nearly 35%, the manual unloading time can be greatly reduced, and the production efficiency can be significantly improved. At the same time, this design can also be applied to other industrial production in the field of material of water washing and processing, which has high value in engineering application and popularization value.

References

1. Baojun, W.: Application of centrifuge in the treatment of pickling wastewater in steel plants. Chem. Equip. Technol. **5**, 5 (2017)
2. Qianyun, W.: Influencing factors analysis of dewatering effect of horizontal centrifuge in coal washing plant. Shandong Ind. Technol. **20** (2018)
3. Wu wei. The Application of vertical scraper discharging centrifuge. Min. Equip. **5** 2018
4. ATMEL Corporation. AT89S52 datesheet. www.atmel.com
5. Lijie, H.: Study on improving the comprehensive efficiency of imported scraper centrifuge. China Plant Eng. **10** (2014)
6. Zhenxiang, Z.: The factors affecting the centrifugal efficiency of siphon automatic discharge centrifuge. Chem. Equip. Technol. **2** (2014)
7. Wittmer, D.: pH sensor introduction (2015)
8. Dallas Semiconductor: DS18B20 datesheet. www.dalsemi.com

Research on Multi-disciplinary Museum Lighting Design's Emotional Response to Visitors: A Case Study of Dalian Modern Museum

Dan Zhu[1], Zhisheng Wang[1,2(✉)], Yukari Nagai[2], Cong Zhang[1], Haiwen Gao[1], and Nianyu Zou[1]

[1] Research Institute of Photonics, Dalian Polytechnic University, Dalian 116034, China
wangzs@dlpu.edu.cn
[2] Japan Advanced Institute of Science and Technology, Nomi 923-1292, Japan

Abstract. The main research direction of this article is to explore the artificial lighting of the museum emotionally by adopting the innovative thinking and design methods of multidisciplinary cross integration. Three main innovations are proposed. 1. The theory of tourists' demand and space is analyzed. 2. We use creative thinking to answer the relationship between optical engineering and the time that tourists stay in the exhibition area. 3. Interactive formulas are used to quantify emotions. This experiment takes the Dalian Modern Museum as the research object. Firstly, for the space division of the museum and the tour route, we analyzed and discussed it according to the needs of tourists and the size of the exhibition space. Under a certain space size condition, the illuminance, color temperature and color rendering index were measured and then we collated and analyzed the data obtained. Under different light environment conditions, we obtained the relationship between the R9, Rf and Rg values in the color rendering index and the residence time of visitors in the exhibition area. Finally, the factor analysis method is used to analyze the experimental data. This experiment provides theoretical support for the emotional design of artificial lighting, and has creative thinking and research significance.

Keywords: Artificial lighting design · Emotional response level · Creativity

1 Introduction

As a carrier of cultural exchange, the museum's lighting environment greatly affects the visitors' viewing experience. At the same time it is also a functional facility, its main purpose is to display, educate and spread culture [1]. The rationality of lighting design is an important indicator to measure the level of museum construction. Different types of museums should have different display themes, which means different lighting technologies [2]. For the museum, the design of the lighting environment should not only consider the protection of cultural relics, but also consider the color performance of the

X. Jiang and P. Li (Eds.): GreeNets 2020, LNICST 333, pp. 84–93, 2020.
https://doi.org/10.1007/978-3-030-62483-5_10

exhibits and the texture clarity of exhibits [3],we must also pay attention to whether the vision conveys the information inside and around the museum's collection [4]. The lighting in the museum not only meets the visual requirements of tourists, but also makes tourists feel comfortable and enjoy during the visit of the museum [5]. The psychological experience of museum visits and the emotional and educational significance to the visitors need to be conveyed. Therefore, a suitable lighting condition needs to be determined by combining space theory and optical indicators in a multi-disciplinary way. According to the purpose of museum visitors, different tour routes and space attributes are divided. We combine it with the color temperature and color rendering index in the light environment of the corresponding area to provide tourists with a high-quality viewing experience while meeting people's psychological and physiological needs.

In order to ensure the reliability and authenticity of the data, we selected Dalian Modern Museum as the research object and conducted a field survey. This experiment mainly explores the factors that affect tourists' emotions through a combination of subjective evaluation and objective evaluation. Objective experimental data mainly test the relevant parameters of the lighting in the display area (such as illuminance, color temperature, color rendering index (hereinafter referred to as CRI) and the size of the exhibition space. In subjective evaluation, we invite tourists to complete a subjective questionnaire survey. The survey object is to randomly select visitors at the museum site. The objective experimental data is combined with the purpose and route of tourists in the subjective questionnaire survey to analyze, so as to determine the most appropriate lighting environment when visiting the museum. The overall survey has innovative practical and research significance.

2   Measurement and Analysis of Museum Light Environment

2.1   Spatial Theory and Psychological Needs

Tourists have different psychological needs and purposes when visiting the museum, such as visiting the museum as a tourist attraction, accompanying their family or friends, alleviating the pressure on life, etc. The relationship between space and psychological needs is complementary. For example, different space shape, height and size, color and texture of wall paint will affect the psychological experience of tourists [5]. We sum up the two points as Fig. 1 and Fig. 2. Figure 1 is the tour route of most tourists. The investigation of Dalian modern museum mainly divides visitors' visiting purposes into six categories. As shown in Fig. 2, most tourists regard the museum as a scenic spot or accompany their families to visit, and these people are in a happy mood. There are relatively few people who are decompressed, and they are calm when they visit the exhibits.

2.2   Measurement and Analysis of Lighting Conditions

Illuminance Measurement
The illumination measurement mainly adopts the central point method, dividing the area into square grids on average. The central point of the grid is the test area [6]. The

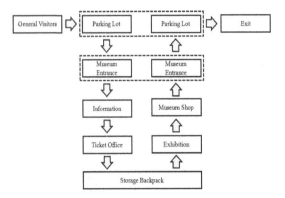

Fig. 1. Museum visit process for general visitors.

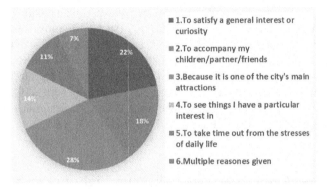

Fig. 2. Visitor's purpose

objective measurement is mainly aimed at the educational exhibition hall, the modern military exhibition hall, the modern life exhibition hall and the street exhibition area. The ground uniformity and the light color test uses the central point method to take the points, taking the average (Eq. 1) and the average of the reflectivity of nine points. The illumination uniformity of the whole environment is obtained by Eq. (2).

Eav is the average illuminance value, the unit is Lux (Lx), Ei is the illuminance value of a point, $Emin$ is the minimum of illumination M is the longitudinal measuring point, and N is the transverse measuring point (Eq. 1 and 2):

$$E\text{av} = \frac{1}{M \cdot N} \sum Ei \tag{1}$$

$$U_0 = \frac{Emin}{Eav} \tag{2}$$

As shown in Fig. 3, Fig. 4 and Fig. 5, the education exhibition hall mainly relies on LED panel lights and natural light for lighting, because the exhibition hall mainly carries out the education function without cultural relics display, so the overall level of illumination is relatively high. The average illumination is 145.3 lx (Eq. 1) for 9 points. The illumination uniformity of the whole environment is 0.7 by Eq. 2.

Fig. 3. Education exhibition hall

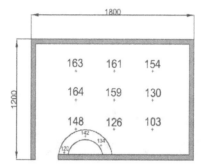

Fig. 4. Illumination distribution of Education exhibition hall

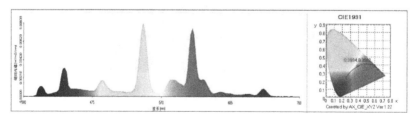

Fig. 5. Education exhibition hall illumination spectrum

Figure 6 is a modern living exhibition hall, which mainly relies on track metal halide lamp for lighting. The test points are 2 * 6, the average is 16 lx, and the illumination uniformity is 0.625. Modern Life Exhibition Hall Illumination Spectrum is shown in Fig. 7.

Fig. 6. Illumination distribution and scene of modern life exhibition hall

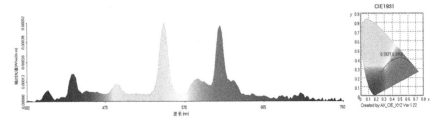

Fig. 7. Modern life exhibition hall illumination spectrum

Fig. 8. Modern military exhibition hall

Figure 8 is a modern military exhibition hall. The LED panel and guide metal halide lamp are used for lighting. The test points (Fig. 9) are 4 * 3, the average illumination is 16 lx, the illumination uniformity is 0.625. The illumination spectrum of modern military exhibition hall is shown in Fig. 10.

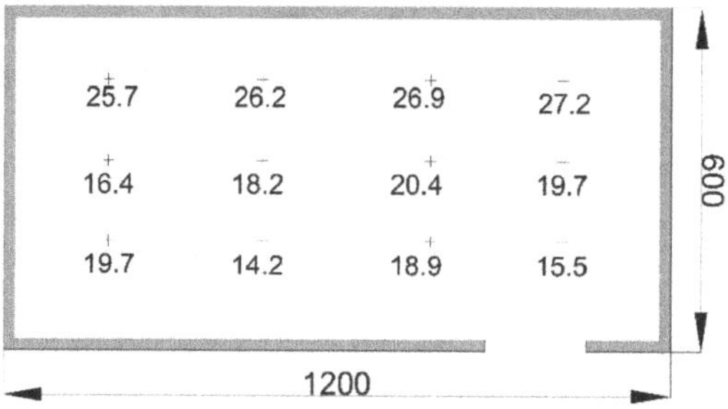

25.7	26.2	26.9	27.2
16.4	18.2	20.4	19.7
19.7	14.2	18.9	15.5

600

1200

Fig. 9. Illumination distribution of modern military exhibition hall

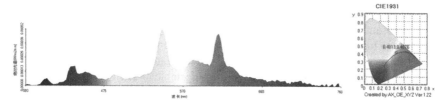

Fig. 10. Illumination spectrum of modern military exhibition hall

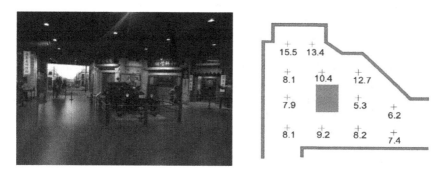

Fig. 11. Illuminance distribution and scene map of street exhibition area

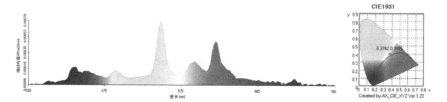

Fig. 12. Illumination spectrum of street display area

Figure 11 is a street display area with LED. The panel lamp is illuminated, and the test point is 4 * 3, the average is 16 lx, and the illumination uniformity is 0.625. the illumination spectrum of street display area is shown in Fig. 12.

The results show that in the four exhibition halls, only the illumination value of the education exhibition hall is relatively standard. The illumination of the other three exhibition halls is usually low. The details, stereoscopic sense and texture clarity of the exhibits cannot be well guaranteed. As a result, tourists spend less time in this environment than expected.

Color Temperature Measurement

Color temperature conditions have a subtle influence on our viewing experience. In the study of interior lighting design, kruithof defined "comfortable area" by comparing the correlation between color temperature (CCT) and illuminance, and proposed a method to achieve "pleasant area" [5]. kruithof's rule shows that the specific area of high (low) CCT corresponds to high (low) illuminance, which makes the view Observers feel happy [6], 3000–4000K is a relatively popular color temperature range, or a relatively high visual comfort range, which makes people feel more comfortable and bright [7], different types of museums have different emotions to convey to people. Flexible use of the relationship between light and color temperature enables visitors to have different sensory experience during the visit. In this experiment, we use spectroradiometer to measure the color temperature in the field. The number of measuring points in each area shall not be less than 9, and each functional area shall have more than three measuring points. The color temperature of each light source or mixed light source under different light sources is measured. The results show that the average color temperature of education exhibition hall is 3643k, modern life exhibition hall is 3273k, modern military exhibition hall is 2953k, and street display area is 3010k. The CCT of the four exhibition halls is in the comfortable area, the modern life exhibition hall, the modern military exhibition hall and the street exhibition area is at a low CCT level, which will make the tourists have a heavy and depressing mood in the process of visiting.

CRI Measurement

Lighting with high CRI can truly restore the original appearance, color and texture of the cultural relics, and can better spread the culture [9]. For the museum, where there are abundant collections, exhibits with different attributes need different light sources to display. The CRI of light sources directly affects the hue and saturation of the color of the exhibits. For displayed paintings, color fabrics and other places with high color identification requirements, The CRI of commonly used light source shall not be less than 90. For places with general color identification requirements, The CRI of commonly used light source shall not be less than 80 [10].

The measurement results of the color rendering index of the modern museum are shown in the Table 1. It suggests that higher color rendering index of education exhibition hall can make people feel clear and relaxed. Modern life exhibition hall has a higher overall color rendering index because of its rich silk fabrics and high color temperature. At the same time, the main lighting fixtures in the exhibition area are metal halide lamps with guide rails, R9 is normal. For the modern military exhibition hall and street display area, the requirements of collection protection and color rendering are not high, and the lighting lamps mainly use LED panel lamp. Because of the inherent shortcomings of LED lamp and the lack of red band, the R9 value and overall CRI are low.

Table 1. Measurement results of the color rendering index of modern museums

Hall	Ra	R9	Rg	Rf	SDCM
Education Exhibition Hall	90.1	81	96	90	3.2
Modern Life Exhibition Hall	92.1	83	97	91	2.7
Modern Military Exhibition Hall	82.3	−7	96	92	4.9
Street Exhibition Area	80.3	−7	79	97	5.1

3   Emotional Response Model

3.1   Sensory Questionnaire

The subjective evaluation is carried out by inviting tourists to fill in a questionnaire on the spot, which is reliable. A total of 117 sets of data were collected from more than 20 individuals. The exhibition area is education exhibition hall, modern life exhibition hall, modern military exhibition hall, painting exhibition hall and street exhibition area, and non exhibition area is corridor. Measurements are made at the following levels: A is 10, a - is 8, B is 7, B - is 6, C is 5, C - is 4, D is 3, D - is 0. 11 indicators, such as the size of the authenticity color, the preference of the light source color, the detail expressiveness of the exhibits, the visual adaptability and the pleasure of viewing the exhibits, can be calculated. According to each evaluation, the corresponding scores are above 80, above 70, above 60 and below 60. It is divided into four levels: excellent, good, medium and poor [11]. Get illumination of the display area. As can be seen from Fig. 12, the lighting conditions in the basic display area are of general grade.

Item\Sample\Score	A+10	A−8	B+7	B−6	C+5	C−4	D+3	D−0	mean value	Secondary weights	Weighting×10
Realistic color of exhibits	2	10	5	4	1	1	0	0	7.30	20%	14.60
Light source color preference	1	11	7	2	0	2	0	0	7.26	5%	3.63
Exhibits' detail expressiveness	2	6	12	1	0	1	0	1	7.04	10%	7.04
Three-dimensional expressive force	1	8	7	4	2	0	1	0	6.95	5%	3.48
Texture clarity of exhibits	1	5	7	8	0	1	0	0	6.82	5%	3.41
Outer Contour Clarity of Exhibits	1	7	9	4	1	1	0	0	7.04	5%	3.52
Brightness Acceptance of Exhibits	2	9	7	5	0	0	0	0	7.43	5%	3.72
Visual adaptability	6	8	5	3	0	1	0	0	7.86	5%	3.93
Psychological pleasure	1	10	5	5	1	1	0	0	7.13	5%	3.57
Lighting artistic preference	2	9	7	4	2	0	0	0	7.30	20%	14.60
Infectivity	2	11	4	3	1	2	0	0	7.26	10%	7.26
Total										100%	68.745

Fig. 13. Basic display subjective assessment results

3.2 Emotional Response

The purpose of this study is to define the relationship between emotional perception and the different combinations of CCT, CRI and various illumination values. The uncertainty of the experimental evaluation data was tested by the root mean square (RMS) on the emotional response level table through the variability between the observer and the observer (Eq. 3). The smaller the RMS value, the greater the consistency between the two data sets [12]. The higher the RMS value, the worse the consistency within or between observers. For variability among observers, y_i is the score of the individual observer for the I stimulus. x_i is the average score of all observers for the I stimulus; N is the total number of stimuli. As shown in Table 2 the root mean square values of variability between and within observers in this experiment.

Table 2. RMS values of variability between and within observers

Emotional Scale	Observer Change	Internal Observer Change
Pleasant	0.92	0.38
Comfort	0.92	0.37
Bright	0.89	0.38
Color	0.92	0.36
Clean	0.88	0.33
Nature	0.96	0.38
Lively	0.92	0.36
Ease	0.90	0.37
Soft	0.83	0.34
Classical	0.85	0.35
Warm	0.84	0.35
Average	0.89	0.36

$$RMS = \sqrt{\frac{\sum_{i1}^{n}(y_i - x_i)^2}{n}} \tag{3}$$

By adding the measured data to the emotional quantification evaluation and substituting it into formula (3), we can get a high consistency with the quantitative value of RMS below. Emotional indicators are 11 phrases in Fig. 13. According to the results of the visitors' evaluation of each index, different experimental conditions of light environment were constructed subjectively. Finally, by changing the objective parameters such as illuminance, color temperature and color rendering index, combined with the different lighting environment constructed according to the subjective, the emotional response of museum lighting can be obtained.

4 Discussion and Summary

Through the experiment of Dalian Modern Museum, we can draw a conclusion that the tourists' visiting purpose is directly related to the size of the space. For different tourists' visiting purposes, most tourists who aim at visiting scenic spots or accompanying family and friends will complete the whole process. The visiting time is shorter and most of them will stay in small space and exhibits-intensive areas for a long time. For pressure relief tourists, they spend longer time visiting places with large space and do not choose to go to Museum stores. For visitors who visit designated exhibits and satisfy their curiosity, they are more excited. They usually go directly to designated exhibits and follow-up visits are slower. At the same time, combined with the relationship between residence time and illumination, color temperature and color rendering index, under the same illumination condition, R9 value of some spatial rendering index using LED as the main light source is lower or negative, which is due to the inherent disadvantage of the ratio of LED light source - less red band, and with the increase of the efficiency of LED light, the reduction of the red part will be more serious. In these places where the R9 value is low, most of the tourists are not very happy, and the sense of ornamental experience will decrease. According to these data and analysis, four kinds of emotional models of museum lighting are determined by interactive formula.

References

1. Wang, L.: Application and Research of Visual Conditions in Museum Exhibition Design Light Environment. Xi'an University of Architecture and Technology, Shanxi (2013)
2. Rui, D., Mingyu, Z., Gang, L., et al.: Investigation and study on Chen lighting of museum exhibition based on cultural relic protection. China Illum. Eng. J. **24**(3), 18–23 (2013)
3. Zhai, Q.Y., Luo, M.R., Liu, X.Y.: The impact of LED lighting parameters on viewing fine art paintings. Light. Res. Technol. **48**(6), 711–725 (2016)
4. Ajmat, R., Sandoval, J., Arana Sema, F., O'Donell, B., Gor, S., Alonso, H.: Lighting design in museums: exhibition vs. Preservation. WIT Transactions on The Built Environment, vol 118 © 2011 WIT Press (2011)
5. Liu, J., Wang, Z., Nagai, Y., et al.: Research on the emotional response level of museum visitors based on lighting design methods and parameters. International Conference on Green Communications and Networking. Springer, Cham, 221–239 (2019)
6. Royer, M.P., Houser, K.W., Wilkerson, A.M.: Color discrimination capability under highly structured spectra. Color Res. Appl. **37**(6), 441–449 (2012)
7. Kruithof, A.A.: Tubular luminescence lamps for general illuminatio. Philips Technical Rev. **6**, 65–96 (1941)
8. Ajmat, R., Sandoval, J., Arana Sema, F., et al.: Lighting design in museums: exhibition vs. preservation. WIT Trans. Built Environ. **118**, 195–206 (2011)
9. Scuello, M., Abramov, I., Gordon, J., et al.: Museum lighting: optimizing the illuminant. Color Res. Appl. **29**(2), 121–127 (2004)
10. Wei, M., Houser, K.W., David, A., et al.: Effect of gamut shape on colour preference. In: Proceedings of CIE 2016 Lighting Quality & Energy Efficiency, pp. 32—41, Melbourne, Australia (2016)
11. Cuttle, C.: Control of damage to museum objects by exposure to optical radiation. In: Proceeding of the 24th Session of the CIE, vol. 1, pp. 324–328(1999)

Anti-collision Device of DSTWR and SFKF Hybrid Positioning Vehicle

AnHua Wang[✉], HongKai Ding, and DongDong Wan

Electronic and Information Engineering Institute, Heilongjiang
University of Science and Technology, Harbin 150022, China
3036361_cn@sina.com

Abstract. In order to suppress the hidden danger brought by the blind area of construction vehicles, a vehicle collision prevention device is designed. The realization method is introduced in the paper. Combined with DSTWR and SFKF algorithm, UWB detection technology is applied to achieve reliable ranging and positioning. It is found through one-to-many ranging experiment that the maximum static absolute error is 0.093 m, the average static error is 0.37 m, and the maximum static relative error is 0.0106; the maximum dynamic absolute error is 0.15 m, the average dynamic error is 0.073 m, and the maximum dynamic relative error is 0.0174. The positioning experiment shows that the maximum errors along the direction of the three axes are 0.17 m, 0.29 m, and 0.45 m, and the average position error of the node is about 0.241 m. The experimental results show that the proposed method satisfies the actual functional requirements, and can ensure stable detection with high speed and accuracy. The large error and track discontinuity under the action of motion and vibration factors can be effectively suppressed by the method, which has application value.

Keywords: Anti-collision of the vehicle · Ultra-wideband · Positioning · Kalman filter

1 Introduction

With the continuous improvement of industrial mechanization, engineering vehicles are widely used in the construction of ports, mines, road, bridge and so on. There are blind spots for many engineering vehicles in the operation process, making a personal safety hazard for on-site construction personnel. Therefore, it is of great significance to develop a wireless ranging and positioning system with automatic detection and early warning functions [1–3].

At present, more mature ranging methods include laser, microwave, infrared, GPS and ultrasonic. The basic principles of laser, ultrasonic and infrared measurement distances are: The target sends laser, ultrasonic or infrared light, and the receiver sends them by the original way. And then the time interval between transmission and reception is obtained to estimate the distance. Laser ranging must be point-to-point ranging, and

© ICST Institute for Computer Sciences, Social Informatics and Telecommunications Engineering 2020
Published by Springer Nature Switzerland AG 2020. All Rights Reserved
X. Jiang and P. Li (Eds.): GreeNets 2020, LNICST 333, pp. 94–107, 2020.
https://doi.org/10.1007/978-3-030-62483-5_11

ultrasonic and infrared ranging are easily affected by multipath environments, resulting in low accuracy [4, 5]. A microwave with a carrier wavelength of 0.8–10 cm is used to achieve microwave ranging, and it is modulated and transmitted by the primary station, received and forwarded via the secondary station. The post-lag phase difference of the modulated wave after mixing processing is measured to estimate the distance. The cost of microwave ranging is high [6, 7]. Based on the global positioning system and the Beidou positioning system, GPS ranging is combined with an electronic map and road data to achieve ranging [8]. Existing GPS navigation systems are affected by complex environments. For example, accurate location services cannot be provided in viaducts and tunnel areas [9, 10]. Affected by various factors, these ranging methods have the defects, such as low immunity, high signal transmission power, channel fading, and low energy efficiency. They are not suitable for application between construction vehicles and construction personnel and construction equipment [11, 12].

In recent years, UWB communication technology has developed rapidly. UWB communication has the advantages of low power consumption, wide bandwidth, low spectral density, high data transmission rate, strong confidentiality, high multi-path immunity and high resolution [13]. How to apply UWB to achieve accurate and reliable ranging has become a research hotspot, where the ranging algorithm and signal collision avoidance mechanism are the keys [14]. TOA, TDOA, AOA, RSSI, and TWR and so on are common ranging algorithms [15–18].

The waveform characteristics of numerous UWB signals are analyzed by Stefano M and Wymeersch H et al. NLOS state identification and error reduction are carried out using support vector machine to effectively eliminate NLOS error and improve ranging accuracy. However, this method is based on a large number of data statistics with a large workload; Huerta J M et al. use particle filter and unscented Kalman filter to locate position and measure speed, which can effectively suppress NLOS error. However, the algorithm is too complicated to implement [19, 20]; in the literature [21], ranging is achieved by integrating RTT (Round Trip Time) and AOA (Angle Of Arrival) to propose a grid-based clustering algorithm that does not need to master the prior conditions such as the indoor environment. However, its hardware part should be based on the antenna array, with a high cost; literature [22] judges the NLOS state based on the prior information, and identifies the NLOS error by INS (Inertial Navigation System) and processes it. Since this method is more complicated, the additional hardware is required.

In order to improve the reliability of measurement information, the theory of robust estimation is proposed by Yang Y X et al. to eliminate measurement error[19]. Combined neural network and Kalman filtering, adaptive suppression state vector and measurement interference method are proposed by Gao Weiguang and others. But the training process of the neural network has a large workload [23, 24]. A new vector robustness factor is proposed by Miao Yuewang et al. using the GNSS/INS combination to improve the system accuracy [25, 26].

Considering the system detection speed, accuracy and reliability requirements under actual working conditions, the combination of DSTWR (Double-sided Two-way Ranging) algorithm and SFKF (Sampling Fitting Robust Kalman Filter) is proposed to achieve ranging and positioning [27].

2 Ranging Method

DSTWR algorithm based on the improvement of TWR does not need to consider the synchronization problem between nodes in the ranging process and has higher precision and speed in the actual measurement [1]. The DSTWR ranging diagram is shown in Fig. 1.

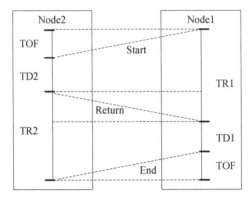

Fig. 1. Ranging principle diagram of DSTWR.

Node 1 sends a Start signal to Node 2. After a T_{D2} time, node 2 sends a Return signal to node 1. After a T_{D1} time, node 1 sends an End signal to node 2. Supposing that time interval of sending and receiving of node 1 is T_{R1}, and that time interval of sending and receiving of node 2 is T_{R2}, and the flight time between the two nodes is TF, it can be obtained that:

$$T_{R1} = 2T_F + T_{D2} \tag{1}$$

$$T_{R2} = 2T_F + T_{D1} \tag{2}$$

Simultaneous formula (1) and (2), the flight time is:

$$
\begin{aligned}
T_F &= \frac{T_F(2T_{D2} + 4T_F + 2T_{D1})}{2T_{D2} + 4T_F + 2T_{D1}} \\
&= \frac{T_F(2T_{D2} + 4T_F + 2T_{D1})}{T_{R2} + T_{R1} + T_{D2} + T_{D1}} \\
&= \frac{(T_{D2} + 2T_F)(T_{D1} + 2T_F) - T_{D2} \times T_{D1}}{T_{R2} + T_{R1} + T_{D2} + T_{D1}} \\
&= \frac{T_{R2} \times T_{R1} - T_{D2} \times T_{D1}}{T_{R2} + T_{R1} + T_{D2} + T_{D1}}
\end{aligned}
\tag{3}
$$

The distance between the two nodes is:

$$d = \frac{T_{R2} \times T_{R1} - T_{D2} \times T_{D1}}{T_{R2} + T_{R1} + T_{D2} + T_{D1}} \times C \tag{4}$$

In the formula (4), C represents a signal transmission speed.

Supposing the offset errors of two node clocks are e_1 and e_2 respectively, and they are brought in the Eq. (4) to obtain the actual distance d' of the two nodes:

$$d' = \frac{T_{R2}(1 + e_2) \times T_{R1}(1 + e_1) - T_{D2}(1 + e_2) \times T_{D1}(1 + e_1)}{T_{R2}(1 + e_2) + T_{R1}(1 + e_1) + T_{D2}(1 + e_2) + T_{D1}(1 + e_1)} \times C \quad (5)$$

The distance error Δd of the simultaneous Eqs. (4) and (5) is:

$$\Delta d = d - d' = (T_F(e_1 - e_2) + T_F \times O(e_1, e_2, T_{D2}, T_{R2})) \times C \quad (6)$$

It can be seen that the DSTWR algorithm does not need to consider the node synchronization problem, with small calculation amount and small error, which can ensure the high measurement accuracy and speed [28, 29].

3   Location Method

Positioning is the extension of point-to-point ranging, which is achieved by measuring the distance from a node to multiple base stations [30]. The distance value during the positioning process is sent to the upper computer, and the upper computer calculates the distance according to the distance from the node to all the base stations, as shown in Fig. 2. Taking the three base stations as an example, the coordinates of the three base stations are known, the coordinates of each base station are taken as the origin point, draw three circles based on the distance from the node to each base stations, and the intersection point of the three circles is node coordinates [31–33].

Due to external noise interference, when the positioning node is in the moving state, the measured node movement trajectory is intermittent [34]. Considering the requirements of algorithm accuracy, speed and complexity in practical applications [35], the Kalman filtering algorithm is introduced, which makes the measurement results smoothed and more in line with the movement trajectory.

Supposing the nonlinear state equation and the observation equation expression are:

$$Y(m + 1) = f[y(m), u(m), w(m)] \quad (7)$$

$$G(m) = h[g(m), u(m), v(m)] \quad (8)$$

In Eqs. (7) and (8), $f[\cdot]$ represents the transition function of the nonlinear state, $w(m)$ represents the noise component, $h[\cdot]$ represents a nonlinear observation function, $v(m)$ represents an observed noise component. Both variable $w(m)$ and variable $v(m)$ are Gaussian white noise, and the variances are respectively Q and R. $Y(m + 1)$ is the state vector, $g(m)$ is the observation vector and $u(m)$ is the control vector. According to the Gaussian distribution, the recursion relationship of the state variable $Y_{m,m}$ and $Y_{m,m-1}$ at time m is:

$$\hat{Y}_{m,m} = \hat{Y}_{m,m-1} + A_m \cdot (G_m - \hat{G}_m) \quad (9)$$

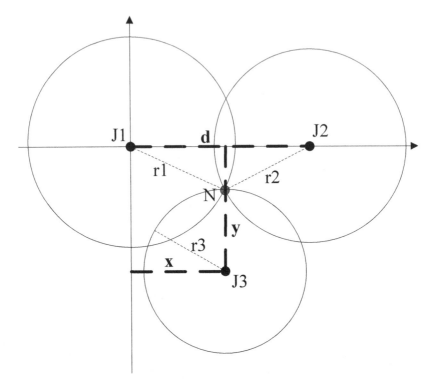

Fig. 2. Three base stations positioning principle.

In Eq. (9), A_m is the denoising gain, $Y_{m,m-1}$ is the previous state value and G_m is the optimal prediction value. System updates and forecasts can be expressed as:

$$
\begin{aligned}
A_m &= P_{y_m g_m} \cdot P_{g_y}^{-1} \\
P_m &= P_{m,m-1} - A_m \cdot P_{g_m} A_m^T \\
\hat{Y} &= \hat{Y}_{m,m-1} + A_m \cdot (G_m - \hat{G}_m)
\end{aligned}
\tag{10}
$$

The Jacobian linearization matrix can be expressed as:

$$
\begin{aligned}
\phi_{m,m-1} &= \frac{\partial f[y_{m-1}, w_{m-1}]}{\partial y_{m-1}} |y_{m-1} = \hat{y}_{m-1} \\
H_m &= \frac{\partial h[y_m, v_m]}{\partial y_m} |y_m = f(\hat{y}_{m-1}, w_{m-1})
\end{aligned}
\tag{11}
$$

The Eq. (10) and (11) are substituted into Eq. (7):

$$
\begin{aligned}
\hat{y}_m &= f(\hat{y}_{m,m-1}, w_{m-1}) + A_m \cdot \{g_m - h_m[f(\hat{y}_{m-1}, w_{m-1}, v_m)]\} \\
A_m &= P_{m,m-1} \cdot H_m^T \cdot (H_m \cdot P_{m,m-1} \cdot H_m^T \cdot R_m)^{-1} \\
P_{m,m-1} &= \phi_{m,m-1} \cdot P_{m-1} \cdot \phi_{m,m-1}^T + Q_{m-1} \\
P_m &= (I - A_m \cdot H_m) \cdot P_{m,m-1} \\
\hat{y}_0 &= E[y_0], \; P_0 = E[(y_0 - \hat{y}_0)(y_0 - \hat{y}_0)^T]
\end{aligned}
\tag{12}
$$

Equation (12) belongs to the state recursion equation, the initial value determines the accuracy of the prediction result, and the high precision can be maintained only when the linear equation is approximated. Therefore, the state vector is first sampled to fit the linear equation, and then filtering is carried out to make the prediction result closer to reality.

Let the state equation of a wireless sensor node be:

$$Y_{m+1} = f_m(Y_m) + W_m \qquad (13)$$

The measurement equation is:

$$G_m = h_m(Y_m) + V_m \qquad (14)$$

In the Eqs. (13) and (14), m represents the time m and g_m represents a measurement vector corresponding to the time m. y_m indicates the state vector at time m, h_m is the measured value, f_m represents the state transition function. w_m and v_m are zero mean white noise and they are not correlated with each other.

The sigma point sample is carried out for the state vector y. The y statistical property is $(\bar{y}, P_y)$. There are $2n + 1$ sampling points, which are respectively $\xi_i (i = 0, 1, 2, \cdots, 2n)$. The expressions of ξ and the weight ω are:

$$\begin{cases} \xi_0 = \bar{Y} \\ \xi_i = \bar{Y} + (\sqrt{(n+\lambda)P_y})_i, & (1 \le i \le n) \\ \xi_i = \bar{Y} - (\sqrt{(n+\lambda)P_y})_i, & (n < i \le 2n) \end{cases} \qquad (15)$$

$$\begin{cases} \omega_0^{(jz)} = \frac{\lambda}{\lambda+n} \\ \omega_i^{(jz)} = \frac{0.5}{\lambda+n}, & (i = 1, 2, \ldots, 2n) \end{cases} \qquad (16)$$

$$\begin{cases} \omega_0^{(fc)} = \frac{\lambda}{\lambda+n} + (1 - \alpha^2 + \beta) \\ \omega_i^{(fc)} = \frac{0.5}{\lambda+n}, & (i = 1, 2, \cdots, 2n) \end{cases} \qquad (17)$$

In Eqs. (15)–(17), λ is the scaling factor of the sampling point and mean spacing. Supposing $\lambda = \alpha^2(n + l) - n$, l generally takes 0. α usually takes 0.1, indicating the degree of dispersion of the sampling points; $(\sqrt{(n+\lambda)P_y})_i$ corresponds to the ith column square root of the matrix; β takes 2 (consistent with Gaussian distribution) and describe the distribution information of y; $\omega^{(jz)}$ represents the mean weight value and $\omega^{(fc)}$ represents the variance weight value.

The prediction function of the node state is:

$$\begin{cases} \xi_{m|m-1}^i = f_m(\xi_{m-1|m-1}^{(i)}), & i = 0, 1, 2, \cdots, 2n \\ \hat{y}_{m|m-1} = \sum_{i=0}^{2n} \omega_i^{(jz)} \xi_{m|m-1}^{(i)} \\ P_{m|m-1}^{(i)} = \sum_{i=0}^{2n} \omega_i^{(fc)} (\xi_{m|m-1}^{(i)} - \hat{y}_{m|m-1})(\xi_{m|m-1}^{(i)} - \hat{y}_{m|m-1})^T + Q_{m-1} \end{cases} \qquad (18)$$

The prediction function of node measurement is:

$$
\begin{cases}
\zeta^{(i)}_{m|m-1} = h_m(\xi^{(i)}_m), i = 0, 1, 2, \cdots, 2n \\
\hat{g}_{m|m-1} = \displaystyle\sum_{i=0}^{2n} \omega^{(jz)}_i \zeta^{(i)}_{m|m-1} \\
P_{\hat{y}_m} = \displaystyle\sum_{i=0}^{2n} \omega^{(fc)}_i (\zeta^{(i)}_{m|m-1} - \hat{y}_{m|m-1})(\zeta^{(i)}_{m|m-1} - \hat{y}_{m|m-1})^T + R_m \\
P_{\hat{y}_m \hat{g}_m} = \displaystyle\sum_{i=0}^{2n} \omega^{(fc)}_i (\zeta^{(i)}_{m|m-1} - \hat{y}_{m|m-1})(\zeta^{(i)}_{m|m-1} - \hat{y}_{m|m-1})^T
\end{cases}
\tag{19}
$$

The update of the covariance matrix and the Kalman coefficient K of the node state prediction can be expressed as:

$$
\begin{cases}
\hat{y}_{m|m} = \hat{y}_{m|m-1} + K_m(g_m - \hat{g}_{m|m-1}) \\
K_m = P_{\bar{y}_m \bar{g}_m} P^{-1}_{\bar{g}_m} \\
P_{m|m} = P_{m|m-1} - K_m P_{\bar{g}_m} K^T_m
\end{cases}
\tag{20}
$$

According to Eqs. (15)–(20), the nodes' position estimates of every moment can be obtained.

Supposing the node coordinate is $y = (x, y)^T$ and state vector is $y_m = (x_m, y_m)^T$, then the state equation is:

$$
y_{m+1} = Dy_m + w_m
\tag{21}
$$

In Eq. (21), w_m represents noise and D represents a state transition matrix (unit matrix).

The quantity measurement model varies with the observed value, base station node is supposed to be $n(n = 1, 2, 3)$. L_m is a three-dimensional distance vector, and v_m is three-dimensional noise. When the observed value is the distance between the base station and a node, the measurement function is:

$$
\begin{cases}
g_m = L_m = h_m(y_m) + v_m \\
L_m = \begin{pmatrix} L_{m1} \\ L_{m2} \\ L_{m3} \end{pmatrix} = \begin{pmatrix} \sqrt{(x_m - x_{n1})^2 + (y_m - y_{n1})^2} \\ \sqrt{(x_m - x_{n2})^2 + (y_m - y_{n2})^2} \\ \sqrt{(x_m - x_{n3})^2 + (y_m - y_{n3})^2} \end{pmatrix}
\end{cases}
\tag{22}
$$

In Eq. (22), $L = (L_1 L_2 L_3)^T$ is the distance from a node to the base station, $L_m = (L_{m1} L_{m2} L_{m3})^T$ is the distance vector at the mth calculation. $x_{n1} = (x_{n1}, y_{n1})^T$, $x_{n2} = (x_{n2}, y_{n2})^T$ and $x_{n3} = (x_{n3}, y_{n3})^T$ are respectively the coordinate vector of the base station $n(n = 1, 2, 3)$.

4 Construction of Hardware Experiment Platform

The wireless ranging and positioning system is a wireless network composed of a plurality of base station devices and a plurality of node devices. The structure of each base station or node is shown in Fig. 3, including the control unit with STM32F103C8T6 as the core, DWM1000 communication unit, alarm unit, and power supply unit. The power supply unit of the base station is composed of USB interface and voltage regulator chip, and the power supply unit of the node is composed of USB interface, voltage conversion module, battery and charging management module.

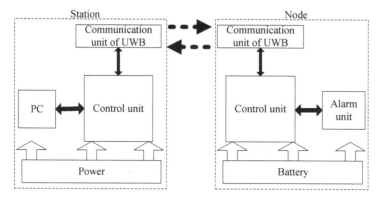

Fig. 3. Structure diagram base station (node).

Each base station and node has a unique code. Each base station and the node can communicate wirelessly. The base station detects and calculates the distance of each node, and the alarm can be triggered when the detected distance value exceeds the safe range. The measurement results of multiple base stations can be referred to locate the node. The control unit mainly completes the work such as timing provision, distance calculation and control instruction release for the UWB unit; with UWM data transceiving function, the communication unit can accurately measure time stamps [36].

5 Test Experiment

Ranging experiment and positioning experiment are carried out respectively according to the anti-collision alarm function, detection accuracy, speed demand, simulation system application environment. First, a base station and four nodes are used to perform basic system function tests, that is distance detection between the base station and each node. The notebook is connected to the base station by using the serial port, and the operation results of the base station at two times are randomly selected as shown in Fig. 4(a) and Fig. 4(b). The selected node numbers are node [1], node [2], node [3], and node [4], respectively. Therefore, dist[1], dist[2], dist[3], and dist[4] in Fig. 4(a) and Fig. 4(b) respectively represent the distance values from the base station to the four nodes, which are in centimeters. So the system is able to provide valid data for test experiments. Before each test, the distance from the base station to the node must be calibrated.

5.1 Static Ranging Experiment

A square is selected as a test site to simulate a working environment such as a dock and a distribution center, and place a base station in the car body. Considering the actual

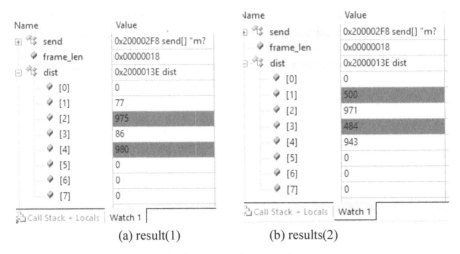

(a) result(1)   (b) results(2)

Fig. 4. Base station operation.

application, 12 m, 10 m and 8 m are respectively set as alarm thresholds, and correspond to different alarm sounds. One node is placed in turn at 12 m, 10 m, and 8 m from the car body, and stayed at each position for enough time to obtain enough experimental data. The photograph of ranging experiment is shown in Fig. 5(a), and the photograph of collision prevention device is shown in Fig. 5(b). The base station calculation results are shown in Fig. 6(a), Fig. 6(b), and Fig. 6(c). The horizontal axis represents the number of tests and the vertical axis represents the distance value. It can be seen that the maximum ranging errors of 12 m, 10 m, and 8 m are 0.089 m, 0.093 m, and 0.085 m respectively, and the average errors are 0.0374 m, 0.0380 m and 0.0349 m respectively.

(a) the ranging experiment   (b) the collision prevention device

Fig. 5. The photos of physical test.

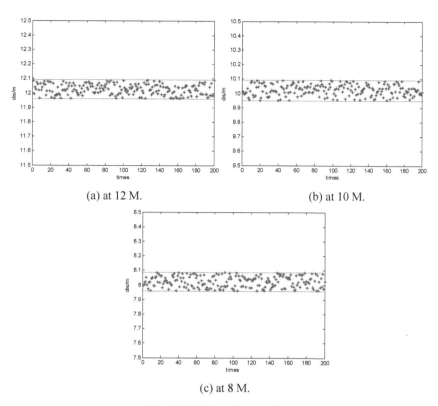

(a) at 12 M.

(b) at 10 M.

(c) at 8 M.

Fig. 6. Static ranging results.

5.2 Dynamic Ranging Experiment

In order to simulate the actual working conditions, multiple vehicles are arranged around the vehicle and nodes at the base station as a signal barrier. Generally, engineering vehicles are not too fast. So the base station vehicle approaches the node at a speed of 25 km/h. The operation results of the base station are shown in Fig. 7(a), Fig. 7(b), and Fig. 7(c). The maximum ranging errors which are visible at 12 m, 10 m, and 8 m is 0.148 m\0.15 m\0.139 m, respectively. The average errors are 0.073 m, 0.075 m, and 0.070 m, respectively.

5.3 Positioning Experiment

The three base stations are respectively placed at fixed positions (at the same height, and within the effective communication distance) to form a triangular structure. The nodes are randomly moved within the effective range, and the observation positioning result and the actual position are as shown in Table 1. It can be seen that the maximum positioning errors in the three coordinate axes of X, Y and Z are 0.17 m, 0.29 m, and 0.45 m, respectively, and the average errors are 0.11 m, 0.10 m, and 0.19 m respectively.

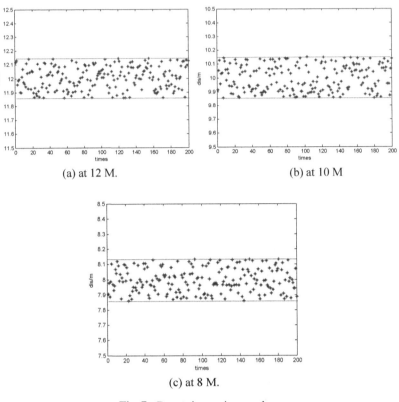

(a) at 12 M.   (b) at 10 M

(c) at 8 M.

Fig. 7. Dynamic ranging results.

Table 1. Locating experimental results.

	Measuring coordinates/m			Actual coordinates/m			Error/m		
	X	Y	Z	X	Y	Z	ΔX	ΔY	ΔZ
1	5.06	4.97	0.98	4.99	5.01	0.85	0.07	0.04	0.13
2	5.63	9.73	0.96	5.58	9.66	0.91	0.05	0.07	0.05
3	10.8	7.54	1.06	10.7	7.52	1.24	0.12	0.02	0.18
4	8.28	7.90	0.66	8.39	8.02	0.21	0.11	0.12	0.45
5	7.35	0.83	0.47	7.18	1.12	0.78	0.17	0.29	0.31
6	3.63	1.46	0.07	3.68	1.55	0.11	0.05	0.09	0.04
7	1.98	10.2	1.86	2.13	10.2	2.09	0.15	0.01	0.23
8	0.84	3.35	0.78	0.69	3.51	0.92	0.15	0.16	0.14

6   Conclusion

The implementing method of vehicle anti-collision device is introduced in this paper. Taking DWM1000 as the core, the system achieves reliable ranging and positioning by combining DSTWR with SFKF algorithms. The test results show that the vehicle and personnel with the detection device perform different distance detection in the static state, with the maximum absolute error of 0.093 m, the average error of 0.037 m, and the maximum relative error of 0.0106; distance detection is carried out in the moving state, with the maximum absolute error of 0.15 m, the average error of 0.073 m, and the maximum relative error of 0.0174; in the positioning experiment, the maximum errors in the three coordinate axes are 0.17 m, 0.29 m, and 0.45 m, respectively, and the average position error of the node is about 0.241 m. The experimental results prove that the method in this paper can effectively eliminate large errors, stabilize the errors in small range, keep the detection track continuous, meet the actual functional requirements, and has application value. It should be noted that in practical applications, the detection algorithm and communication mechanism should be further optimized for the number of nodes.

Acknowledgment. This work was supported by Harbin Science and Technology Innovation Talents Special Project (NO. 2017RAQXJ031); Heilongjiang Fundamental Research Foundation for the Local Universities in 2018 (2018KYYWF1189); Key project Task of Public Safety Risk Control and Emergency Technical Equipment of National Key R&D Program (NO. 2017YFC0805208); 2017 National Nature Fund, (NO. 51674109).

References

1. Angelis, G.D., Moschitta, A., Carbone, P.: Positioning techniques in indoor environments based on stochastic modeling of UWB round-trip-time measurements. IEEE Trans. Intell. Transp. Syst. **17**(8), 2272–2281 (2016)
2. Abdulrahman, A., et al.: Ultra wideband indoor positioning technologies: analysis and recent advances. Sensors **16**(5), 1–36 (2016)
3. Mikhaylov, K., Petäjäjärvi, J., Hämäläinen, M., Tikanmäki, A., Kohno, R.: Impact of IEEE 802.15.4 communication settings on performance in asynchronous two way UWB ranging. Int. J. Wirel. Inf. Netw. **24**(2), 124–139 (2017). https://doi.org/10.1007/s10776-017-0340-9
4. Luo, Q., Han, B.: Distance and Azimuth testing system based on ultrasonic and infrared detecting technology. Comput. Autom. Meas. Control. **13**(4), 304–306+334 (2005)
5. Bartoletti, S., Giorgetti, A., Win, M.Z., Conti, A.: Blind selection of representative observations for sensor radar networks. IEEE Trans. Veh. Technol. **64**(4), 1388–1400 (2015)
6. Wang, D.Y., Zhao, C.F., Peng, J.Y., Li, M.: Error factors impact analysis of microwave range finder ranging accuracy. Electron. Meas. Technol. **4**, 42–44 (2010)
7. Bartoletti, S., Dai, W., Conti, A., Win, M.Z.: A mathematical model for wideband ranging. IEEE J. Sel. Top. Sig. Process. **9**(2), 216–228 (2015)
8. Yu, H.L., Guo, A.H., Luo, W.: Method of ranging based on GPS positioning data. Electron. Meas. Technol. **7**, 95–98 (2011)
9. Sangodoyin, S., Niranjan, S., Molisch, A.F.: A measurement-based model for outdoor near-ground ultrawideband channels. IEEE Trans. Antennas Propag. **64**(2), 740–751 (2016)

10. Wang, L.X., Huang, Z.G., Zhao, Y.: Term design and analysis of user range accuracy in GPS navigation message. J. Nanjing Univ. Sci. Technol. **38**(5), 620–625 (2014)
11. Zhao, X.M., Hui, F., Shi, X., Ma, J.Y., Yang, L.: Concept, architecture and challenging technologies of ubiquitous traffic information service system. J. Traffic Transp. Eng. **14**(4), 105–115 (2014)
12. Movassaghi, S., Abolhasan, M., Lipman, J., Smith, D., Jamalipour, A.: Wireless body area networks: a survey. IEEE Commun. Surv. Tutor. **16**(3), 1658–1686 (2014)
13. Xu, H.T., Chen, B., Lv, Z.G.: Modulation technology and application prospects of UWB. Inf. Technol. **4**, 175–178 (2011)
14. Wen, F., Wan, Q.: Time delay estimation based on entropy estimation. J. Electron. Sci. Technol. **11**(3), 258–263 (2013)
15. Giorgetti, A., Chiani, M.: Time-of-arrival estimation based on information theoretic criteria. IEEE Trans. Sig. Process. **61**(5–8), 1869–1879 (2013)
16. Wu, S., Li, J., Liu, S.: Single threshold optimization and a novel double threshold scheme for non-line-of-sight identification. Int. J. Commun. Syst. **27**(10), 2156–2165 (2014)
17. Fu, S.C., Li, Y.M., Zong, K., Zhang, M.J., Wu, M.: Accuracy analysis of UWB pose detection system for road-header. Chin. J. Sci. Instrument. **38**(8), 1978–1987 (2017)
18. Han, Z., He, J., Geng, Y., Xu, C., Xu, L., Duan, S.: CRLB for TOA based near-ground swarm robotic localization. In: Ubiquitous Intelligence and Computing and 2015 IEEE 12th Intl Conf on Autonomic and Trusted Computing and 2015 IEEE 15th Intl Conf on Scalable Computing and Communications and Its Associated Workshops (UIC-ATC-ScalCom), pp. 1853–1858 (2015)
19. Liu, T., Xu, A.G., Sui, X.: Adaptive robust Kalman Filtering for UWB indoor positioning. Chin. J. Sens. Actuators **31**(4), 678–686 (2018)
20. Peng, S., Chen, C., Shi, H., et al.: State of charge estimation of battery energy storage systems based on adaptive unscented Kalman Filter with a noise statistics estimator. IEEE Access **5**, 13202–13212 (2017)
21. Mao, K.J., Wu, J.B., Jin, H.B., Miao, C.Y., Xia, M., Chen, Q.Z.: Indoor localization algorithm for NLOS environment. Acta Electronica Sinica **44**(5), 1174–1179 (2016)
22. Liu, T., Xu, A.G., Sui, X.: Application of UWB/INS combination in indoor navigation and positioning. Sci. Surv. Mapp. **41**(12), 162–166 (2016)
23. Gao, W., Chen, G.: Integrated GNSS/INS navigation algorithms combining adaptive filter with neural network. Geomat. Inf. Sci. Wuhan Univ. **39**(11), 1323–1328 (2014)
24. Taran, S., Nitnaware, D.: Performance evaluation of 802.15.3a UWB channel model with antipodal, orthogonal and DPSK modulation scheme. Int. J. Microw. Wirel. Technol. **6**(1), 34–42 (2015)
25. Miao, Y.W., Zhou, W., Tian, L., et al.: Extended robust Calman filtering detection and its application based on innovation x ~ 2. Geomat. Inf. Sci. Wuhan Univ. **41**(2), 269–273 (2016)
26. Kok, M., Hol, J.D., Schön, T.B.: Indoor positioning using ultrawideband and inertial measurements. IEEE Trans. Veh. Technol. **64**(4), 1293–1303 (2015)
27. Zhang, S., Pedersen, G.F.: Mutual coupling reduction for UWB MIMO antennas with a wideband neutralization line. IEEE Antennas Wirel. Propag. Lett. **15**, 166–169 (2016)
28. Ding, H., Liu, W., Huang, X., Zheng, L.: First path detection using rank test in IR UWB ranging with energy detection receiver under harsh environments. IEEE Commun. Lett. **17**(4), 761–764 (2013)
29. Yu, K., Dutkiewicz, E.: NLOS identification and mitigation for mobile tracking. IEEE Trans. Aerosp. Electron. Syst. **49**(3), 1438–1452 (2013)
30. Angelis, A.D., Dwivedi, S., Händel, P.: Characterization of a flexible UWB sensor for indoor localization. IEEE Trans. Instrum. Meas. **62**(5), 905–913 (2013)

31. Okamoto, E., Horiba, M., Nakashima, K., Shinohara, T., Matsumura, K.: Particle swarm optimization-based low-complexity three-dimensional UWB localization scheme. In: Sixth International Conference on Ubiquitous and Future Networks. pp. 120–124 (2014)
32. Ershadh, M.: Study of the design evolution of an antenna and its performance for UWB communications. Microw. Opt. Technol. Lett. **57**(1), 81–84 (2015)
33. Zhang, Z., Zhu, J., Ruan, J., Song, G.: Distance measurement for the indoor WSN nodes using WTR method. Int. J. Distrib. Sens. Netw. **1**, 1–13 (2014)
34. Ojaroudi, N.: Compact UWB monopole antenna with enhanced bandwidth using rotated L-shaped slots and parasitic structures. Microw. Opt. Technol. Lett. **56**(1), 175–178 (2014)
35. Demir, U., Bas, C.U., Ergen, S.C.: Engine compartment UWB channel model for intravehicular wireless sensor networks. IEEE Trans. Veh. Technol. **63**(6), 2497–2505 (2014)
36. Richardson, P., Xiang, W., Shan, D.: UWB outdoor channel environments: analysis of experimental data collection and comparison to IEEE 802.15.4a UWB channel model. Int. J. Ultra Wideband Commun. Syst. **3**(1), 1–7 (2014)

Nonlinear Resistance Circuit Curve Intersection Method Algorithm Research

Wang Haiyue[1,2(✉)] and Su Xunwen[1]

[1] Heilongjiang University of Science and Technology, Harbin 150022, China
2509614001@qq.com
[2] Shang Du Power Plant, Inner Mongolia, China

Abstract. The volt-ampere relation (VCR) of nonlinear resistance elements is a nonlinear function relation which does not satisfy ohmic law. In this paper, the curve intersection method of nonlinear resistance circuits is studied and analyzed. Firstly, the existence and uniqueness of the solution of nonlinear resistance circuit are discussed, and then the feasibility of the algorithm is verified by computer simulation. Then the actual operation is verified by Multisim software. The experiment is an algorithm used when the volt-ampere characteristics of nonlinear resistance are given by the actual test data. This is an algorithm for finding the intersection point of arbitrary straight line and polynomial curve function. Compared with the actual measured results, the results obtained by the algorithm are almost resulting in. It can be said that the algorithm is an efficient and simple method to solve the problem.

Keywords: Non-linear · Curve intersection method · Cubic polynomial · Simulation simulation

1 Introduction

Nonlinear circuit theory has always been one of the important research directions of circuit theory. Because all real circuits have nonlinear circuit properties from a realistic perspective. Therefore, strictly speaking, all circuits belong to nonlinear circuits in reality.

Linear circuit is a special form in nonlinear circuit. If the difficulties facing the nonlinear circuit are overcome, the difficulty that the linear circuit needs to solve is naturally solved.

In this paper, the nonlinear resistance circuit is solved by the method of curve intersection in MATLAB calculation software. The measured data points are fitted with the three-time polynomial fitting through the MATLAB software, the obtained function is used instead of the nonlinear resistance characteristic curve equation, and the intersection point of the load line and the fitting curve is obtained. The intersection of two volt-ampere characteristic curves is the static working point of the circuit.

A set of experimental data are generated by using Multisim software to simulate and test, and then the coordinates of intersection points are obtained by using interpolation and fitting algorithm in MATLAB software.

X. Jiang and P. Li (Eds.): GreeNets 2020, LNICST 333, pp. 108–116, 2020.
https://doi.org/10.1007/978-3-030-62483-5_12

2   Solution Method of Nonlinear Resistance Circuit

In linear circuits, the characteristic of linear components is that their parameters do not change with voltage or current. If the parameter of the circuit element varies with voltage or current, it is referred to as a non-linear element. A circuit comprising a non-linear element is referred to as a non-linear circuit.

All of the actual circuits in the reality are non-linear circuits. In general, those elements whose non-linear degree changes are not very obvious are treated as linear elements, and the accuracy and accuracy of various parameters calculation can be ensured while the circuit analysis is simplified. However, for the circuit whose nonlinear characteristics can not be ignored, it must be treated as a nonlinear circuit. Avoid the difference between the calculated value and the actual value without significance.

The method of analyzing the non-linear circuit has curvilinear intersection algorithm, piecewise linear method, small signal analysis. These methods can solve the nonlinear circuit with simple circuit diagram, but it is difficult to find the approximate solution satisfying the accuracy for the complex nonlinear circuit.

With the rapid development of science and technology. The computer algorithm replaces the traditional manual calculation. It not only can improve the computational efficiency, but also get the exact value.

In the following article, the non-linear resistance circuit is simply introduced, and several commonly used methods for solving the non-linear resistance circuit are mentioned.

2.1   Curve Intersection Method

The advantage of curve intersection is that it can depict the content by chart.

A simple nonlinear Resistance circuit by voltage source U_0 and linear resistance R_0 nonlinear resistance R composition. Figure 1(a) is Nonlinear Resistance Circuit.

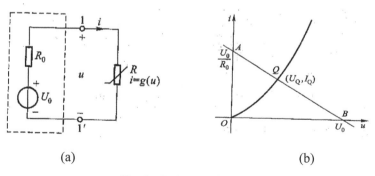

(a)   (b)

Fig. 1. Static operation point

The electrical network is consisted of the voltage source U_0 and the linear resistor R_0. The nonlinear resistance element R is an external circuit. The intersection of the volt-ampere characteristic curve of the two-part circuit is the static working point $Q(U_Q, I_Q)$. Figure 1(b) is the resulting graph.

2.2 Small Signal Analysis Method

A nonlinear circuit that is usually encountered in an electronic circuit It not only has a DC power supply, but also has an input voltage that varies over time. Suppose in any case, there's an $|u_s(t)| \ll U_0 \cdot u_s(t)$ is called a small signal voltage. Small signal analysis can be used to analyze this kind of circuit.

The method of the small signal analysis method comprises the following steps of:

1. Solving the static working point Q of the non-linear circuit.
2. Solving the dynamic resistance of nonlinear circuits.
3. Make a small signal equivalent circuit for a given nonlinear resistance at a static operating point. Figure 2 is the equivalent circuit

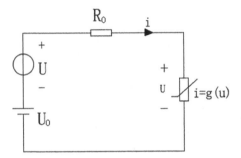

Fig. 2. Small signal equivalent circuit

4. The Solution of the Equivalent Circuit Based on the Small Signal.

3 Existence of Solutions for Nonlinear Resistance Circuits

3.1 Solution of Nonlinear Resistance Circuit

Unique solution: In the voltage-controlled resistance circuit, if the volt-ampere characteristic curve is monotonous, and this set of voltage and current values satisfies the KCL, KVL at the same time. Figure 3(a) is a curve with only one solution.

Multiple solutions: The electrical network is consisted of the voltage source and the linear resistor. The nonlinear resistance element is an external circuit. In practice, the volt-ampere characteristic curves of some nonlinear resistance are not monotone. The load line and the non-linear resistance volt-ampere characteristic curve have a plurality of intersection points. The volt-ampere characteristics of nonlinear resistance curves with multiple solutions is shown in Fig. 3(b).

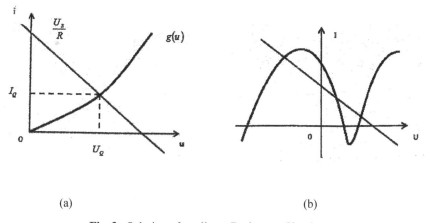

(a)   (b)

Fig. 3. Solution of nonlinear Resistance Circuit

4   MATLAB Programming

4.1   Introduction of MATLAB Drawing Function

In MATLAB, the most basic and widely used drawing function is plot, which can draw different curves on two-dimensional plane, while the plot drawing statement used in this design is used in this design.

Plot function is used to draw the graph of linear coordinate on two - dimensional plane.

For example, when you draw a function image of $y = \cos(x)$, we can use the plot statement. Figure 4 is the function graph.

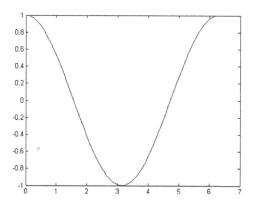

Fig. 4. $y = \cos(x)$ function graph

Enter the command in the command window:

$$\gg x = 0 : pi/100 : 2 * pi;$$
$$\gg y = \cos(x);$$
$$\gg plot(x, y); \ \text{drawing} \ y = \cos(x) \ \text{images}$$

4.2 Cubic Spline Interpolation

With the rapid development of modern scientific and technological means. People often encounter that function expressions are too complex to be worked out in a short period of time. Therefore, it is necessary to study a new interpolation method-Spline interpolation method.

The term spline comes from life, and a set of data actually measured is expressed in plane coordinates. All the points in this set of data are connected smoothly to form a smooth curve that connects all the experimental data points. Such a curve is called a spline. It is actually connected by the cubic polynomial curve, and the so-called spline interpolation is obtained mathematically.

5 Multisim Simulation

5.1 Simulation Circuit Design and Measured Data

The simulation is simulated according to the schematic Fig. 5. Select a linear sliding rheostat with an adjustment range of 0–100 Ω. The regulated range of the DC voltage source is 1–10 V. Select model 1N4149 Diode as nonlinear resistance. Adjust the size of the DC power supply and record the voltage u_i and current i_i of the nonlinear resistance. Table 1 is the actual measured data.

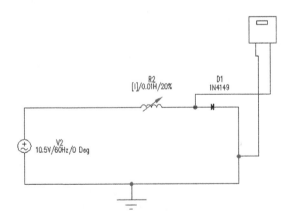

Fig. 5. Wiring diagram of the simulation circuit

Table 1. Measured data of simulation circuit

$i_i(A)$	1.452	1.555	1.773	1.804	2.110	2.318	2.553	4.295
$u_i(V)$	0.108	0.124	0.159	0.164	0.218	0.257	0.304	0.773

5.2 Volt-Ampere Characteristic Curve of Nonlinear Resistance

Enter directly in the command window.

The command is executed as shown in Fig. 6.

>>x=[1.452 1.555 1.773 1.804 2.110 2.318 2.553 4.295];

>>y=[0.108 0.124 0.159 0.164 0.218 0.257 0.304 0.773];

>>xi=1.4:0.01:4.3;

    yi=spline(x,y,xi);

    plot(x,y,'o',xi,yi);

    $p = polyfit(xi, yi, 3);$

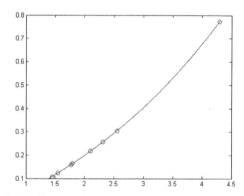

Fig. 6. Nonlinear resistance volt-ampere characteristic curve

5.3 The Principle of the Algorithm for Finding the Intersection Point

In this paper, the one-dimensional interpolation method is used. The plane volt-ampere characteristic curve of nonlinear resistance $g(u)$ is drawn by one-dimensional interpolation method and the analytical expression is obtained. In this method, the test data points are used as interpolation nodes and described by cubic splines interpolation function.

Use $s(u)$ if the allowable error is close to the wireless small Gradually approach $g(u)$. According to the derivative in the course of higher mathematics, $s(u)$ must be a convergence function and must have a second-order smoothness condition.

The principle of $l(u)$ and $s(u)$ intersection is shown in Fig. 7. The coordinate axis of independent variables is divided into infinitely many small intervals, and it is necessary to ensure that the interval length of each new interval is less than the length of the original interval. Until the coordinates of the intersection point within the allowable error range can be obtained, which is the solution of the circuit. In essence, the iterative process is gradually approaching to the real value, and the solution within the allowable range of the error is obtained. The specific algorithms are as follows:

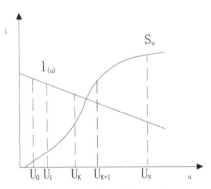

Fig. 7. Intersection of isometric cut

First average segmentation $\Delta_1 : a = u_0 < u_1 < \ldots u_j < \ldots u_N = b$, $j = 0, 1, 2, \ldots N$, $u_j = u_0 + j[(b - a)/N]$, Interval length $\delta_1 = (b - a)/N$. Subtract each dividing point u_j, just as $f(u_j) = l(u_j) - s(u_j)$. If there is a unique intersection Q within the area [a, b], the two sides $f(u_j)$ and $f(u_k)$ of the Q point are opposite to each other.

Then the subinterval of the first isometric segmentation $\delta_1 = (b - a)/N$ is replaced by the total length of the original interval for the second isometric segmentation.

It is determined that the coordinates of the intersection point Q are in a narrower range of area, the width of which is $1/N$ of the length $[a_1, b_1]$ of the small section of the last isometric division. This cycle iterates until an interval length is obtained and within the allowable error range. The iterative process is ended and the intermediate value of the area is taken as the intersection coordinate Q.

5.4 Calculation of Static Working Point

The load line and the non-linear characteristic curve are both represented in the software, and the intersection points of the two curves are calculated by using the assembler language. This is the very important static working point we need to ask for. The programming content is as follows:

```
>> x = [1.452 1.555 1.773 1.804 2.110 2.318 2.553 4.295];
>> y = [0.108 0.124 0.159 0.164 0.218 0.257 0.304 0.773];
>> xi = 1.4 : 0.01 : 4.3;
```

$$yi = spline(x, y, xi);$$

$$plot(x, y, o, xi, yi);$$

```
>>hold on;
>>x2=0:4.5;
>>y2=-0.05*x2+0.435;
>>plot(x2,y2);
>>[x,y]=solve('0.003*x^3+0.0069*x^2+0.1132*x-0.0801','y2=-0.005*x2+0
```

.435');

```
x=

    2.671

y=

    0.319
```

Figure 8 is intersection curve. By using the algorithm, the coordinates of the intersection point are obtained to be (2.671, 0.319). That is the static operating point of the
diode.

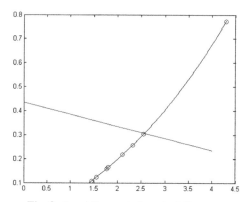

Fig. 8. Load line and characteristic curve

6 Conclusion

The static working point of the circuit is obtained by the curve intersection method,
which has higher accuracy than the traditional graphic method. This practical solution
can be applied to any monotonous nonlinear resistance circuit.

It is found that the curve intersection method is the best method with fast calculation,
low complexity and high accuracy.

Acknowledgment. This work was financially supported by the National Science Foundation of China under Grant (51677057), Local University Support plan for R & D, Cultivation and Transformation of Scientific and Technological Achievements by Heilongjiang Educational Commission (TSTAU-R2018005) and Key Laboratory of Modern Power System Simulation and Control & Renewable Energy Technology, Ministry of Education (MPSS2019-05).

References

1. Martins: Validation of fixed speed wind turbine dynamic models with measured data. IEEE Trans. Power Electron. **28**(11), 5100–5110 (2012)
2. Bottrell, N., Prodanovic, M., Green, T.C.: Dynamic stability of a microgrid with an active load. IEEE Trans. Power Electron. **28**(11), 5107–5119 (2013)
3. Piyasinghe, L., Miao, Z., Khazaei, J., Fan, L.: Impedance model-based SSR analysis for TCSC compensated type-3 wind energy delivery systems. IEEE Trans. **6**(1), 179–187 (2015)
4. Huang, L., Li, J., Liu, X., et al.: Realization of complex torque coefficient method in time domain based on PSCAD/EMTDC and research on the Influence on Electrical Damping. Sensor World **24**(2), 13–18 (2018)
5. Li, G., Zhang, S., Zhang, Z.: Analysis of the possibility of subsynchronous resonance under the high series compensation level of the 330 Kv main network in Northwest China by frequency scanning method. Proc. CSEE **12**(1), 58–61 (1992)
6. Yang, X.: Study on Network Topology Analysis and Operation Modes Combination in Power System, vol. 27. Huazhong University of Science and Technology, Wuhan (2007)
7. Wen, B., Boroyevich, D., Burgos, R., et al.: Small-signal stability analysis of three-phase AC system in the presence of constant power loads based on measured d-q frame impedances. IEEE Trans. Power Electron. **30**(10), 5952–5963 (2015)
8. Wang, K., Xu, Z., Du, N., et al.: Subsynchronous resonance by thyristor controlled series compensation. High Volt. Eng. **42**(1), 321–329 (2016)
9. García-Gracia, M., Comech, M.P., Sallán, J.: Modelling wind farms for grid disturbance studies. IEEE Trans. Power Electron. **30**(10), 4952–4963 (2013)
10. Blaabjerg, F., Teodorescu, R., Liserre, M., Timbus, A.V.: Overview of control and grid synchronization for distributed power generation systems. IEEE Trans. Ind. Electron. **53**(5), 1398–1409 (2006)

New Analytical Formulas for Coupling Coefficient of Two Inductively Coupled Ring Coils in Inductive Wireless Power Transfer System

Mohammed Al-Saadi[1(⊠)], Stanimir Valtchev[1], José Gonçalves[1],
and Aurelian Crăciunescu[2]

[1] UNINOVA-CTS, Department of Electrical and Computer Engineering, Faculdade de Ciências e Tecnologia, Universidade Nova de Lisboa, Lisbon, Portugal
mohamedshihab91@gmail.com
[2] Facultatea de Inginerie Electrică, Universitatea POLITEHNICA din București (UPB), Bucharest, Romania

Abstract. In this paper, an analytical formula for the coupling coefficient (k) was introduced for two inductively coupled coils of ring configuration. The response surface methodology (RSM) was used as a tool to develop this formula. The k was tested as a function of the geometrical parameters which include the followings parameters: an air-gap (d) between inductively coupled coils; coils dimensions which include the inner (r_1) and outer (r_2) radii of the transmitter coil, inner (R_1) and outer (R_2) radii of the receiver coil; and misalignment parameters. Therefore, the introduced k formula is facilitating of a ring coil design, performance optimization of an IPT system, and prediction of system behaviour at normal or misalignment cases. The percentage effect of each parameter on the k was calculated. It was found that the d has the most considerable impact on the k among other geometrical parameters.

Keywords: Wireless power transfer · Inductive power transfer · Battery charging · Electric vehicles · Ring coil · Coupling coefficient · Box-Behnken design · Response surface methodology

1 Introduction

The The inductive power transfer (IPT) technology has a pivotal role in the widely spreading of contactless battery charging, such as electric vehicles (EVs), mobile phones, and laptops etc. This technology plays major role in addressing the issues of convenience and safety which are causes by the wired charging systems. The IPT technology based on the electromagnetic induction, where the electric power is transferred from the transmitter coil to the receiver coil wirelessly. These coils could take many geometrical configurations, such as ring, rectangular, square etc. It has been found that the geometrical

X. Jiang and P. Li (Eds.): GreeNets 2020, LNICST 333, pp. 117–127, 2020.
https://doi.org/10.1007/978-3-030-62483-5_13

configuration of the coils and their dimensions have a significant influence on the power transfer level and distance (air-gap) [1].

Recently, much researches have been focused on enhancing the IPT power transfer capability in terms of optimization the geometrical coils parameters of the inductively coupled coils, or using resonance topologies [2, 3]. However, the power transfer capability and efficiency of an IPT system are mainly affected by the magnetic coupling coefficient (k) between the inductively coupled coils; this parameter is geometrical. The optimizing of this parameter results in an optimization of the whole system performance. For example, Hiroya T. et al. have found that the optimum k of the H-shaped core transformer obtained at 1.4 of the winding width to the air-gap ratio [4]. On the other hand, it was observed that as the receiver area increases, the k increases [5]. Also, it has been noted that a misalignment between the inductively coupled coils has a negative effect on the IPT system performance [6].

One way to optimize a ring coil performance is to find a mathematical expression of k. For example, an analytical formula of mutual inductance (M) and self-inductances of the transmitter (L_1) and receiver (L_2) coils in [7] was proposed based on the Neumann's expression; then, k can be calculated due to Eq. (1). However, the ferrite plates effect is not considered; the ferrite plates are usually used behind each coil to enhance the magnetic coupling. In [8], it was mentioned an analytical formula for k. However, this formula does not considered the inner radius of the coils, the ferrite effect, and the misalignment parameters. In this paper, a new analytical formula of k has introduced for ring coil optimization, design, and performance prediction in an IPT system. The response surface methodology (RSM) is used as a tool to develop a formula for the k of two inductively coupled ring coils. Based on this approach, the k can be explored as function of geometrical ring coil parameters which include the coils dimensions, air-gap, and the misalignments parameters. The effect of the ferrite plate behind each coil was considered.

$$k = \frac{M}{\sqrt{L_1 L_2}}. \tag{1}$$

This paper is organized as follows: Sect. 2 presents the IPT system structure, coils configurations and misalignment between the inductively coupled coils. Section 3 describes the proposed the proposed methodology to find the optimization parameters of a ring coil. Section 4 presents the simulation results. Section 5 examines the validation of the introduced formulas. Finally, Sect. 6 shows the conclusion.

2   Fundamentals of the Inductive Per Transfer Technology

2.1   System Structure

The structure of the typical IPT system for contactless batteries charging is illustrated in Fig. 1. It composes of two unattached sides, which are the transmitter and receiver. The transmitter side is include the following components: a rectifier; an inverter is utilized to excite the transmitter coil (L_1) with a high AC voltage of high frequency through a resonance circuit topology. The receiver side is compose of the following components:

a receiver coil (L_2) to pick up the electromagnetic field from the transmitter coil and convert it to electric power with the same frequency of the inverter; a rectifier for purpose of AC/DC conversion to supply the battery load with a DC output voltage (V_L) [8].

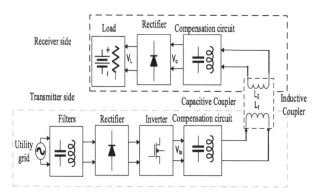

Fig. 1. Typical IPT structure for contactless charging system.

The inductive coupler is described by the k, M, L_1, and L_2 parameters. The k parameter is usually used to predict an IPT system performance, efficiency and power capability that could be transferred [2, 12, 13]. The power delivered to the load (P_L) in a compensated system is expressed by Eq. (2) [1]. The quality factor of the receiver coil is given by Eq. (3) [7].

$$P_L = \omega I_1^2 \frac{M^2}{L_2} Q_2 = V_{in} I_1 k^2 Q_2 \tag{2}$$

$$Q_2 = \begin{cases} \frac{\omega L_2}{R_L} & \text{for series compensation} \\ \frac{R_L}{\omega L_2} & \text{for parallel compensation} \end{cases} \tag{3}$$

2.2 Inductively Coupled Coils

Ithis paper, a ring configuration has been used to represent the inductively coupled coils. A disc of ferrite plate was added behind each coil. The ferrite plates are commonly added to the inductively coupled coils for the purpose of improving magnetic coupling [11]. However, the key aspects of the geometrical parameters of the inductively coupled ring coils are involve: inner radius (r_1) and outer radius (r_2) of the transmitter coil; inner radius (R_1) and outer radius (R_2) of receiver coil, (see Fig. 2a); and air-gap (d) between the coils (see Fig. 2b).

2.3 Misalignments

The misalignment is a deviation of a moveable receiver coil from its normal position with respect to the transmitter coil which is fixed. In this research, two misalignment cases were considered: lateral misalignment (L) and angular misalignment (α) (see Fig. 3a and b).

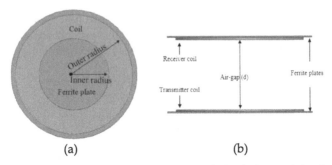

Fig. 2. (a) A ring coil with ferrite plate; (b) inductively coupled coils.

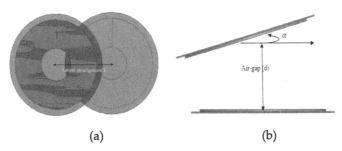

Fig. 3. Misalignment: (a) lateral; (b) angular.

3   Methods

This paper involves analysing two inductively coupled coils of ring configuration to develop an analytical formula for k. This parameter was explored as a function of the geometrical parameters. The geometrical parameters are including the following parameters: dimensions of the inductively coupled coils, air-gap, and misalignment.

The finite-element analysis (FEA) as well as the response surface analysis (RSA) were applied to explore the k model. Firstly, series of FEA simulations were carried out in the ANSYS Maxwell software to analysis the inductively coupled coils at systematic variation of the geometrical parameters based on a design technique that utilized in the RSM. Secondly, the FEA results were inserted into JMP software to conduct the RSA. The RSA outputs include the quadratic k model; this model is function of the geometrical parameters [12].

3.1   Finite Element Analysis (FEA)

This study was conducted with two ring coils of Litz copper wire. A ferrite plate behind each coil was considered with a radius of 4 cm larger than the outer radius of transmitter/receiver coil. The FEA was carried out in the ANSYS Maxwell software to calculate the k.

3.2 Response Surface Analysis (RSA)

The RSM utilizes quantitative data from the pertinent experiment to define the regression model to optimize the output model which is affected by several input parameters. The RSM includes a group of statistical and mathematical techniques that are used to fit the empirical model to the experimental outcomes acquired from the relative to experimental design. A quadratic surface fitting is implied by the RSM, this helps to optimize the output model with a minimum number of experiments, furthermore, analysing the interaction effect between the input parameters [13]. Consequently, several linear and/or square polynomial terms are utilized to describe the behaviour of the studied system, and then explore the conditions of an experimental system until its optimization [14]. In this work, the FEA simulation data were utilized instead of experimental data.

The RSA of a model involves the followings three basics steps; multivariate design of experiment (DOE); design technique which utilized for prediction of the coefficients in the output quadratic model. Finally, analysis of variance (ANOVA) which is used for purpose of the analysis and evaluation of the resulted quadratic models. The RSA was achieved with aid of the utilizing the JMP statistical software package.

3.2.1 Multivariate Design of Experiment (DOE)

The multivariate DOE based on the RSM was performed with the following aspects: k is the output response; the r_1, r_2, R_1, R_2, d, L, and α are the independent input parameters. In this research, these input parameters are varied with the ranges mentioned in Table 1. These ranges are usually used in electric vehicles applications. However, the maximum, average and minimum values are coded in the JMP software as $+$, 0, and $-$, respectively.

Table 1. The ranges of the geometrical parameters of two ring coils.

Studied points	r_1 (cm)	r_2 (cm)	R_1 (cm)	R_2 (cm)	d (cm)	L (cm)	α (deg)
Minimum	2	30	2	30	10	0	0
Maximum	20	60	20	60	50	20	7
Average	11	45	11	45	30	20	3.5

3.2.2 Design Technique

There are many experimental design techniques based on the RSM [14]. We have used Box–Behnken to find the coefficients of the k analytical model. Box-Behnken has an efficient estimation of the coefficients in the mathematical models. This technique is more economical among the others, especially for a large number of input parameters [14]. However, based on the Box-Behnken design, a total of 62 simulated data are required for the geometrical seven input parameters [13]. These data were obtained using FEA.

3.2.3 Analysis of Variance (ANOVA)

The graphical analyses are used to define the interaction between the input variables and the output responses to estimate the statistical parameters influence. For this purpose, ANOVA was implemented by using the JMP statistical software package design for the regression analysis. The output model involves linear, quadratic and the interactions terms between every two input parameters. For example, for two input parameters (x_1 and x_2) the output model (y) can be described by Eq. (4) [14].

$$y = b_0 + b_1 x_1 + b_2 x_2 + b_{11} x_1^2 + b_{22} x_2^2 + b_{12} x_1 x_2 \tag{4}$$

Where b_0 is a constant coefficient, b_i is a coefficient represents the linear effect of x_i on y, b_{ii} is a coefficient reveals the quadratic impact of xi on y, and b_{ij} is the effect factor due to interaction between x_i and x_j on y.

In the ANOVA, the significance of the statistical analysis can be examined. The parameters standard deviation (SD) and R^2 are an indication of the degree of the model fit. However, the significance of each parameter/term in the polynomial model can be evaluated based on the F-value or probability value (P-value) at a confidence interval of 95%.

The ANOVA table contains the following columns: source, the degree of freedom (DF), the sum of squares, mean square, F-value, and P-value. The F-value and P-value are utilized for assessment the response surface of the output model. The significant factor is described by F-value or P-value with a 95% confidence level. This means only the parameters/terms of P-value less than 5% or 0.05 have a significant effect on the output model. A large F-value or the smaller P-value of a parameter/term points out the significance effect of the corresponding parameter/term [12].

4 Results and Discussion

In this section, an analytical formula for k of inductively coupled two ring coils is introduced. The k was tested as a function of seven parameters with their ranges as were mentioned earlier in Sect. 3.2.1. According to the Box-Behnken design technique, a total of 62 experimental data are needed for the seven input parameters [12]. In this work, these data were acquired utilizing the FEA in ANSYS Maxwell software, then entered into the JMP software, as listed in Table 2.

4.1 ANOVA for the k Model

The terms that have P-value greater than 0.05 were removed from the ANOVA table of the k model because of their negligible effects. The ANOVA table of the miniature k model is detailed in Table 3. For this model, the F-value is as much as 558.023. This points out that this model is statistically significant. The following parameters terms are considered significance in the k model since they have a P-value less than 0.05: r_2, R_2, d, L, $r_2 * R_2$, $r_2 * d$, $d * L$, r_2^2, R_2^2, d^2, and L_2.

Table 2. Output response, k for seven input variables.

No.	Pattern	Independent input variables							Output responses due FEA
		r_1 (cm)	r_2 (cm)	R_1 (cm)	R_2 (cm)	d (cm)	L (cm)	α (deg)	k
1	− −0−000	2	30	11	30	30	10	3.5	0.12726
2	− −0+000	2	30	11	60	30	10	3.5	0.18519
3	−0−000−	2	45	2	45	30	10	0	0.27112
4	−0−000+	2	45	2	45	30	10	7	0.25728
.	.	.	.	.	.	.	.	.	.
.	.	.	.	.	.	.	.	.	.
.	.	.	.	.	.	.	.	.	.
61	+0+000+	20	45	20	45	30	10	7	0.25789
62	++0−000	20	60	11	30	30	10	3.5	0.19057

Table 3. ANOVA details for the reduced k model.

Source	DF	Sum of Squares	F-value	P-value (Prob > F)
Model	11	1.9056906	558.0230	
r_2	1	0.1118922	360.4062	$<.0001*$
R_2	1	0.0882226	284.1661	$<.0001*$
d	1	1.4131795	4551.869	$<.0001*$
L	1	0.0625464	201.4626	$<.0001*$
$r_2 * R_2$	1	0.0094160	30.3292	$<.0001*$
$r_2 * d$	1	0.0077811	25.0630	$<.0001*$
$d * L$	1	0.0332267	107.0236	$<.0001*$
r_2^2	1	0.0062627	20.1724	$<.0001*$
R_2^2	1	0.0056681	18.2570	$<.0001*$
d^2	1	0.1482140	477.3992	$<.0001*$
L^2	1	0.0027515	8.8627	$0.0045*$
Error	50	0.0155231		
Lack of fit	13	0.01255379	13.1078	
Pure error	37	0.00276928		
Correlated total	61	1.9212136		

As can be seen from Table 3, the d has the most significant effect on k among other parameters, since it has the largest F-value. The quadratic terms that have a significant effect on k are: d, r_2, R_2, and L. On the other hand, the parameters that have a negligible effect on k are: r_1, R_1, and α. However, only the parameters that have a major impact on k are appeared in its analytical model which can be written by Eq. (5). The absolute brackets were added to avoid the negative value of k due to fitting error. However, the SD for the k model is only 0.177469. This means a good accuracy of the k analytical model was obtained with a R_2 as much as 0.99.

$$k = \left| 0.265045 + 0.068280125.\left(\frac{r_2 - 45}{15}\right) + 0.06062954.\left(\frac{R_2 - 45}{15}\right) - 0.242657.\left(\frac{d - 30}{20}\right) \right.$$
$$- 0.05104995.\left(\frac{L - 10}{10}\right) + 0.034307.\left(\frac{r_2 - 45}{15}\right).\left(\frac{R_2 - 45}{15}\right)$$
$$- 0.0311871.\left(\frac{r_2 - 45}{15}\right).\left(\frac{d - 30}{20}\right) + 0.0644463.\left(\frac{d - 30}{20}\right).\left(\frac{L - 10}{10}\right)$$
$$- 0.0209289.\left(\frac{r_2 - 45}{15}\right)^2 - 0.0199105.\left(\frac{R_2 - 45}{15}\right)^2$$
$$\left. + 0.1018143.\left(\frac{d - 30}{20}\right)^2 - 0.0138724.\left(\frac{L - 10}{10}\right)^2 \right| \tag{5}$$

4.2 Impact of the Input Parameters on the k Model

The graphical analyses are utilized to determine the impact of the interaction between the input variables on the output model (i.e. k) to estimate the statistical impact of the input parameters based on Eq. (5). Figure 4(a) details the interaction impact of (r_2 and R_2) on k; Fig. 4(b) explains the interaction effect of the (d and L) on the k.

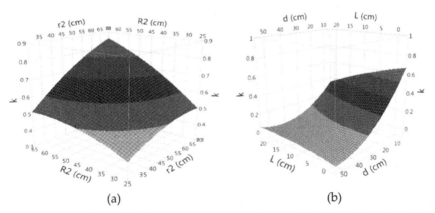

(a)   (b)

Fig. 4. Interaction effect of the input parameters on k: (a) r_2 and R_2 at d = 10 cm, L = 10 cm; (b) L and d at R_2 = 30 cm, r_2 = 30 cm.

As can be seen from Fig. 4, the k is considerably declining as d increases. Likewise, as the L increases, the k is slightly diminishing within the studied range of L. However, the increasing of d or L have a negative impact on the k value. On the other hand, as the r2 and/or R2 increasing this result in slightly increasing of k. However, the effect of the d and L on the k slightly decreases as the r_2 and/or R_2 increasing.

The numerical representation was applied to measure the percentage impact of the input parameters on k, the outcomes are presented in Fig. 5.

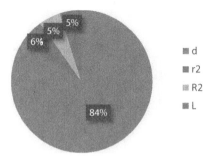

Fig. 5. The percentage impact of the significant input parameters on k.

According to Fig. 5, the d has a considerable impact on k which is as many as 84%, the r_2 and R_2 have a close effect on k i.e. only 6, and 5, respectively. The L has a slight effect on k i.e. 5%. The maximum value of k i.e. 0.864 occurs at maximum r_2 (i.e. 60 cm), minimum d (i.e. 10 cm), maximum R_2 (i.e. 60 cm), and minimum L (i.e. 0 cm).

5 Validation

To verify the analytical formula of k, its results data is compared with the FEA data. Since d has the most impact on the k among all other input parameters, it was chosen as a variable parameter to validate the analytical k formula with the following considerations: $r_2 = R_2 = 30$ cm, and L = 0 cm. The results due to the presented analytical formula and FEA are explained in Fig. 6.

As can be seen from Fig. 6, the data obtained using the introduced formula of k are in the line with that obtained by FEA especially when the d within the range specified in Table 1, at d = 5 cm the k formula is still working but the error increases. This points out that the introduced k formula has a very good accuracy with specified range, which listed in Table 1. However, the error increasing gradually as the input parameters are get far from the investigated ranges.

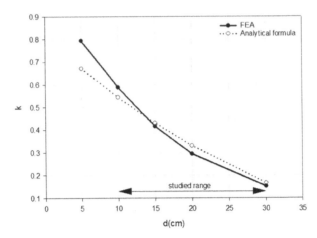

Fig. 6. Validation of the analytical k formula with FEA data.

6 Conclusion

In this paper, an analytical formula for magnetic coupling coefficient (k) of the two inductively coupled ring coils was introduced. The RSM has been used as a tool to find this formula as function of the geometrical parameters.

In this work, due to our knowledge, it was for the first time to calculate the percentage impact level of the geometrical parameters (i.e. air-gap, each design and misalignment parameter) on the k. This calculation has an importance in an IPT system design and optimization; which helps to specify which parameter has thoroughly to be designed or has a priority to be optimized. We found out that the d has the most impact on k among other geometrical parameters. On the other hand, the k is slightly sensitive to lateral misalignment and has low sensitivity to angular misalignment. This indicates that a ring coil has good misalignment tolerance; this is a desirable feature in the batteries charging of the electric vehicles.

The outcomes due to the introduced analytical model were consistent with the FEA data. This is conclusive evidence that the introduced k model is applicable for any two inductively coupled ring coils. However, the most we can talk about that k model has a very good accuracy within the investigated ranges of the input parameters. This model is still working outside these ranges but the error increases gradually as the parameters get far from their specified range, as given in Table 1.

Acknowledgments. This work funded by CTS – Centre of Technology and Systems of the UNL and the Portuguese Foundation of Science and Technology FCT (Project No. funds this work UID/EEA/00066/2019).

References

1. Chatterjee, S., Iyer, A., Bharatiraja, C., Vaghasia, I., Rajesh, V.: Design optimisation for an efficient wireless power transfer system for electric vehicles. Energy Proc. **117**, 1015–1023 (2017)
2. Nguyen, M.Q., Woods, P., Hughes, Z., Seo, Y., Rao, S., Chiao, J.: A mutual inductance approach for optimization of wireless energy transmission, pp. 4–7 (2014)
3. Hackl, S., Lanschutzer, C., Raggam, P., Randeu, W.L.: A novel method for determining the mutual inductance for 13.56 MHz RFID systems. In: 2008 6th International Symposium on Wireless and Microwave Circuits and Systems, pp. 297–300 (2008)
4. Takanashi, H., Sato, Y., Kaneko, Y., Abe, S., Yasuda, T.: A large air gap 3 kW wireless power transfer system for electric vehicles. In: IEEE Energy Conversion Congress and Exposition ECCE 2012, pp. 269–274 (2012)
5. Lu, F., Zhang, H., Hofmann, H., Mi, C.C.: A dynamic charging system with reduced output power pulsation for electric vehicles. IEEE Trans. Ind. Electron. **63**(10), 6580–6590 (2016)
6. 208126373-Wireless-Power-Transfer-Johnson-I-Agbinya.pdf
7. Sallan, J., Villa, J.L., Llombart, A., Sanz, J.F.: Optimal design of ICPT systems applied to electric vehicle battery charge. IEEE Trans. Ind. Electron. **56**(6), 2140–2149 (2009)
8. Mur-miranda, O., et al.: Wireless Power Transfer Using Weakly Coupled Magnetostatic Resonators, pp. 4179–4186 (2010)
9. Nayak, P.S.R., Kishan, D.: Performance analysis of series/parallel and dual side LCC compensation topologies of inductive power transfer for EV battery charging system. Front. Energy **16**, 166–179 (2018)
10. Zhang, W., et al.: Analysis and comparison of secondary series- and parallel-compensated inductive power transfer systems operating for optimal efficiency and load-independent voltage-transfer ratio. Power Electr. IEEE Trans. **29**(6), 2979–2990 (2014)
11. Chmela, O., Sadílek, J., Vallejos, S., Hubálek, J.: Microelectrode array systems for their use in single nanowire-based gas sensor platforms. J. Eletr. Eng. **68**(NO2), 158–162 (2017)
12. Al-Saadi, M., Ibrahim, A., Al-Qaisi, M., Crăciunescu, A.: New analytical formulas for self-inductances of inductively coupled ring coils in wireless power transfer system. Univ. Politeh. Bucharest Sci. Bull. (2019)
13. Kumar, S., Meena, H., Chakraborty, S., Meikap, B.C.: Application of response surface methodology (RSM) for optimization of leaching parameters for ash reduction from low-grade coal. Int. J. Min. Sci. Technol. **28**(4), 621–629 (2018)
14. Bezerra, M.A., Santelli, R.E., Oliveira, E.P., Villar, L.S., Escaleira, L.A.: Response surface methodology (RSM) as a tool for optimization in analytical chemistry. Talanta **76**(5), 965–977 (2008)

Green Communications

Path Planning of Mobile Robot Based on Simulated Annealing Particle Swarm Optimization Algorithm

Jie Zhao[✉], Xuesong Sheng, and Jianghao Shi

Heilongjiang University of Science and Technology, Harbin, China
zhao_xxsc@163.com

Abstract. In view of the problem of premature convergence of traditional particle swarm optimization (PSO) algorithm in path planning, which is easy to converge to local optimal solution and poor path quality, some theory about the corresponding PSO algorithm of simulated annealing optimization is studied in this paper. While planning the moving path of the robot, it analyzes the effect of initial temperature and cooling coefficient on path length and iteration times from the major contributing factors of simulated annealing algorithm. Thus deduce the law of its change and seek the optimal parameter matching. Simulated annealing algorithm can not only move the updated particle position on the basis of the particle swarm optimization formula, but also select the updated position with a certain probability. The method is used to avoid the particle converging into the local optimal solution in the whole iterative process. The capabilities of the global optimization is strengthened. Compared with the traditional PSO algorithm, the simulated annealing PSO in complex environment has better optimization ability, shorter path and fewer iterations in the simulation results.

Keywords: Particle swarm · Path planning · Simulated annealing · Linear inertia weight

1 Introduction

The path planning of mobile robot has always been the focus of the research in the development of intelligent robot. It is a major task for a moving robot that its travel path has the lower energy consumption, shorter distance and less time. It is necessary to avoiding all obstacles based on the origin and destination point coordinates in the work environment with obstacles.

There are many existing path finding methods, including artificial potential field, visual graph, etc. But these methods have some limitations, such as artificial potential

This work is supported by basic scientific research projects of Heilongjiang University of Science and Technology "The research on visual servo control of mine driving section forming based on image".

field method can not find the path in front of the near obstacles, visual graph method has complicated search path and low efficiency [1].

In recent years, many experts and scholars have studied some new algorithms, such as genetic algorithm, ant colony, bee colony and so on. Compared with some traditional algorithms, an intelligent bionic algorithm is proposed by simulating one or several behaviors of natural creatures, which provides a new way to the path ending in complex environment [2].

PSO has the benefits of fast searching speed, easy accomplishment and simple programming language. Due to its own limitations, such as being very possible to shrink to local optimum solution and being greatly affected by complex constraints in some problems, its search results are not ideal. To prevent premature maturation to the point of convergence of population, an improved PSO algorithm is proposed in this research. PSO uses linear adaptive inertia weight factor and simulated annealing algorithm to optimize, which improves the convergence of PSO. PSO combined with simulated annealing algorithm can make particle swarm jump out of the local optimal point. So the global optimal or approximate optimal point independent of the initial point selection can be accomplished [3]. Through the simulation of robot path planning, the result is analyzed and compared, which shows the effectiveness and rationality of PSO algorithm.

2 PSO Path Planning

2.1 The Basic Principle of PSO

Based on the study of predatory behavior of birds, the path planning of PSO algorithm is proposed. When solving the optimization value, the solution to algorithm corresponds to the flight location of a bird in the searching space, so it becomes a "particle". Every particle lies in a different place and at a different velocity. The fitness value of the function is determined to judge quality of a particle [4]. Each particle continuously updates itself by tracking the influence of the individual optimal location I_{best} and the global optimal location B_{best} of the whole population, so as to generate the next round of new population [5]. The velocity and position update formulas of particles are as follows:

$$T_i^{k+1} = wT_i^k + c_1 rand(I_{best} - X_i^k) + c_2 rand(B_{best} - X_i^k) \tag{1}$$

$$X_i^{k+1} = X_i^k + T_i^{k+1} \tag{2}$$

Where: T_i^k is the kth dimension speed in the ith iteration; X_i^k is the kth dimension location in the ith iteration; w is the inertia weight; c_1 and c_2 are learning factors; 'rand' is the random number distributed between 0 and 1.

Particle path planning is to search for the shortest path without collision from start to end. Its function can be expressed

$$F = \sum_{i=0}^{m} \sqrt{(x_i - x_{i-1})^2 + (y_i - y_{i-1})^2} \tag{3}$$

In the basic PSO algorithm, w is a fixed value. When it is large, the particle's global search ability is strong, which may make particle fly over the lowest point. When it is small, it can make the local search ability of particles stronger. In this way, particles tend to convergence to local optimum and lose the global search ability. PSO algorithm usually consists of three parts: the velocity of particles in the last time, the cognitive part of particles, and the social cognitive part of the whole population. The value of inertia weight affects the flying velocity of particles. Learning factors can regulate self cognition and social cognition. Therefore, there is an important relationship between the inertia weight and the change of learning factors. In view of the above defects, a group of scholars represented by Kennedy proposed an algorithm to change the inertia weight through the linear decreasing law [6], that is, with the iterative process, the value of the inertia weight is constantly reduced. The algorithm has a strong exploration ability in the early stage and can search in a larger space range. In the later stage, the reduction of weight value makes it possible to convergence to a local region for more meticulous search. The weight variation formula is as follows [7]:

$$w = w_{max} - \frac{(w_{max} - w_{min})}{iter_{max}} i \tag{4}$$

Where: w_{max} is maximum; w_{min} is minimum; $iter_{max}$ is the maximum value; i is the current value of iterations of particles. The program flow is similar to the basic PSO, but the final step determines whether the algorithm meets the termination conditions. If not, adjust the value w and replace the particle speed and location to continue the cycle. However, with the different optimization problems, the adjustment strategies of linear inertia weight are also different. So the linear inertia weight have great limitations in application.

2.2 The General PSO Process

The PSO executing steps are as follows:

Step 1: set initialization scale, learning factor, maximum number of iterations and other related parameters;
Step 2: Initialize the velocity and location of the individual, record the optimal location of each particle currently searched as I_{best}, and define the optimal location of particle global searched as B_{best};
Step 3: calculate the objective function of each particle, such as fitness, and save its particles and fitness values;
Step 4: update the particle's velocity, replace the particle's location according to the constraint conditions, initialize the particles beyond the boundary again, and then calculate its fitness value according to the optimal positioning in the process of historical optimization and replace the previous position;
Step 5: check whether the particle can meet the stop condition. If it meets the condition, it will stop and output the optimal value result of the algorithm program. If it does not meet the condition, it will return to the third step to continue to update the particle velocity and position.

3   PSO Path Planning Based on Simulated Annealing

3.1   The Basic Principle of Simulated Annealing

The algorithm of simulated annealing was first proposed by Metropoli. The idea of the algorithm comes from the annealing process, that is, first heating the solid to a higher temperature, and then gradually making the temperature lower. When heating, the internal particles are in a disordered state. When cooling, the particles become orderly and reach equilibrium at each temperature. At room temperature, the energy can reach the minimum value. Therefore, the algorithm has good global search ability, high efficiency and easy to understand program.

The simulated annealing algorithm can generate new solutions randomly between the given solution and its local domain. Through the metropolis acceptance criterion, the solution with better adaptability can be accepted, or the solution with worse adaptability can be accepted within a certain probability. In the process of particle swarm optimization, because the local optimal solution generated by each iteration of the algorithm does not necessarily meet the constraint conditions of the problem, so by introducing the local optimal solution for optimization search, we expect to get more local optimal solutions that meet the constraint conditions and have better adaptive values. A better global optimal solution is generated, which leads to the evolution of the population [8].

The solution process of simulated annealing algorithm is as follows [9]. The initial control parameters are set as follows: enough initial temperature, initial iterative solution, cycle counter, maximum number of iterations of algorithm, decay function of temperature and end criterion of program; The first random disturbance is carried out to generate a new solution X_N. According to the objective function value of the new solution, and then the comparison, if $f(X_N)$ less than $f(X_0)$, the new solution X_N is accepted. Otherwise calculate the selection probability, if it meets the requirements, it will also accept the new solution. According to the desuperheating function, carry out the desuperheating operation, and reach the end criterion to end the desuperheating operation, otherwise compare again.

3.2   The PSO Algorithm with Simulated Annealing

In the original PSO in order to prevent the particles from generating large offset, the flight velocity of each particle will be controlled within a range. If a particle flies to a better location than the current position, then later iterations will be carried out at that position. At this time, the simulated annealing algorithm is added to make particle jump out of the local optimal solution. After each particle flies to the latest position after iteration, the fitness value of the particle is calculated. If the fitness value is greater than the previous position, the particle moves to the new position. If it is not better than the current position, calculate the change value of fitness to judge whether it is greater than the annealing value. If it is greater than the annealing value, the particle moves to the new location to complete annealing operation [10]. The algorithm can make every particle be annealed. The updated particle position can not only move according to the optimization formula of PSO, but also select the updated position with a certain probability. This characteristic of accepting the new solution with a certain probability can effectively avoid the local search, thus greatly improving the search performance of particle swarm algorithm [11].

Here its detailed steps is.

Step 1: Initialize the following parameters: the scale of the initial population, the learning factors c_1 and c_2, the position x_i and velocity v_i of particles, and the initial temperature T_0 of the maximum value of iterations $iter_{max}$;

Step 2: Store location and fitness value in the individual extremum, and calculate the fitness value according to the objective function;

Step 3: Update corresponding information of each particle

$$T_i^{k+1} = wT_i^k + c_1 rand(I_{best} - X_i^k) + c_2 rand(B_{best} - X_i^k) \tag{5}$$

$$X_i^{k+1} = X_i^k + T_i^{k+1} \tag{6}$$

$$w = w_{max} - \frac{(w_{max} - w_{min})}{iter_{max}} i \tag{7}$$

Step 4: Calculate the updated fitness value. If it is more superior than the previous particles, update the values of particles P_{id} and the values of groups P_{gd}. Otherwise, carry out the next step;

Step 5: Add a disturbance near the particle to make the particle generate a new solution X_N.

Step 6: Carry out desuperheating operation and renewal formula of temperature, $T(k + 1) = \alpha * T(k)$, $k = 0, 1, 2 \ldots\ldots$, where α is a number close to 1 and the end temperature is set to 0.2;

Step 7: Judge whether the function reaches the number of termination iterations. If it reaches the number, stop the program and output the last iteration result, which is the optimal extreme point and optimal value. Otherwise, skip to step 3 to continue the iteration.

4   Study on the Influence of Simulated Annealing Parameters

The optimal configuration of important parameters in the simulated annealing algorithm is very important for path planning, including the initial temperature and annealing coefficient. The setting of these two parameters plays a decisive role in whether algorithm can achieve ideal results.

4.1   Initial Temperature

The initial temperature may influence the global optimal solution of the system. The pros and cons of the algorithm are determined by the selection probability, the denominator of which is the initial temperature. An important step of simulated annealing algorithm is to design a reasonable initial temperature. If its probability is not reasonable, it is likely that it will get a global optimal solution. Generally, a large initial temperature is set at the beginning of the algorithm, but too high temperature will make the calculation time longer.

This paper selects the initial temperature as low temperature, medium temperature and high temperature to carry out simulation experiments and compares the effects of different initial temperature on the results. Here, when the temperature is 800, 4000, 25000, 35000 to carry out five simulation experiments under the complex environment map with annealing coefficient of 0.95, the data is recorded and the average value of each group of data is curve fitted (Fig. 1).

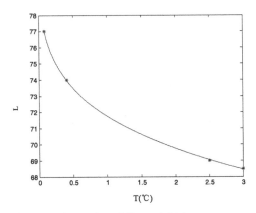

Fig. 1. Path length at different initial temperatures

It can be seen that at the initial temperature of high temperature, the average path length can get better value than at low temperature and medium temperature, and the result at 25000 °C is basically consistent with that at 35000 °C. So when the initial temperature is set to high temperature, better simulation results can be obtained.

Most of the reasons for this effect come from the most important judgment basis of simulated annealing algorithm $\exp(f(X_0) - f(X_N)/T) < randm$. The difference between the new solution and the original solution will be adopted within 0.69 times of the initial temperature. And because the initial temperature will become smaller with the iteration, the annealing is realized.

4.2 The Cooling Coefficient

The cooling coefficient is also an important parameter. Only by setting a reasonable desuperheating coefficient, the algorithm can slowly cool down like the solid annealing, so that the balance can be reached at each temperature. Therefore, the setting of the cooling coefficient is usually between 0.90 and 0.99. The smaller the value is, the faster the cooling is. But the algorithm can not fully optimize.

However, when the initial temperature is set high, the temperature is still large when reaching the limit of iterations times, which may not be conducive to the later local search. Set the initial temperature as 25000 °C, and the coefficient of desuperheating as

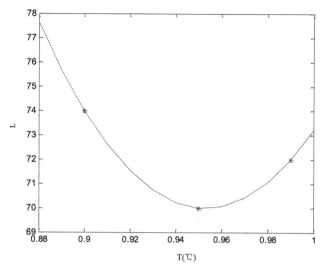

Fig. 2. Path length under different cooling coefficient

0.90, 0.95 and 0.99 to carry out five simulation experiments and make the average value of the data into a fitting curve as shown in Fig. 2.

It can be seen that in the case of selecting high temperature, with the increase of annealing coefficient, the path length of simulated annealing particle swarm optimization algorithm first decreases and then increases. Therefore, we consider adding an exponential annealing coefficient $\alpha = \exp(-CK^{1/N})$, where C is constant set to 2, K is the maximum value of iterations, and N is quantity of parameter. In this way, annealing coefficient has a large value in the beginning, and the slow cooling makes the annealing temperature have a better search for the whole at a high temperature. In the later stage, the value is smaller, and the annealing speed is faster, which can reach a smaller temperature and strengthen the local search.

It can avoid the situation that the simulation result is not ideal when setting a higher desuperheating coefficient at high temperature. Finally, through the simulation experiment, the path length obtained by using this exponential type of temperature coefficient is better than that obtained by using constant temperature coefficient.

In conclusion, the particle swarm optimization algorithm of simulated annealing chooses the initial temperature as 25000 and the annealing coefficient as index.

5　The Global Path Planning Simulation and Result Analysis

The grid method is used to build the environment map. In the environment, the above PSO and the simulated annealing PSO after parameter matching are used to plan the path, and the ability of the two methods in the complex environment is compared and analyzed.

The initialization settings of the parameters are m $= 200$, $iter_{max} = 500$, $c_1 = 1.7$, $c_2 = 1.8$, $T_0 = 25000$, $\alpha = \exp(-CK^{1/N})$. The simulation results are shown in Figs. 3 and 4.

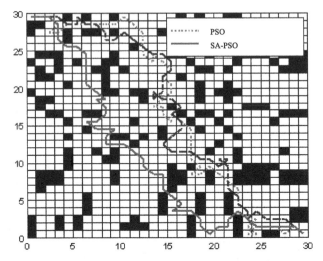

Fig. 3. Simulation of each algorithm in complex environment

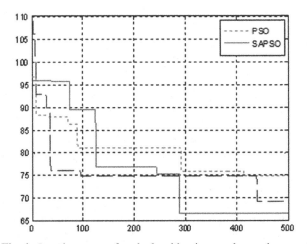

Fig. 4. Iterative curve of each algorithm in complex environment

After five times of simulation, the simulated annealing particle swarm optimization algorithm can finally get a smaller objective function, namely the path length, which is 66.4975.

In the complex case, the simulated annealing PSO iterates to the optimal value as soon as possible, and the objective function value of the path length is the minimum.

From the above simulation chart and convergence curve, we can know that in the complex environment, the basic particle swarm optimization algorithm after iterating 421 times falls into the local optimal solution, the optimal path fitness function is 74.8112, and the inflection point is 39. The PSO algorithm of linear inertia weight gets the local optimal solution in 442 iterations, the function of path is 69.1534, and the inflection point is 41.

When the simulated annealing PSO iterates 290 times, it jumps out of the local optimal value. Under the current maximum number of iterations, the most effective path fitness function is 66.4975, and the inflection point is 35. The turning points involve the turning times of the robot. It can be seen from the path above that the simulated annealing PSO algorithm is more smooth than the other two algorithms.

6 Conclusion

Traditional PSO has the advantages of simple modeling, simple calculation process, fast convergence and less parameters for robot path planning, but it is also easy to fall into the local optimal solution, so it can not get the global optimal solution in some complex environments, which makes the optimization result not ideal. The PSO algorithm combined with simulated annealing can make up for the above shortcomings. In conclusion, the simulated annealing PSO algorithm has stronger search ability in complex environment, better path and shorter path length and the algorithm is effective and reasonable.

References

1. Zhao, Z., Lin, Y.: Mean particle swarm optimization algorithm based on adaptive inertia weight. Comput. Eng. Sci. **38**(2), 501–505 (2016)
2. Lozano-Perez, T., Wesley, M.: An algorithm for planning collision free paths among polyhedral obstacles. Commun. ACM (5), 436–450 (1975)
3. Li, W., Zhao, D.: Whole area robot based on grid method and neuron covering algorithm. Mech. Des. Manuf. **08**, 232–238 (2017)
4. Montiel, O., Sepúlveda, R., Orozco-Rosas, U.: Optimal path planning generation for mobile robots using parallel evolutionary artificial potential field. J. Intell. Robot. Syst. **79**(2), 237–257 (2015)
5. Liu, X.: Particle swarm optimization and its application development. Sci. Technol. Econ. Guide **26**(04), 138–139 (2018)
6. Shi, Y., Eberhart, R.C.: A modified particle swarm optimize. In: Proceedings of IEEE Congress on Evolutionary Computation, Anchorage, pp. 69–73 (1998)
7. Erdem, H.A.: Solving container loading problem with genetic algorithm. In: 15th International Symposium on Computational Intelligence and Informatics, pp. 391–396. IEEE, Budapest (2014)
8. Wang, R., Xiao, H., Mo, Q., Huang, G.: Multi base station cooperative power allocation based on simulated annealing algorithm. J. Guilin Univ. Electron. Sci. Technol. **37**(05), 350–354 (2017)
9. Li, J., Xu, R.: Research on the combination of annealing algorithm and neural network algorithm in path planning. Autom. Instrum. **32**(11), 6–31 (2017)
10. Sun, B., Zhu, D., Yang, S.X.: An optimized fuzzy control algorithm for three-dimensional AUV path planning. Int. J. Fuzzy Syst. **20**(2), 1–14 (2018)
11. Guo, X., Zhang, B.Y., Feng, G.H.: Multi-objective optimization of low-speed and high-torque direct-drive permanent magnet synchronous motor based on hybrid particle swarm optimization. J. Mech. Electr. Eng. **35**(11), 1214–1219 (2018)

Modulation Recognition of Digital Signal Based on Decision Tree and Support Vector Machine

Fugang Liu[✉], Ziwei Zhang, Shuang Zheng, and Zhaoju Jia

Heilongjiang University of Science and Technology, Harbin 150022, China
liufugang_36@163.com

Abstract. The modulation recognition of digital signal is widely used in the field of communication. In this paper, a decision tree modulation recognition algorithm based on feature extraction and a conventional classifier recognition based on SVM are proposed. 9 kinds of common digital signals are identified and simulated. The results show that the recognition rate of SVM classifier and decision tree is high at low SNR.

Keywords: Modulation recognition · Feature extraction · SVM

1 Introduction

Modulation recognition and classification of wireless communication signals is vital when the electromagnetic spectrum is shared among civilian, government, and military to improve spectrum efficiency and shortage problem. Fast recognition and classification of a wireless signal is a significant process for accurately learning and reliably sharing the spectrum to improve spectrum utilization efficiency [1].

Early modulation parameters were mainly identified by using some external instruments and the experience of artificial individuals. This method of manual analysis and judgment is inefficient, expensive and its accuracy is not guaranteed. With the development of science and technology, modulation recognition technology has been rapidly improved, and automatic modulation recognition (AMR) technology has emerged. As shown in Fig. 1, one of AMR algorithms is maximum likelihood hypothesis recognition method based on Bayesian decision theory. There are some common algorithms such as Average likelihood ratio rest (ALRT) [2–4], Generalized Likelihood Ratio Test (GLRT) [5], Hybrid LRT [6, 7], Quasi-Hybrid LRT [8, 9] and some improved algorithms [10–13]. Maximum likelihood algorithm can achieve optimal recognition performance theoretically by minimizing the probability of misjudgment. This method can ideally obtain high recognition rate by using short message. However, the algorithm requires more prior knowledge (such as carrier rate, channel response information, signal and noise power) and huge computation, which is not conducive to the application of recognition technology.

X. Jiang and P. Li (Eds.): GreeNets 2020, LNICST 333, pp. 140–150, 2020.
https://doi.org/10.1007/978-3-030-62483-5_15

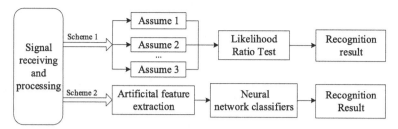

Fig. 1. Two AMR schemes

The scheme 2 in Fig. 1 is based on feature extraction. It has received increasing interest for its application in spectrum sensing and modulation recognition. In this scheme, there is no threshold setting, and the computation is less than likelihood ratio test scheme.

The modulation recognition technique based on feature extraction can be divided into three steps: signal preprocessing, feature extraction and classification (recognition).

(1) Signal preprocessing: This step mainly completes the necessary pre-processing of received signal data. It includes carrier synchronization, frequency down conversion, noise suppression, estimation and processing of parameters such as signal-to-noise ratio, symbol period and carrier frequency.
(2) Feature extraction: According to different schemes, this step can be divided into two categories: manual feature extraction and automatic feature extraction. The main task is to use signal processing tools such as wavelet, cyclostationary and cumulant to extract the characteristic parameters of signal in time domain, frequency domain or the other transform domain.
(3) Classification and recognition: Different algorithms correspond to different classifiers. According to the extracted feature parameters and classification requirements, the applicable decision rules and recognition classifiers are selected and determined.

The accuracy of feature data has a great impact on the performance of learning algorithm and recognition results. Reviewing the literature in recent years, these algorithms for modulation recognition mainly include decision tree, the k-nearest neighbor[14], support vector machine [15], artificial neural network and some hybrid algorithms [16].

In this paper, we study the recent research work and study the modulation recognition method based on feature extraction. We focus on two methods: decision tree method and support vector machine method. In this paper, the feature parameters of MPSK, MFSK, MASK and MQAM are extracted from the features of high-order cumulant, instantaneous and wavelet transform. Then, the modulation recognition of communication signals is carried out by using decision tree and support vector machine respectively. Finally, the simulation experiment is carried out.

2   Decision Tree and Support Vector Machine

Decision Tree is a decision analysis method, which is a graphical method to evaluate project risk and judge the feasibility of net present value by constructing a Decision

Tree to obtain the probability that the expected value of net present value is greater than or equal to zero on the basis of knowing the probability of various situations. Because this kind of decision branch is drawn like the trunk of a tree, it is called decision tree. In machine learning, decision tree is a predictive model, which represents a mapping relationship between object properties and object values.

In SVM regression, the input samples are not linear in general, we need to use a non-linear relationship to map each input sample into a high-dimensional feature space, to make it as linear as possible in the high-dimensional space. This makes it easier for us to use linear relations to deal with them, and then to get the Nonlinear regression of the source space of the sample.

The regression function is

$$f(x, w) = w\psi(x) + b = (w, \psi(x)) + b \tag{1}$$

w is the weight vector and b is a constant. The SVM regression problem is generally solved by introducing a loss function. Coefficients w and b in equation.

The minimization of is estimated by:

$$\min Q = \frac{1}{2}\|w\|^2 + C \sum_{i=1}^{m} (\xi_i' + \xi_i) \tag{2}$$

The constraints are:

$$\begin{cases} wx_i + b - y_i \leq \varepsilon + \xi_i \\ y_i - wx_i - b \leq \varepsilon + \xi_i' \\ \varepsilon, \xi_i' \geq 0; i = 1, \ldots \ldots m \end{cases} \tag{3}$$

Where C is the error penalty factor, ξ_i and ξ_i' relaxation factors, and ε is the loss function.

Since the feature space has a high dimension and the objective function is not differentiable, the Lagrangian multiplier method is used to calculate the feature space.

Quadratic programming problems with linear inequality constraints:

$$W(a_i, b_i) = \sum_{i=1}^{m} y_i(a_i - b_i) - \varepsilon \sum_{i=1}^{m} (a_i + a_i') - \frac{1}{2} \sum_{i,j=1}^{m} (a_i - a_i')(a_j - a_i')x_i x_j' \tag{4}$$

The constraints are:

$$\begin{cases} \sum_{i=1}^{m} (a_i - a_i') = 0 \\ a_i \geq 0, a_i \leq C \end{cases} \quad i = 1, 2, \ldots, m \tag{5}$$

x_i, x_j' is the input, y_i is the output, a_i and b_i are Lagrangian multipliers. Penalty Factor C is used to control the complexity of regression model and the precision of regression estimation. The larger the C is, the better the fitting degree of the data, that is, the higher the accuracy of the regression estimation, but not unlimitedly large, which may make the model can't predict normally, and also used to control the accuracy of the regression estimation, but also control the generalization ability of the model.

3   Feature Parameter Analysis and Extraction of Signal

3.1   Features of Higher Order Cumulants

Because the noise and signal are independent each other, the cumulative amount of Gaussian noise at the fourth order and above is zero according to the nature of high-order accumulation, the effect of Gaussian noise can be ignored. Therefore, the high-order accumulation has a strong noise suppression ability. If the signal received by the receiver has been Carrier synchronization, code element timing, matching filtering, channel noise is Gaussian white noise, then the sequence of code element synchronous sampling and complex signals obtained at the output is:

$$\sum_k a_k \sqrt{E} p(t - kT_s) \exp(j\theta_c) + n(t) \tag{6}$$

Where, k is 1,2,3, ..., N; N is the length of the send element sequence, a_k represents the code element sequence, $p(t)$ is the sender element waveform, the T_s is the code element width, the fc is the carrier frequency, the θ_c is the carrier phase, The E in the energy of the signal, $n(t)$ is the compound white noise with a zero with a zero mean. For the smooth random process $x(t)$ of the zero mean, its P-level mixing moment is defined as:

$$M_{pq} = E\{X(K)^{p-q}[X^*(K)]^q\} \tag{7}$$

The second-order cumulant is:

$$C_{20} = Cum(X, X) = M_{20} \tag{8}$$

$$C_{21} = Cum(X, X^*) = M_{21} \tag{9}$$

The fourth-order cumulant is:

$$C_{40} = Cum(X, X, X, X) = M_{40} - 3M_{20}^2 \tag{10}$$

$$C_{41} = Cum(X, X, X, X^*) = M_{41} - 3M_{20}M_{21} \tag{11}$$

$$C_{42} = Cum(X, X, X^*, X^*) = M_{42} - |M_{20}|^2 - 2M_{21}^2 \tag{12}$$

The sixth-order cumulant is:

$$C_{60} = Cum(X, X, X, X, X, X) = M_{60} - 15M_{40}M_{20} + 30M_{21}^3$$
$$C_{63} = Cum(X, X, X, X^*, X^*, X^*) = M_{63} - 9M_{42}M_{21} + 9|M_{20}|^2M_{21} + 12M_{21}^3 \tag{14}$$

The three feature parameters extracted based on higher-order cumulants are F_1, F_2 and F_3:

$$F_1 = |C_{40}|/C_{42}, \quad F_2 = |C_{41}|/C_{42}, \quad F_3 = |C_{63}|^2/|C_{42}|^3 \tag{15}$$

As can be seen from Fig. 2, The three features of higher-order cumulants have better performance Robustness at low SNR. F_1 can be used to divide the nine signals into {8PSK, 2FSK, 4FSK} and {2ASK, 2PSK, 16QAM, 64QAM, 4ASK and 4PSK}, {2ASK, 2PSK, 16QAM, 64QAM, 4ASK, 4PSK} can be further divided into {2ASK, 2PSK, 4PSK} and {16QAM, 64QAM, 4ASK} by F_2. Finally, {2ASK, 2PSK, 4PSK} and {16QAM, 64QAM, 4ASK} can be divided into {2ASK, 2PSK} and 4ASK, {16QAM, 64QAM}and 4PSK by F_3. In other words, 4ASK and 4PSK can be recognized by extracting high-order cumulant feature parameters and setting appropriate threshold, and the remaining signals can be divided into three sets: {2ASK, 2PSK}, {16QAM, 64QAM} and {8PSK, 2FSK, 4FSK}.

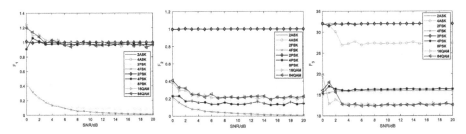

Fig. 2. Specific parameter for F_1, F_2, F_3

3.2 Transient Characteristics

For the real signal $x(t)$, if we do the David Hilbert transform on $x(t)$, then the David Hilbert transform signal is $y(t)$ where the David Hilbert transform formula is:

$$y(t) = \frac{1}{\pi t} \otimes x(t) = \frac{1}{\pi} \int_{-\infty}^{+\infty} \frac{x(\tau)}{t - \tau} d\tau \tag{16}$$

$$s(t) = x(t) + jy(t) \tag{17}$$

The instantaneous amplitude is

$$A(i) = \left(x^2(i) + y^2(i)\right)^{\frac{1}{2}} \tag{18}$$

The instantaneous phase is

$$\varnothing(i) = \varnothing(i) + C(i) \tag{19}$$

Because $\theta(i)$ has phase collapse, it is modified by Modulo 2

$$C(i) \begin{cases} C(i-1) - 2\pi, \theta(i+1) - \theta(i+1) > \pi \\ C(i-1) + 2\pi, \theta(i) - \theta(i+1) > \pi \\ C(i-1), \text{else} \end{cases} \tag{20}$$

$$\theta(i) = \begin{cases} \tan^{-1}\left[y(i)\big/x(i)\right], \ x(i) > 0 \\ \tan^{-1}\left[y(i)\big/x(i)\right] - \pi, \ x(i) < 0, y(i) \geq 0 \\ \tan^{-1}\left[y(i)\big/x(i)\right] + \pi, \ x(i) < 0, y(i) \geq 0 \\ \pi/2, \ x(i) = 0, y(i) \leq 0 \\ -\pi/2, \ x(i) = 0, y(i) > 0 \end{cases} \tag{21}$$

The instantaneous frequency is

$$f(t) = \frac{1}{2\pi T}[\varnothing(i) + \varnothing(i-1)] \tag{22}$$

(1) The fluctuation degree of instantaneous amplitude can be reflected by the Maximum value of center normalized instantaneous amplitude spectral density γ_{max}, whose function is to distinguish the envelope time-varying signal from the envelope constant signal, that is, to separate the MFSK signal from the MPSK signal.

$$\gamma_{max} = max|DFT(a_{cn}(i))|^2\Big/N_S \tag{23}$$

As shown in the Fig. 3 (a), 2FSK and 4FSK are envelope stable, 2FSK and 4FSK are equal to zero. However, 2FSK and 4FSK are small constants due to noise, while 8PSK is a constant that is not zero. Therefore, the system can be used the appropriate threshold $T5$ to divide 8PSK, 2FSK and 4FSK into 8PSK and 2FSK, 4FSK.

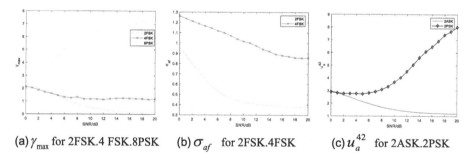

(a) γ_{max} for 2FSK.4 FSK.8PSK   (b) σ_{af} for 2FSK.4FSK   (c) u_a^{42} for 2ASK.2PSK

Fig. 3. Instantaneous characteristics of different signals

(2) The function of the Standard deviation of absolute value of zero center normalized non-weak signal instantaneous frequency σ_{af} is to distinguish the signal with absolute and direct frequency information of normalized center frequency from the signal whose absolute value is constant.

$$\sigma_{af} = \sqrt{\frac{1}{C}\left(\sum_{a_n(i)>a_i} f_N^2(i)\right) - \left(\frac{1}{C}\sum_{a_n(i)>a_i} \left|f_N^2(i)\right|\right)^2} \tag{24}$$

The absolute value of the zero center normalized instantaneous frequency of 2FSK is a constant, and its standard deviation is zero, but this system is carried out in the noisy environment, so the σ_{af} of 2FSK is a relatively small constant But 4FSK instantaneous frequency absolute value is not zero, therefore 4FSK's σ_{af} is bigger than 2FSK'S σ_{af}. As shown in the Fig. 3 (b), the system can identify 2FSK and 4FSK with appropriate threshold $T7$.

(3) Amplitude information can be represented by the Compactness of the zero-center normalized instantaneous amplitude u_a^{42}.

$$u_a^{42} = \left(E\left[f_{CN}^4(i)\right] \middle/ \left\{E\left[f_{CN}^2(i)\right]\right\}^2 \right) - 1 \tag{25}$$

The function of u_a^{42} is to distinguish the signals with higher instantaneous amplitude distribution from those with higher instantaneous amplitude distribution. As shown in the Fig. 3 (C), there are two levels in the 2ASK, so the compactness of u_a^{42} is relatively small. The 2ASK and 2PSK are separated by appropriate threshold $T4$.

3.3 Wavelet Transform Feature

The modulated signal is transformed by wavelet. Because there are many burrs in the signal after wavelet transform, the median filter is used to remove the burrs of wavelet transform amplitude. The variance of the obtained wavelet transform amplitude after median filtering. This feature reflects the stability of wavelet transform amplitude of various signals. As shown in Fig. 4, since the amplitude stability of 64QAM signal is obviously different from that of 16QAM signal, the system uses proper threshold $T6$ to recognize the two signals in 16QAM and 64QAM

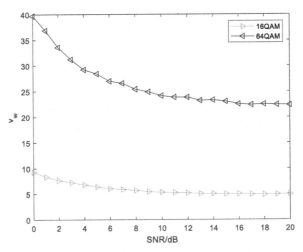

Fig. 4. V_w for 16QAM and 64QAM signals

4 Classification and Recognition of Modulation Signals Based on Decision Tree and SVM

4.1 Flow Chart

The recognition flow of the algorithm is shown in Fig. 5.

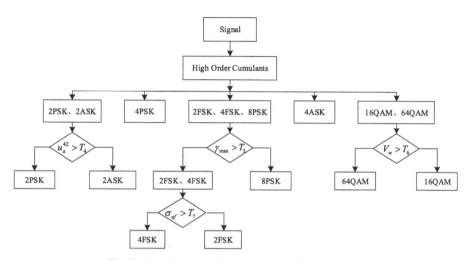

Fig. 5. Classification flowchart of modulation signals

According to the feature parameters extracted above, nine common digital signal modulation modes are identified by using binary tree structures. The 4PSK and 4ASK signals are identified, and divided into three parts:{2ASK, 2PSK}, {16QAM, 64QAM} and {8PSK, 2FSK, 4FSK}; Then{2ASK and 2PSK}, {2FSK, 4FSK, 8PSK}and{16QAM, 64QAM} are identified by u_a^{42} , γmax , V_w, and 2FSK and 4FSK are identified by σ_{af}.

4.2 Identification Simulation and Performance Analysis

In this paper, MATLAB software is used to verify the algorithm, after the down conversion to baseband signal, carrier synchronization, symbol timing, matching filter, nine common digital signal modulation mode is identified. The simulated signal is 125 symbol length, 4000 Hz carrier frequency, 16000 Hz sampling frequency, 2000 bps symbol rate, 0–20 dB signal-to-noise ratio, and the extracted characteristic parameters are averaged after 100 calculations.

Experiment 1 was a decision tree based statistical recognition model:

Figure 6 (a) is the recognition rate of 1000 times for 9 kinds of signals in 0–20 dB SNR. From Fig. 6 (a), the recognition rate of 2PSK and 2FSK is poor in 0–2 dB SNR, and most of the recognition rate is over 90% when the SNR is 5 dB. Figure 6 (b) is the total recognition rate of 9 kinds of signals under the SNR of 0 to 20 dB. From Fig. 6 (a), the overall recognition rate is below 90% under the SNR of 0 to 5 dB, and the total recognition rate is 95% and the highest is 99% under the SNR of 6 dB.

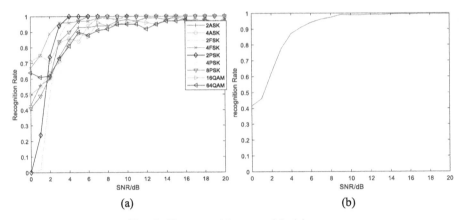

Fig. 6. The recognition rate of decision tree

Experiment 2 was Support Vector Machine based recognition:

Under the condition of signal-to-noise ratio from 0 dB to 20 dB, 500 data samples are used as training set and 500 data samples are used as test set for each feature. The effect of recognition is shown in Fig. 7 (a) and Fig. 7 (b).

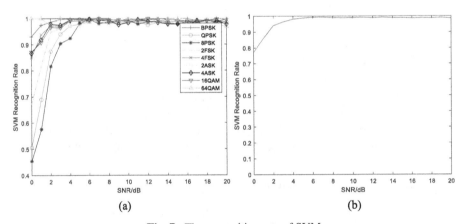

Fig. 7. The recognition rate of SVM

Figure 7 (a) shows the recognition rate of 9 signals at 0–20 dB SNR. From Fig. 7 (a), the recognition rate of 4PSK and 8PSK is worse than other signals at 0–2 dB SNR, and the recognition rate of 64QAM and 2PSK signals is better at low SNR. The recognition rate of the 9 signals is over 99% when the signal-to-noise ratio is 6 dB. Figure 7.(b) is the total recognition rate of 9 kinds of signals under the signal-to-noise ratio of 0 to 20 dB. From Fig. 7 (b), the overall recognition rate is above 90% when the signal-to-noise ratio is 0 to 2 dB, and the total recognition rate is 99% when the signal-to-noise ratio is 5 dB, and the highest is 99.8%.

5   Conclusion

In this paper, the higher-order cumulants of 9 common digital signals are calculated, then the features of higher-order cumulants, 3 instantaneous features and wavelet transform are extracted, 9 kinds of signals are identified by Decision Tree Algorithm and SVM on MATLAB Platform. The experimental results show that the recognition rate of the method based on decision tree is poor at low snr. After the SNR is 6 dB, the recognition rate reaches over 90% and the highest is 99% The recognition rate of SVM method at 0 dB has reached 78%, when the SNR is 2 dB, the recognition rate is more than 90%, the recognition rate is stable at 99% when the SNR is 4 dB, and the highest recognition rate is 99.8%. Therefore, the algorithm has a good recognition rate in low SNR, which proves the effectiveness and reliability of the Algorithm.

Acknowledgements. This work has been partially supported by the National Natural Science Foundation of China project (51674109) and 2017 scientific research project of basic scientific research expenses of provincial colleges and universities in Heilongjiang Province.

References

1. Zhou, R., Liu, F., Gravelle, C.W.: Deep learning for modulation recognition: a survey with a demonstration. IEEE Access **8**, 1–12 (2019)
2. Kim, K., Polydoros, A.: Digital modulation classification: the BPSK versus QPSK case. In: MILCOM 88, Century Military Communications–What's Possible?. Conference Record. Military Communications Conference, pp. 431–436, San Diego, CA. IEEE (1988)
3. Polydoros, A., Kim, K.: On the detection and classification of quadrature digital modulations in broad-band noise. IEEE Trans. Commun. **38**(8), 1199–1211 (1990)
4. Hwang, C., Polydoros, A.: Advanced methods for digital quadrature and offset modulation classification. In: MILCOM 91- Conference Record, pp. 841–845, McLean, VA. IEEE (1991)
5. Boiteau, D., Le Martret, C.: A general maximum likelihood framework for modulation classification. In: Proceedings of the 1998 IEEE International Conference on Acoustics, Speech and Signal Processing, pp. 2165–2168, Seattle, WA. IEEE (1998)
6. Panagiotou, P., Anastasopoulos, A., Polydoros, A.: Likelihood ratio tests for modulation classification. In: MILCOM 2000 Proceedings, pp. 670–674, Los Angeles, CA. IEEE (2000)
7. Tadaion, A., Derakhtian, M., Gazor, S., Aref, M.: Likelihood ratio tests for PSK modulation classification in unknown noise environment. In: Canadian Conference on Electrical and Computer Engineering, pp. 151–154, Saskatoon, Sask. IEEE (2005)
8. Abdi, A., Dobre, O., Choudhry, R., Bar-Ness, Y., Su, W.: Modulation classification in fading channels using antenna arrays. IEEE MILCOM 2004. Military Communications Conference, pp. 211–217. IEEE, Monterey, CA (2004)
9. Li, H., Dobre, O., Bar-Ness, Y., Su, W.: Quasi-hybrid likelihood modulation classification with nonlinear carrier frequency offsets estimation using antenna arrays. In: MILCOM 2005 – 2005 IEEE Military Communications Conference, pp. 570–575, Atlantic City, NJ. IEEE (2005)
10. Xu, J.L., Su, W., Zhou, M.: Likelihood function-based modulation classification in bandwidth constrained sensor networks. 2010 International Conference on Networking. Sensing and Control (ICNSC), pp. 530–533. IEEE, Chicago, IL (2010)
11. Phukan, G.J., Bora, P.K.: Parameter estimation for blind classification of digital modulations. IET Sig. Process. **10**(7), 758–769 (2016)

12. Liu, F., Xu, J., Hu, F., Wang, C., Wu, J.: Lightweight trusted security for emergency communication networks of small groups. Tsinghua Sci. Technol. **23**(2), 195–202 (2018)
13. Arya, S., Yadav, S.S., Patra, S.K.: WSN assisted modulation detection with maximum likelihood approach, suitable for non-identical Rayleigh channels. In: 2017 International Conference on Recent Innovations in Signal processing and Embedded Systems (RISE), pp. 49–54, Bhopal, India. IEEE (2017)
14. Aslam, M.W., Zhu, Z., Nandi, A.K.: Automatic modulation classification using combination of genetic programming and KNN. IEEE Trans. Wirel. Commun. **11**(8), 2742–2750 (2012)
15. Zhang, W.: Automatic modulation classification based on statistical features and support vector machine. In: 2014 XXXI-th URSI General Assembly and Scientific Symposium (URSI GASS), pp. 1–4, Beijing, China. IEEE (2014)
16. Sun, X., Su, S., Huang, Z., Zuo, Z., Guo, X., Wei, J.: Blind modulation format identification using decision tree twin support vector machine in optical communication system. Opt. Commun. **438**, 67–77 (2019)

Text Mining and Analysis of Meituan User Review Text

Yong-juan Wang$^{(\boxtimes)}$, Guang-hua Yu, Li-nan Sun, and Pei-ge Liu

Heihe College, Heihe 164300, China
641917127@qq.com

Abstract. Based on the current situation of the market, this paper obtains related data of online user reviews on Meituan the online food delivery platform by soft wares and implements preprocessing and mining by language correlation function, and finally draws a conclusion that judging from the mining results of the featured words and the emotions in the reviews, the Meituan platform and its food delivery service have been evaluated by users as being cheap, economical, convenient and fast, and the key elements that users concern regarding to the merchant rating are the merchant's attitude, the delivery man's attitude, the food taste and the food security respectively.

Keywords: Online user reviews · Text segmentation · Keyword extraction

1 Research Design

Based on the empirical study of Meituan food delivery and from the perspective of social psychology, this paper finds out that the basic requirement of users is the key factor for their choices on products to meet their expectations, while the exposure, sales, rating and specific review content of a merchant help them to decide whether to choose it. After a research into the text mining theory applied in the paper, we mined and processed the data as shown in the Fig. 1 below. Firstly, we preprocessed the data source to achieve a denoise effect. We created a user customized dictionary and removed stop words to guarantee the accuracy of the word segmentation, and it demanded repetitions for the best effect. And finally we carried out mining and analysis on the review content after the word segmentation.

X. Jiang and P. Li (Eds.): GreeNets 2020, LNICST 333, pp. 151–155, 2020.
https://doi.org/10.1007/978-3-030-62483-5_16

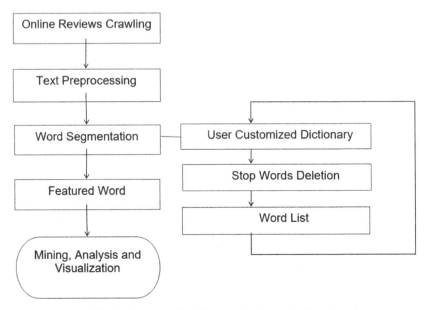

Fig. 1. The analytical framework of empirical study

2 The Empirical Analysis

2.1 Data Crawling and Process

We have crawled 3013 reviews of the biggest food delivery platform in China Meituan integrated on the website Baidu Koubei (https://koubei.baidu.com/, a special website integrating users' reviews and merchants' catalogs), and a part of merchants' user reviews on Meituan itself by using a third party software Octopus Crawler. And finally fixed the merchant Xuji Pie as the object of this text mining research considering its high rating and quantity of users' reviews.

There are altogether 597 valid users' reviews crawled by Octopus Crawler on Xuji, a restaurant mainly selling pie in Heihe city. The first step is to take a denoise processing on the text data obtained, and preprocess the data by keeping the useful content while deleting some irrelevant fields including the nicknames, review times and irrelevant reviews. And then deleting poor quality reviews including reviews in default templates and texts in a length no more than 2 characters.

2.2 Keyword Extraction of the Review Texts

The original crawled online reviews should be processed in the form of the structured data for analysis. A part of the review data is set as the training library, and then the review content is processed in the steps of word segmentation and featured word extraction. Partial results are shown as in Fig. 2.

```
> wk<-worker()

> words<-"好吃!米饭有点硬 但是锅包肉超级好吃!海带丝有点咸!不太好吃地三

+     鲜也好吃!"

> segment(words,wk)

[1] "好吃"  "米饭"  "有点"  "硬"   "但是"  "锅包肉" "超级"

[9] "海带丝" "有点咸" "不太"  "好吃"  "地"   "三鲜"  "也"   "好吃"
```

Fig. 2. Training result of review text segmentation

The training result of word segmentation reveals that some proper nouns like "地三鲜" (means 3 fresh vegetables growing underground, a dish of stewed fried potatoes, eggplants and peppers) has been separated into two words "地" (ground) and "三鲜" (3 fresh), and the phrase "不太好吃" (not very good) has been separated into two words "不太" (not very) and "好吃" (yummy). Those segmentation results will affect the accuracy of the text feature extraction, so it is necessary to create a user customized dictionary on the basis of merchants' information for accurate segmentation of review content. And we can also see a large quantity of noise words like "too" or "but" where we still need to denoise the segmented texts by the steps of setting down filter criterias, making a stop word list and deleting useless words accordingly. The stop words are mined by following the principle that only words with the useful information will be adopted, and then added into the stop word list for a further deletion.

Basically, we create a user customized dictionary based on the product, cut reviews into sentences, finish segmentation by using the customized dictionary, and then mark the part of speech of each word. Because the product characteristic values are mainly nouns, we filter the nouns by filtering criteria when extracting featured keywords, and also delete the stop words in the list of marked texts.

All the crawled review texts are processed by word segmentation in terms of the training results. The processed texts are used for word frequency statistics and keyword extraction. After processing the featured words by algorithm, we finally selected 20 key words for the use of Meituan's user review analysis. The 20 key words selected from Xuji Pie's user reviews for analysis are listed in Fig. 3 below.

```
> keys<='mydata.txt'
31226.3   1668.3 986.093 682.193 526 399.133  363.915 356.982
   " "  "好吃"   "吃"  "味道"  "不错"   "菜" "地三鲜" "骑手"
328.698 302.484 293.48   292.16 278.563 270.002 270.002 246.523
   "肉"  "馅饼"    "量"  "米饭"  "砂锅" "锅包肉"  "盖饭"   "饼"
246.523 245.773 234.784   228.49
   "做"  "小哥"    "点"  "实惠"
```

Fig. 3. Keyword extraction result of Xuji Pie's user reviews

3 Review Mining and Result Analysis

A further visualization is conducted on those segmentation words and extracted key words. The cloud picture as shown in Fig. 4 made by the words can directly show the core content of the user reviews.

Fig. 4. The cloud picture of Meituan's user review words on Baidu Koubei

The word "food delivery" has been mentioned many times among all reviews, which can prove that the food delivery is the featured business of Meituan. The key words standing for the comprehensive assessment are "economical", "cheap" and so on. And we can also infer from Fig. 4 that quite a large number of user's order food delivery by using Meituan, and the most commonly words to describe their experience are "not bad", "cheap", "economical" and "love it". And another word "Islamic" is also a high frequent word, but according to the research, the specific reviews related to "Islamic" topic are almost passive reviews such as "discrimination", "rubbish".

Considering that the words reflecting the comprehensive assessment are "economical", "cheap" and so on, Meituan and its food delivery team should aim to provide low prices and convenient and fast service in order to acquire more users, which can also help to improve users' satisfaction on both the service and the food, and hence to increase their dependence and tendency.

Fig. 5. The cloud picture of review words of Xuji Pie

Figure 5 finds that most users prefer to describe their user experience by mentioning the words "deliveryman", "deliveryman buddy", "attitude", "volume", "salty", "on time", "taste" and "clean" in their reviews, which means the factors that users concern in their experience lie in the punctuality of deliveryman, the attitude of deliveryman, the attitude of the merchant, food taste, food volume and food security.

For merchants, to attract users or improve their satisfaction degree, they need to serve with a good attitude and meanwhile guarantee food security and taste. For delivery men, they should remember to keep a good attitude and deliver the food on time in safety so that they can help to improve the user loyalty to Meituan, and hence to increase both the user number and business sales of Meituan food delivery.

References

1. Li, T.: Data Mining Where Theory Meets Practice. Xiamen University Press, Hainan (2013)
2. Li, Z.: Research on the Influence of O2O Take out Catering Platform Online Comments. E-commerce Press (2017)
3. Dong, S., Wang, X., Ge, Z.: Analysis of content characteristics of B2C shopping website online comments based on text mining. Library Theory and Practice (2017)
4. Li, X.: Feature word extraction in Chinese text classification. Comput. Eng. Des. (2019)
5. Tao, Q., Ge, T.: Research on big data mining technology based on THDS. J. Mudanjiang Norm. Univ. (2017)
6. Zhang, L.: Language Data Analysis and Mining Practice. Mechanical Industry Press, China (2015)
7. Hao, Y.: An Empirical Study on the Influence of Online Reviews on Consumers' Perception and Purchase Behavior. Harbin Institute of Technology, Harbin (2010)
8. Fang, N., Liu, X.: Business process change domain analysis based on data flow constraint of petri net. J. Mudanjiang Norm. Univ. (2018)

Extraction of Baseline Based
on Second-Generation Wavelet Transform

Jiancai Wang[✉]

Office of Academic Affairs, Hei Longjiang University of Science and Technology,
Harbin 150022, China
154539860@qq.com

Abstract. In the analysis of signals processing, due to the various kinds of interference in the transformation and sampling of the analytical instruments, the baseline of the signals is presented in the upper and lower drift. The upper and lower baseline could affect the accuracy of quantitative calculation, analysis and evaluation. In the study, the principle of second-generation wavelet is discussed and introduced to extract the baseline. The features of signals are analyzed and the quantitative accuracy of components has been significantly improved by the baseline extraction. The second-generation wavelet method successfully realizes the split of baseline from the signal peak with high efficiency and is easy to be implemented.

Keywords: Second-generation wavelet transform · Signals processing · Baseline extraction

1 Introduction

In the process of collecting signals, due to there are the various kinds of interference in the transformation and sampling of the analytical instruments, the baseline of the signals is presented in the upper and lower drift. The baseline shift can change the shape of the signal peak, affecting peak height and peak area calculation, which has a bad effect on the components calculation, quantitative analysis and evaluation [1–3].

There are various kinds of methods for baseline shift correction, such as Digital filtering, Baseline fitting, Adaptive filtering and Wavelet transform [4] etc. Baseline fitting is by means of mathematics, a function model (usually n order polynomial) is established to describe the baseline and then to correct, but it's difficult to deal with in technology and real-time processing is not high. In addition, the curve fitting method is also used to correct baseline shift, but the TLC baseline spectrum is a slowly varying frequency signal, when the signal is weak, the difference was very difficult to extract.

Recently, the method of wavelet transform is widely applied in traditional Chinese medicine fingerprint and high performance liquid chromatography fingerprint to correct baseline shift. In 1995, Sweldens proposed a new wavelet construction method, which does not depend on Fourier transform, the lifting scheme (Scheme Lifting), which is called the second generation wavelet transform (SGWT) [5]. In the transformation process, through the design of predictor and updater, the wavelet function is obtained with

X. Jiang and P. Li (Eds.): GreeNets 2020, LNICST 333, pp. 156–161, 2020.
https://doi.org/10.1007/978-3-030-62483-5_17

expected properties [6]. Using the lifting wavelet to carry out baseline shift correction, signal denoising, it has the characteristics of simple structure, fast operation, saving storage space and realizing the integer wavelet transform.

In the study, the D4 wavelet is used in the process of exacting baseline, which will make a correction effect based on the discussion of the theory of second generation wavelet transform, with the advantages of simple coefficients and a few floating-point calculations, the method has a good inhibiting effect on baseline shift.

2   Methods

There are three steps to realize the second generation wavelet transform (SGWT) constituted by the lifting process: Split, Predict and Update. The reconstruction process is the inverse [7]. The specific implementation is shown in Fig. 1.

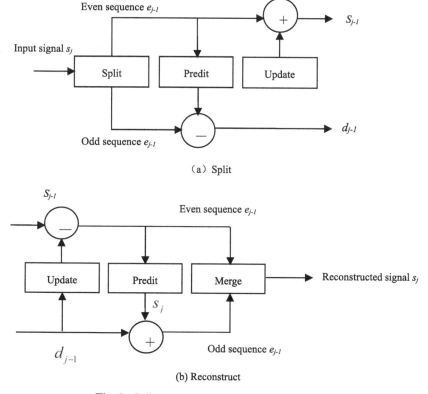

(a) Split

(b) Reconstruct

Fig. 1. Split and reconstruction process of SGWT

2.1 Split

The original signal is divided into two subsets, which is usually divided into even sequences and odd sequences. The length of each subset is half of the atom sets and does not intersect each other. That is:

$$split(s_j) = (e_{j-1}, o_{j-1}) \qquad (1)$$

In the formula, $e_{j-1} = \{e_{j-1,k} = s_{j,\,2k}\}$, $o_{j-1} = \{o_{j-1,\,k} = s_{j,\,2k+1}\}$

2.2 Predict

According to the correlation between the sequences, generally, the odd sequence is predicted by the even sequence. In the prediction, there is a deviation dj − 1 between the actual value Oj − 1 and the predicted value p (ej − 1), the difference is called the detail coefficient or the wavelet coefficient, which reflects the degree of approximation. The amplitude change of wavelet coefficient is inversely proportional to the correlation of the data. The prediction process as follows:

$$d_{j-1} = o_{j-1} - P(e_{j-1}) \qquad (2)$$

In fact, though it is impossible from the subset e_{j-1} to predict the subset o_{j-1} accurately, p (e_{j-1}) may be close to o_{j-1}. In this way, d_{j-1} contains less information than the original. Repeat the process of split and predict, after n steps, the original signal sets can be represented by $\{e_n, o_n, \ldots\ldots, e_1, o_1\}$.

2.3 Update

After the procedure of split, some of the overall characteristics (such as the mean) of the subsets may not be consistent with the original data. At this time, the procedure of update is needed to maintain the overall characteristics of the original data. The process as follows:

$$s_{j-1} = e_{j-1} + U(d_{j-1}) \qquad (3)$$

P and U take different functions, can be constructed by different wavelet transforms.
Sweldens has been proved to make Integers be set to an integer wavelet transform based on the lifting algorithm [8]. An integer set is obtained by SGWT and it also proves that any traditional wavelet can be converted into the corresponding second generation integer wavelet. Select the appropriate algorithms of prediction and lifting for discrete data samples to be processed, the required frequency band is obtained [9]. For the signal data, DB wavelet as the prediction and lifting algorithm is chosen in the study, the second generation of discrete wavelet construction scheme is used to extract the baseline from the original signal.

3 Results and Discussion

3.1 Description of the Algorithm

There are many commonly used wavelet functions such as Haar wavelet, Shannon wavelet and Daubechies wavelet. Due to the frequency spectrum of TLC baseline signal is low, the trend is relatively flat, and the detection effect of Daubechies wavelet to non-stationary signals is better, D4 wavelet function is used to split and reconstruct TLC signal to realization the extraction of the baseline.

The wavelet function of Daubechies wavelet is:

$$\phi(t) = \sqrt{2} \sum_{n=0}^{2N-1} g_n \phi(2t - n) \tag{4}$$

The scaling function of Daubechies wavelet is:

$$f(t) = \sqrt{2} \sum_{n=0}^{2N-1} h_n f(2t - n) \tag{5}$$

Thereinto,the coefficient of High-pass filter is g_n, the coefficient of Low-pass filter is h_n. The z transoform of D4 wavelet is $h(z) = h_0 + h_1 z^{-1} + h_2 z^{-2} + h_3 z^{-3}$, $g(z) = -h_3 z^{-2} + h_2 z^1 - h_1 + h_0 z^{-1}$. Thereinto, the coefficients of Low-pass filter are $h_0 = \frac{1+\sqrt{3}}{4\sqrt{2}}$, $h_1 = \frac{3+\sqrt{3}}{4\sqrt{2}}$, $h_2 = \frac{3-\sqrt{3}}{4\sqrt{2}}$, $h_3 = \frac{1-\sqrt{3}}{4\sqrt{2}}$. The improved Laurent polynomial Euclidean algorithm is used to factorize the multi-phase matrix p (z) of D4 wavelet filter.

$$p(z) = \begin{pmatrix} h_0 + h_2 z^{-1} & -h_3 z^1 - h_1 \\ h_1 + h_3 z^{-1} & h_2 z^1 + h_0 \end{pmatrix}$$

$$= \begin{pmatrix} 1 & -\sqrt{3} \\ 0 & 1 \end{pmatrix} \begin{pmatrix} 1 & 0 \\ \frac{\sqrt{3}-2}{4} z^{-1} & 1 \end{pmatrix} \begin{pmatrix} 1 & \frac{2}{\sqrt{3}} \\ 0 & 1 \end{pmatrix} \begin{pmatrix} 1 & \frac{1}{4} z - \frac{1}{\sqrt{3}} \\ 0 & 1 \end{pmatrix} \begin{pmatrix} \frac{1+\sqrt{3}}{2\sqrt{2}} & 0 \\ 0 & \frac{2\sqrt{2}}{1+\sqrt{3}} \end{pmatrix}$$

Therefore, the procedures of baseline extraction are described as follows:

(1) The input signal is split into even and odd. The original signal x with length N is split into even sequence s_l^0 and odd sequence d_l^0 respectively. $s_l^0 = x_{2l}$, $d_l^0 = x_{2l+1}$,$l = 0,1,\ldots,N/2\text{-}1$.

(2) The steps of predict and update are carried out alternately. The predictive value of the P dual signal of the filter is used as the odd signal. There are four steps to enhance and dual lifting obtained:

$$s_l^1 = s_l^0 + \sqrt{3} d_l^0, \ d_l^1 = d_l^0 - \frac{\sqrt{3}-2}{4} s_{l-1}^1, \ s_l^2 = s_l^1 - \frac{2}{\sqrt{3}} d_l^1, \ d_l^2 = d_l^1 - \frac{\sqrt{3}}{2} s_l^2,$$
$$s_l^3 = s_l^2 - \frac{1}{4} d_{l+1}^2 + \frac{1}{\sqrt{3}} d_l^2.$$

(3) Finally, scale transformation is carried out. Scale transform is used to get $s_l = \frac{\sqrt{3}+1}{2\sqrt{2}} s_l^3, d_l = \frac{2\sqrt{2}}{\sqrt{3}+1} d_l^3$, here s and d are the low frequency and the

high frequency of the wavelet decomposition respectively. Among them, $s = \{s_0, s_1, \ldots, s_{N/2-1}\}$, $d = \{d_0, d_1, \ldots, d_{N/2-1}\}$, $l = 0, 1, \ldots, N/2-1$.

(4) The reconstruction process of the second generation wavelet decomposition is basically the inverse process. First performed scaling, and then update and predict steps, finally carry out the parity inversion to reconstruct the signal.

3.2 Extraction of the Baseline

Before extraction of the baseline, the signal denosing is carried out first. The separation of the high-frequency signals and the low-frequency signals is achieved by Wavelet transform. After separating the high frequency and low frequency signals, the high frequency part of the corresponding peak is zero, and the reconstructed data contains low frequency data so that the baseline can be obtained from the reconstructed data. The method is applied in the TLC. The sample of TLC is obtained by Thin layer chromatography scanner, and then light density detection. The original signal and the result after baseline extraction are shown in Fig. 2.

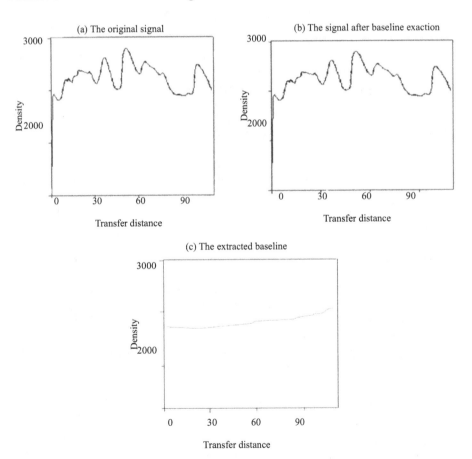

Fig. 2. The original signal and the result after baseline extraction

4   Conclusion

Experiments show that, with the basic principles of the second generation wavelet transform the second generation D4 wavelet is used to extract the baseline based on the lifting algorithm, which not only to speed up the processing speed, but also effectively suppress baseline shift. After extraction, the quantitative accuracy of crude oil components has been significantly improved. In the study, the method of baseline extraction on TLC signal has a certain practical significance and application values, but the general applicability and stability of this method is verified by a large number of real sample analysis.

Acknowledgements. The work was supported by Hei longjiang Fundamental Research Funds for the Local Universities in 2018 (No. 2018-KYYWF-1189).

References

1. Jiangtie, L.,Wei, L.,Zhixia Z, Xianchun L. Application of thin layer chromatography fingerprint in the quality evaluation of Salvia miltiorrhiza. J. Xiamen Univ. (Nat. Sci.) **44**(6), 801–805 (2005)
2. Yun, Y., Xuefeng, Z.: Baseline spliting of traditional Chinese medicine fingerprint based on the second generation wavelet transform, J. South China University of Technol. (Nat. Sci. Ed.) **35**, 49–52 (2007)
3. Jianxin, C.: Study on the method of visible pattern analysis of fingerprint of traditional Chinese medicine based on the representation of multiple maps. Yanshan University, Hebei (2012)
4. Xiangkui, W.: ECG baseline drift depress algorithm based on multi-resolution analysis. Comput. Eng. Des. **29**(13), 3482–3484 (2008)
5. Sweldens, W.: The lifting scheme: a construction of second generation wavelets SIAM. J. Math. Aanal. **29**(2), 511–546 (1997)
6. Daubechies, Ingrid: Ten Lectures on Wavelets. National Defence Industry Press, Beijing (2004)
7. Xinjun, L.: An method of improved wavelet threshold denoising. Sci. Technol. Vis. **25**, 62–63 (2014)
8. Daubechinces, C.R., Seldens, W.: Wavelet transforms that ma Pinter-gers to intergers. Applied and computational hamonic analysis (1998)
9. Yongqiang, W., Haifeng, Y.: ECG signal denoising algorithm based on second generation wavelet transform. J. China Sci. Technol. Expo. **1**, 95–99 (2012)

The Crawl and Analysis of Recruitment Data Based on the Distributed Crawler

Jiancai Wang[1(✉)] and Jianting Shi[2]

[1] Office of Academic Affairs, Hei Longjiang University of Science and Technology,
Harbin 150022, China
154539860@qq.com
[2] School of Computer and Information Engineering, Hei Longjiang
University of Science and Technology, Harbin 150022, China

Abstract. Because of the rapid development of Internet, how to efficiently and quickly obtain useful data has become an importance. In this paper, a distributed crawler crawling system is designed and implemented to capture the recruitment data of online recruitment websites. The architecture and operation workflow of the Scrapy crawler framework is combined with Python, the composition and functions of Scrapy-Redis and the concept of data visualization. Echarts is applied on crawlers, which describes the characteristics of the web page where the employer publishes recruitment information. In the base of Scrapy framework, the middleware, proxy IP and dynamic UA are used to prevent crawlers from being blocked by websites. Data cleaning and encoding conversion is used to make data processing.

Keywords: Distributed crawler · Scrapy framework · Data processing

1 Introduction

With the widespread of modern network, especially after the application of 5G, people are spending more time in searching useful information through piles of data. Therefore the distributed web crawler is adopted to search and obtain Internet data, which can greatly improve the search efficiency. The Internet has been thriving rapidly and changing people's life greatly for decades. According to China Internet development report 2019 issued by The sixth World Internet Conference held in Wuzhen, Zhejiang Province in Oct.20, 2019, China has 0.89 billion netizens and Internet penetration mounted 59.6% [1–4]. There are 5.23 million websites and 281.6 billion web pages. What's more, with the promotion of the commercial process of 5G, there will be a dramatic development in Big data, cloud computing, IOT and data size. The traditional way to find the relevant information is to use Internet search engine, but the efficiency is low if search from Big data, and it is also not conductive to data processing and analysis. Web Crawler, also called Web Spider, was originally developed by Matthew Grey from MIT in 1993 [5]. It is a contemporary to World Wide Web and it can't survive without the Internet by nature. If compare the Internet to a spider net, the web crawler is the crawling spider on the net. By

© ICST Institute for Computer Sciences, Social Informatics and Telecommunications Engineering 2020
Published by Springer Nature Switzerland AG 2020. All Rights Reserved
X. Jiang and P. Li (Eds.): GreeNets 2020, LNICST 333, pp. 162–168, 2020.
https://doi.org/10.1007/978-3-030-62483-5_18

Requesting URL address, the crawlers collect and analyze data by responded contents. For example, if the responded content is html, a DOM structure will be analyzed, parse it and adopt regular match; If the responded content is xml/json, a data objects will be converted and make further analysis.

The distributed crawler uses many computers and many crawlers to coincide with many urls. The distributed system can greatly improve program grabbing efficiency, fault tolerance, and its own Message Queue ensuring the disposable web node to url [6]. There are a great many web crawler researches up to now, and the first crawler is Wanderers [7]. Mercator, based on Java, was developed in 1999 with good scalability. The Apache Foundation published a Java crawler program Nutch [8]. People like Mehdi Bahrami applied distributed crawler to cloud platform. Every crawler, capable of storing a lot of unstructured data, stores results in Cloud Azure, even in Azure Blob. The studies of web crawler are relatively less and late in China. Shanghai University studied distributed crawler based on P2P in 2005. A study of Shanghai Jiaotong University called Igloo, further optimized the performance of distributed crawler system. Therefore, a distributed crawler with a high speed and a high level needs to be devised and applied in the actual situation.

2   Methods

Scrapy is a framework for web scraping developed by Python for scraping web sites and extracting structured data from pages. This framework is more efficient and does not need to consider the problems of multi-process and multi-threading, and the module classification is relatively detailed, which is widely used for data mining and automated testing. Scrapy uses Twisted which is an asynchronous framework to handle network communication. The architecture is clear and includes various middleware interfaces, which can send multiple requests at the same time. Scrapy is a popular Python event-driven networking framework written by Twisted.

2.1   Scrapy Framework

The overall architecture is roughly as follows in Fig. 1.

First, the program gives the initial URL in Spiders, and the engine will pass the URL to the Scheduler; Next, the URL is taken from the scheduler and passed to the Downloader through Requests; After downloading from the Internet, the downloader returns the corresponding response to the Spiders; Spiders will parse and generate two parts, one of which is data and the other is the new URL; The data part is pushed to the project Pipeline for processing data and then the new URL is assigned to the Scheduler and the cycle continues.

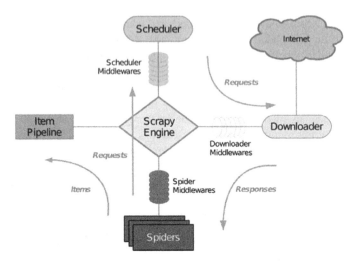

Fig. 1. Scrapy framework

2.2 Distributed Crawler Crawling Data

Open a command line window in PyCharm with Python 3.5, and set the crawler name in the project folder specified. Note: Generally, it is named after the website domain name when creating the crawler file Then there is the spiders folder and its directory structure. A projetc is designed in the crawler file. Define the crawled fields in the parse_item module, as shown in Fig. 2, the fields are necessary for crawling Python-related companies located in Nanjing in a website called "the future worry-free".

```
yield {
    "title" : title,
    "company_name" : company_name,
    "area" : area,
    "exp" : exp,
    "xue_li" : xue_li,
    "ren_shu" : ren_shu,
    "shi_jian" : shi_jian,
    "min_exp" : min_exp,
    "max_exp" : max_exp,
    "min_money" : min_money,
    "max_money" : max_money,
    "address" : address,
}
```

Fig. 2. Defined fields

2.3 Data Cleansing Processing

Data cleaning is the process of re-examining and verifying data, with the goal of removing duplicate information, correcting existing errors, and providing data consistency. The data in each web page is not necessarily the same format. There are many spaces, line breaks, tabs, fields may be misplaced, and the value is empty. Therefore, regular expressions are used to uniformly format the data. Data cleaning is a prerequisite for future data analysis. The specific methods such as using the 'split' method to split a piece of information, the 'strip' method to remove spaces, and then using the regular expression 'findall' to extract the specified content. Finally, these methods replace different data types with the same data type and unify the units. After processing the data, use the 'parse' method to extract the URL of the web page, use the 'for' loop to crawl from the first page, call the parse_item method each time to clean the data, and then parse the source code of the web page until it reaches the specified end. Up to one page. After all the data has been crawled, push them to the pipeline module, and open the database in the pipeline module to achieve the storage of the data. The format after the data cleaning process is completed is shown in Fig. 3.

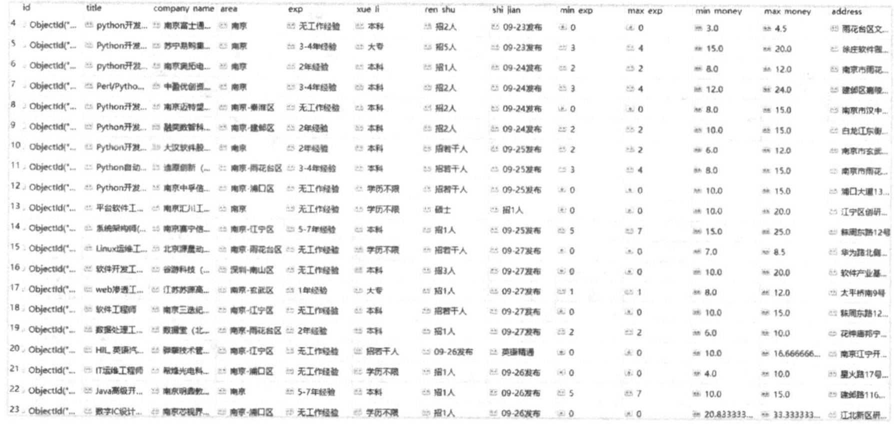

id	title	company name	area	exp	xue li	ren shu	shi jian	min exp	max exp	min money	max money	address
4	Objectid("... python开发...	南京富士通...	南京	无工作经验	本科	招2人	09-23发布	0	0	3.0	4.5	雨花台区文...
5	Objectid("... Python开发...	苏宁易购重...	南京	3-4年经验	大专	招5人	09-23发布	3	4	15.0	20.0	徐庄软件园...
6	Objectid("... python开发...	南京奥拓电...	南京	2年经验	本科	招1人	09-24发布	2	2	8.0	12.0	南京市雨花区...
7	Objectid("... Perl/Pytho...	中国电信股...	南京	3-4年经验	本科	招2人	09-24发布	3	4	12.0	24.0	建邺区奥体...
8	Objectid("... Python开发...	南京迈特曼...	南京·秦淮区	无工作经验	本科	招2人	09-24发布	0	0	8.0	15.0	南京市汉中...
9	Objectid("... Python开发...	融测敏智科...	南京·建邺区	2年经验	本科	招2人	09-24发布	2	2	10.0	15.0	白地江东南...
10	Objectid("... Python开发...	大汉软件股...	南京	2年经验	本科	招若干人	09-25发布	2	2	6.0	12.0	南京市玄武...
11	Objectid("... Python自动...	南京·雨花台区	南京·雨花台区	3-4年经验	本科	招若干人	09-25发布	3	4	8.0	15.0	南京市雨花...
12	Objectid("... Python开发...	南京中孚信...	南京·浦口区	无工作经验	学历不限	招若干人	09-25发布	0	0	10.0	15.0	江口大道13...
13	Objectid("... 平台软件工...	南京江川工...	南京	无工作经验	硕士	招1人	09-25发布	0	0	10.0	20.0	江宁区信研...
14	Objectid("... 系统架构师(...	南京嘉宇信...	南京·江宁区	5-7年经验	本科	招1人	09-25发布	5	7	15.0	25.0	科圈东路12号
15	Objectid("... Linux运维工...	北京源晨和...	南京·雨花台区	无工作经验	学历不限	招若干人	09-27发布	0	0	7.0	8.5	华为基北园...
16	Objectid("... Python开发工...	容游科技 (...	深圳·南山区	无工作经验	本科	招3人	09-27发布	0	0	10.0	20.0	软件产业基...
17	Objectid("... web渗透工...	江苏茅源高...	南京·栖霞区	无工作经验	本科	招若干人	09-27发布	0	0	8.0	12.0	太平桥街9号
18	Objectid("... 软件工程师...	南京三迭纪...	南京·江宁区	无工作经验	本科	招若干人	09-27发布	0	0	10.0	15.0	科圈东路12...
19	Objectid("... 数据处理工...	数据堂 (北...	南京·雨花台区	2年经验	本科	招1人	09-27发布	2	2	6.0	10.0	花神庙南宁...
20	Objectid("... HL 调用气...	骏馒技术管...	南京·江宁区	无工作经验	招若干人		09-26发布	0	0	10.0	16.666666...	南京江宁开...
21	Objectid("... IT运维工程师...	精捷光电科...	南京·浦口区	无工作经验	学历不限	招1人	09-26发布	0	0	4.0	10.0	星大路17号...
22	Objectid("... Java高级开...	南京锐鼎数...	南京	5-7年经验	本科	招1人	09-26发布	5	7	10.0	15.0	建邺路116...
23	Objectid("... 数字IC设计...	南京芯视界...	南京·浦口区	无工作经验	学历不限	招1人	09-26发布	0	0	20.833333...	33.333333...	江北新区研...

Fig. 3. Data cleansing processing

3 Results and Discussion

3.1 The Deployment and Implementation of Scrapy-Redis

The Scrapy framework does not support distribution. In order to achieve distributed crawling, foreign software engineers have developed a distributed crawler framework based on redis, allowing crawlers to have distributed crawling capabilities. The principle of this module is somewhat similar to big data, that is, distributed work, multiple machines work together to complete a goal based on the same URL. The machine that stores the URL list and uniformly manages the URL is called the master server, and other machines running crawlers become slaves.

Due to the queue mechanism of scrapy-redis, the links obtained by slaves node will not conflict with each other. In this way, after each slave completes the fetching task, the obtained results are summarized on the server. Experiments can be performed on the cluster. Without so many machines, multiple Linux virtual machines can be deployed on one physical machine. First, write a configuration file on the master to connect to the redis database, and copy the previously written crawler programs to the Linux virtual machine slaves. The crawling out of order is started and the initial URL of the website is run to crawl in redis-cli. Note that the URLs of the crawled web pages should be staggered. Finally, the data in redis is imported into mongodb. Figure 4 shows the process of distributed crawling. Figure 5 shows the data stored in the database after the crawling is completed.

Fig. 4. Distributed crawling process

3.2 Data Visualization Analysis

Data visualization technology converts data into graphs and charts to provide a basis for decision making. The research of data visualization technology has developed rapidly and achieved corresponding achievements. An analysis of the salary situation of Nanjing IT companies, and analysis of the range of the highest and lowest wages are available. In the IT industry, the minimum wage is generally taken when starting a job, and most can get nearly 10 k. After working for a few years, most of the maximum salary can

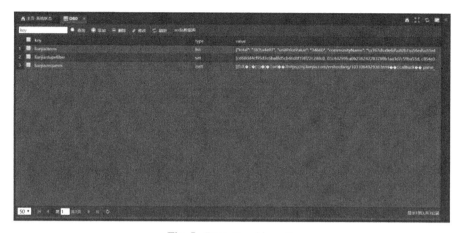

Fig. 5. Data stored in redis

reach 20 k, which is in line with the current average salary status. The data visualization analysis of salary is shown in Fig. 6.

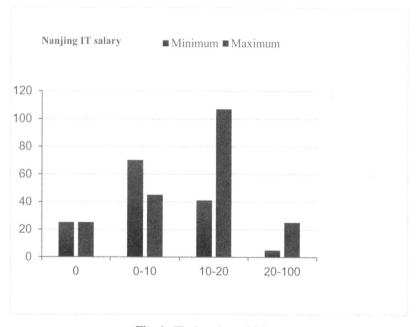

Fig. 6. The bar chart of Salary

The data visualization analysis of education is shown in Fig. 7. Undergraduates in the IT industry account for more than 50% of the undergraduates, with the highest proportion, while there are almost no PhDs, indicating that the IT industry is an industry for the youth only and attaches great importance to the application of technology.

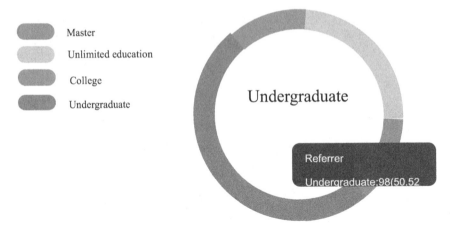

Fig. 7. The pie chart of Education

4   Conclusion

With the development of network technology, how to use crawler and data visualization technology to better understand users and their intentions is a key area in the network era. This paper studies how to collect relevant data from the Internet. Distributed technology makes data collection more efficient. It cleans and filters the collected data, and then displays the useful data and visual analysis to analyze and mine valuable information. There will make full use of the potential value of big data.

Acknowledgements. The work was supported by Hei longjiang Fundamental Research Funds for the Local Universities in 2018 (No. 2018-KYYWF-1189).

References

1. Jie, H.: China's Internet users reached 829 million and 5 G industrialization achieved preliminary results. CNNIC released a report (2019)
2. Biling, G.: 5G key technology and its impact on the internet of things. J. Wirel. Internet Technol. **4**(7), 30–32 (2019)
3. Yuezhong, S., Chao, C., Bin, S.: Analysis of communication technology and challenge of Internet of things in 5G network. J. Inf. Syst. Eng. **7**, 37–38 (2019)
4. Jingke, Z.: 5G key technology research. Inf. Commun. **3**, 261–263 (2018)
5. Sunguo, C.: Research on the implementation of web crawler in subject search engine. Comput. Knowl. Technol. **12**(17), 23–25 (2016)
6. Xiaohui, W.: Improvement of URL de duplication strategy for distributed web crawler. J. Pingdingshan Univ. **24**(5), 116–119 (2009)
7. Yuhao, F.: Design and implementation of distributed web crawler system based on scratch. Chengdu:University of Electronic Science and Technology of China (2018)
8. Hengfei, Z., Yuexiang, Y., Hong, F.: Research and optimization of nutch distributed network crawler. J. Front. Comput. Sci. Technol. **5**(1), 68–74 (2011)

Research upon the Smart Diving Suit Based on Visible Light Communication

Yang Zhou, Jinpeng Wang$^{(\boxtimes)}$, Xin Guan, Ailing Zou, and Nianyu Zou

Dalian Polytechnic University, Dalian 116034, China
wangjp@dlpu.edu.cn

Abstract. Due to the poor anti-noise performance and high delay of the traditional underwater communication technology in the signal transmission process, the safety of divers is potentially threatened in the complex underwater environment. To solve this problem, this paper designed a real-time monitoring of physiological data smart diving suit. Based on the addition of sensor functions to diving suits, this paper researched the overall architecture of visible light communication model. Taking advantage of the high transmission rate and strong stability of visible light communication, blue LED light source is used for signal transmission, and simulation experiments and tests are carried out. The results show that the basic requirements of underwater data transmission are satisfied within a certain communication distance. This provides a reference for further research on underwater communication technology.

Keywords: Visible light communication · The sensor · The smart clothing

1 Introduction

At present, with the increasing research and exploration of various waters, the demand for underwater activities has become increasingly diversified. However, the traditional methods of underwater activities still have some limitations. On the one hand, due to the complex water environment, the irregular operation of divers and the poor diving equipment, the health and safety of divers cannot be guaranteed [1]. On the other hand, traditional wireless communication is difficult to effectively use underwater. The current mainstream underwater wireless communication methods are underwater acoustic communication technology, underwater radio frequency communication technology and underwater quantum communication technology [2, 3]. Underwater acoustic communication has the problems of being greatly affected by complex underwater channels, serious multipath effects, and low communication rates; Although the traditional underwater electromagnetic wave communication is stable, the attenuation of electromagnetic waves in water is very serious. Experiments conducted by Lloret J. in 2016 showed that RF signals can be transmitted at a rate of up to 100 Mbps, but the transmission distance is extremely short, the effective communication distance is only about 15 cm [4]. Visible light communication has developed rapidly in recent years, which provides a new way

© ICST Institute for Computer Sciences, Social Informatics and Telecommunications Engineering 2020
Published by Springer Nature Switzerland AG 2020. All Rights Reserved
X. Jiang and P. Li (Eds.): GreeNets 2020, LNICST 333, pp. 169–181, 2020.
https://doi.org/10.1007/978-3-030-62483-5_19

for underwater communication. For the development of visible light communication, Japan has started its related technology research earlier and matured. Japan's Uema H team has developed a portable underwater communication system to address the drawbacks of traditional underwater word board communication, it meets the communication needs of underwater entertainment [5, 6]. The basic principle of blue-green visible light communication is that the part of the blue-green light in the range of 450–550 nm is in the transmission window [7, 8], it has strong anti-fading ability in water and can achieve the purpose of underwater communication.

This paper proposes a research on an intelligent diving suit based on underwater blue light LED communication. The sensors is implanted in the diving suit to achieve the purpose of dynamically monitoring the physiological data and underwater environmental data of the diver. The blue-green optical communication method has the characteristics of low loss and high transmission rate, which can be used to improve the effectiveness and reliability of the whole communication system [9, 10].

2   System Design Principle and Scheme

The system is mainly composed of a sensor data collection module and a blue visible light communication module. Its system block diagram is shown in Fig. 1. The function of the sensor data collection module is to embed the temperature sensor MLX90614, the heart rate sensor MAX30102, the pressure sensor MS5803 and the MCU into the diving suit through embedded technology. The function of the sensor data collection module is to embed the temperature sensor MLX90614, the heart rate sensor MAX30102, and the MCU into the diving suit through embedded technology, collect the physiological data information of the diver, and then use the blue visible light communication module to modulate the physiological data information after modulation processing. Send to the host computer to complete the physiological data collection and monitoring of the diver.

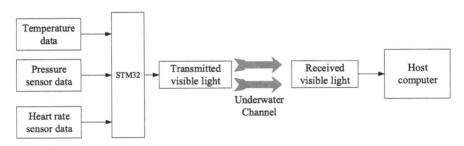

Fig. 1. System block diagram (Color figure online)

In order to meet the needs of practical application, the design of this system should meet the requirements of low power consumption, lightweight, low cost and fast transmission rate.

3   Sensor Data Acquisition Module

3.1   Sensor Module of MLX90614

The MLX90614 sensor can be powered by 3 V or 5 V, and the measurement range reaches $-70\,°C + 380\,°C$. In order to meet the requirements of medical applications, the MLX90614 can also meet the high accuracy requirements ($\pm 0.1\,°C$) in the human body temperature range. In addition, its non-contact temperature measurement method, which uses the internal infrared induction thermopile chip to convert the target's thermal radiation signal into an electrical signal [11], it has the advantages of short response time and easy to realize dynamic measurement. The bottom view of the MLX90614 chip structure and function is shown in Fig. 2. Among them, the function of pin 1-SCL/Vz is to realize serial clock input, and the function of pin 2-SDA/PWM is to perform digital input and output, and can be read Measure the temperature of the object.

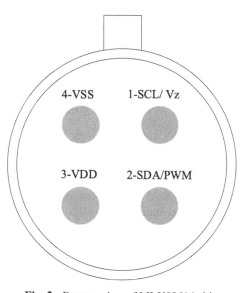

Fig. 2. Bottom view of MLX90614 chip

3.2   Sensor Module of MAX30102

The working voltage of the MAX30102 sensor can also be 3.3 V or 5 V. Its communication interface can use the standard I2C interface for data transmission. Its data output speed is fast, the sampling rate is high, and the operating temperature range is $-40\,°C$ to $+85\,°C$. At the same time, the sensor can be controlled by software to close or open, the standby current is close to zero, and the power supply is always maintained [12, 13]. The light source working inside the chip is red light at 660 nm and infrared light at 880 nm. The optical signal is received by a high-sensitivity phototransistor. The received signal is amplified to support both analog filtering and digital filtering. Finally, the processed data is stored in the internal memory for external MCU to read. The pin diagram of MAX30102 is shown in Fig. 3.

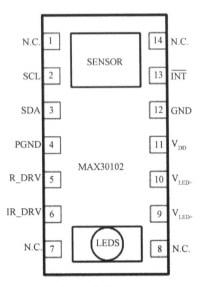

Fig. 3. MAX30102 pin diagram

3.3 Sensor Module of MS5803

MS5803-01BA is a high resolution professional measuring pressure sensor, its outer package is made of ceramic material. It includes an ultra-low-power 24-bit ADC and high linearity sensing module. The communication protocol is simple and does not require programming in the device's internal registers. What's more, it provides a variety of different operating modes, allowing the user to select the conversion speed and sampling frequency.

3.4 Application Principles of Sensor Embedding

The application of sensors in diving suit must fully consider the characteristics of safety, stability, comfort and miniaturization [14]. The details are as follows.

(1) Security. While ensuring the normal operation of the components, it is also necessary to consider whether the electronic components will adversely affect the human body. To ensure safety, for example, a flexible circuit board can be used for packaging.
(2) Stability. The overall package of the system should be firm and electronic components should not be placed in the body's joints and other active parts.
(3) Comfort. The way of wearing must not hinder the normal underwater activities of divers, and it embodies the "people-oriented" principle.

(4) Micromation technology. Components should meet the requirements of small size and light weight, to avoid the appearance of foreign body feeling when wearing, and do not affect the structure of the diving suit [15].

3.5 The Safety Range of Related Physiological Indicators

Heart rate and body temperature are important physiological parameters for monitoring underwater activities of personnel. It is of great significance to maintain the optimal heart rate and normal body temperature to ensure the safety of personnel [16]. In exercise physiology, Table 1 shows the relationship between heart rate and human tolerance. In Physiology, Table 2 shows the relationship between body surface temperature and human subjective feelings. As can be seen from the table, the exercise heart rate of divers is guaranteed to be below 160 BPM, and the body temperature is maintained at 31.5 °C to 34.5 °C, which is relatively appropriate.

Table 1. The relationship of heart rate and body limits

Heart rate (BPM)	Exercise intensity
120–140	Low amount of exercise
140	Normal load
141–160	Medium amount of exercise
161–180	Sub-limit load
More than 180	Ultimate load

Table 2. The relationship of surface temperature and subjective feeling

Surface temperature (°C)	Subjective feeling
Less than 27	Extremely cold
28–29	Shiver
30–31	Cool
31.5–34.5	The most suitable

3.6 The Combination of Wearable Devices

Figure 4 is a schematic diagram of the arrangement of electronic components in a smart diving suit. Place the LED light on the chest to transmit the signal, fix the heart rate sensor on the inside of the cuff with Velcro, and place the temperature sensor on the back. Pressure sensor placed on the neck of the suit. Each sensor is packaged in a soft, comfortable fabric in the form of a pouch, which is secured with Velcro. These devices can be applied to human skin for a long time and reduce friction, so as to reduce the foreign body sensation caused by wearing skin. In addition, the wires are arranged to ensure that the overall circuit is smooth, and fixed with adhesive tapes to prevent short circuits. Finally, after analyzing the underwater activities of the divers, we know that the frequency and amplitude of the activity on the back of the human body is relatively small, so the controller is placed in a patch pocket on the back to ensure the stability of the system work.

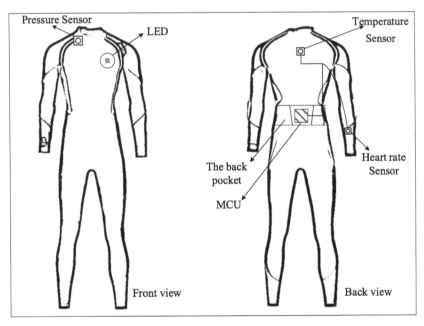

Fig. 4. Schematic diagram of arrangement of sensors and other components

4 Blue Visible Light Communication Module

4.1 Overview of Visible Light Communication System

The schematic diagram of the communication system is shown in Fig. 5. The transmitting end is mainly composed of blue LED and signal modulation drive circuit. Its main function is to modulate the data of the sensor, drive and light the blue LED lamp, and modulate the obtained electrical signal into the driving current of the LED lamp to realize the transmission of blue visible light. In order to effectively increase the bandwidth of the system, Osram LB25 series high-power LED lamps with a shorter carrier lifetime were selected [17], and its wavelength range is 455 nm–470 nm. At the signal receiving end, the photoelectric detector is used to convert the optical signal received from the transmitting end into an electrical signal, and then the operational signal is amplified by the operational amplifier circuit. The amplified electrical signal is restored to the original after demodulation and A/D conversion data. The photodetector chooses PIN photodiode, which has a high response frequency and a much cheaper price than APD avalanche photodiode. Moreover, the use of PIN photodiode to transmit signals has the characteristics of high signal-to-noise ratio, stable performance and small noise.

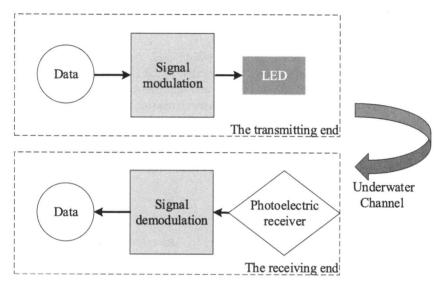

Fig. 5. Schematic diagram of VLC (Color figure online)

4.2 Modulation of Blue LED Signals

This article uses CAP (Carrier-Free Amplitude/Phase Modulation) modulation technology to modulate the optical signal. CAP modulation technology is developed based on QAM modulation technology, they are all single carrier modulation [18]. Compared with OOK modulation, CAP modulation is less disturbed by the pulse noise with abundant low frequency energy and the near-end crosstalk of high frequency. There is no low frequency delay distortion, and the inter-code interference caused by group delay distortion is also smaller.

The synthesis and modulation of CAP signal can be realized in the digital domain, and because of its characteristics of modulation without carrier, even if it is used in the visible light communication system which is more sensitive to the system nonlinearity, the peak-to-mean power ratio of the system will not be too high, so this is a major advantage of the modulation technology applied in the visible light communication [19]. The block diagram of CAP modulation principle is shown in Fig. 6.

4.3 Blue LED Driving Circuit

After the sensor signal is modulated, the working system needs to drive the LED light source to convert it into a light signal and pass it to the receiving end. However, the driving

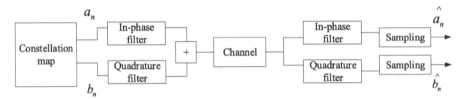

Fig. 6. CAP modulation principle block diagram

capability of the single chip microcomputer is insufficient, so an additional transistor is used to improve the driving efficiency. The function of the capacitor C1 is to increase the switching speed of the LED driving circuit and improve the frequency performance of the entire circuit. Among them, R1 is an adjustable resistor, and R2 is used to balance the impedance of the input terminal and balance the voltage. R3 and R4 play the role of protection circuit to prevent short to ground. R5 is a DC bias resistor. By adjusting the size of resistor R5, the current through the blue LED light source can be adjusted to meet different requirements for actual communication. The schematic diagram of LED drive circuit is shown in Fig. 7.

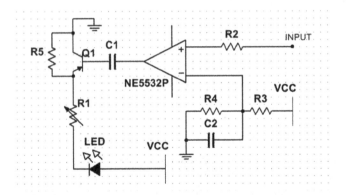

Fig. 7. LED drive circuit block diagram

4.4 Visible Light Receiving Circuit

The signal receiving end is mainly composed of a photodetector, a preamplifier circuit, a controller, and a power supply. The function of the capacitor C1 is to filter out the AC component, so that the DC stabilized power supply VCC outputs a stable DC. Capacitor C2 can filter out the noise superimposed on the signal to achieve the effect of noise reduction. The photodiode D1 converts the light signal emitted by the receiving end into an electric signal, but at this time the signal is relatively small, it needs to be amplified by a preamplifier circuit, and then the controller demodulates it to obtain sensor data. Finally, the diver's body data information is sent to a host computer, and the monitoring party analyzes the dive's physiological indicators to ensure the safety of the diver's underwater activities. The schematic diagram of the photoelectric receiving circuit is shown in Fig. 8.

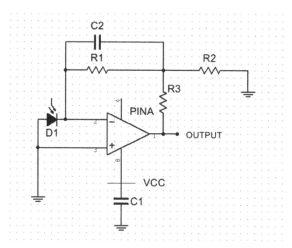

Fig. 8. Visible light receiving circuit block diagram

5 System Test

5.1 Indoor Environment Test

The test location is Lab 103, Teaching Building B, Dalian Polytechnic University. The underwater test chamber with a size of 70 cm * 30 cm * 40 cm is used to simulate the underwater channel. As shown in Fig. 9.

The test in this paper first sets the effect of sending and receiving sensor data under ideal conditions through programming and software simulation, as shown in Fig. 10. The upper square wave is the ideal waveform output by the blue LED driving circuit, and the lower square wave is the ideal waveform of the visible light receiving circuit.

The second step is to use an oscilloscope to detect the waveform sent by the LED in the actual communication process and the waveform received by the circuit at the receiving end, and to compare it to verify the communication effect of the system. After the data information of the sensor is modulated, it is transmitted through the drive circuit of blue light LED. The photodiode of the receiver module receives the optical signal, the signal is amplified by the amplifier circuit, and finally sent to the upper computer through the demodulation of MCU. Figure 11 is the waveform generated by the LED driver, and Fig. 12 is the waveform generated by the photoelectric receiver. The test results show that the waveform of the output signal of the photoelectric receiving end of the communication system is basically the same as the signal of the transmitting end, which indicates that the communication scheme of the system is feasible.

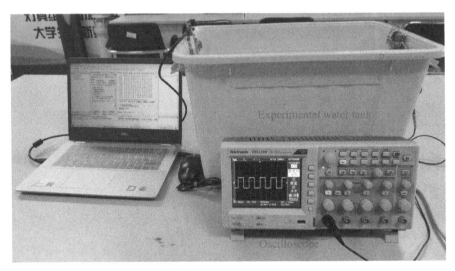

Fig. 9. System experimental environment

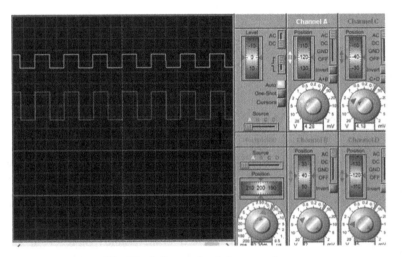

Fig. 10. Software simulation waveform

5.2 Outdoor Environmental Test

The outdoor experiment was conducted in a large artificial lake on campus. Compared with the indoor experimental communication system, the outdoor experimental communication system adds a condenser lens at the emitting end of the LED to reduce the emission Angle of visible light, so as to increase the intensity and distance of visible light and greatly improve the transmission quality of visible light signals. As shown in Fig. 13, when the illumination direction of the blue light LED is aligned with the direction of the photoelectric sensor at the receiving end with the power supply of 5 V, the length of the beam directly observed is about 5.4 m.

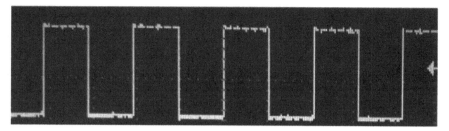

Fig. 11. Waveform output from the driver

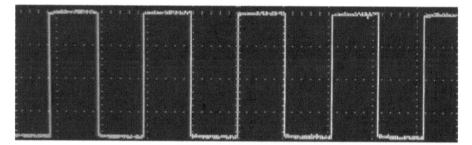

Fig. 12. Waveform received by the receiver

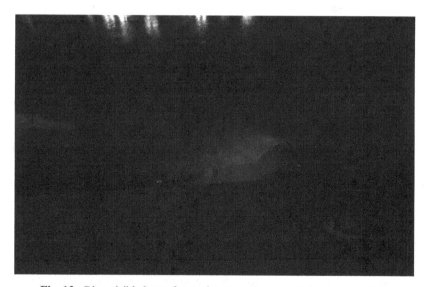

Fig. 13. Blue visible beam for outdoor experiments. (Color figure online)

The experimental data obtained after calculation and arrangement of the experimental data are shown in Table 3. The experimental results show that the system can meet the daily data transmission requirements in outdoor natural underwater channels. In addition, error-free data transmission is achieved when the communication distance between the transmitting end and the receiving end is 5 m and 8 m.

Table 3. The experimental results

Communication distance (m)	Number of error bits	Bit error rate (%)
5	0	0.000
8	0	0.000
15	2	0.277
20	5	0.694

6 Conclusion

Aiming at the disadvantages of traditional underwater equipment and communication methods for divers, this paper designs a smart diving suit that can measure human physiological data based on the underwater visible light communication system. This paper introduces the design idea of the sensor information acquisition module and the overall structure of the visible light communication system. The blue LED light source was used to transmit signals in the underwater channel, and the sensor data was successfully received, thus achieving the purpose of monitoring and warning human physiological indicators. This study also provides reference for future underwater communication activities. Based on the above design, the scheme of the system can still be further optimized, for example, by adjusting the power of the LED light source to meet the communication needs in different environment, there by improving the effectiveness and reliability of data transmission.

Acknowledgements. This research was financially supported via Project of the National Natural Science Foundation of China (61402069), the 2017 Project of the Natural Science Foundation of Liaoning province (20170540059), the General project of Liaoning education department in 2016 (2016J205).

References

1. Zhong, W.: Status and reform suggestions for maintaining navigation safety of local maritime water and underwater activities. China Water Transp. (11), 42–43 (2018)
2. Hu, F.C., Chi, N.: The principle, key technology and applications of underwater visible light communication. China Light Light. (1), 6–13 (2018)
3. Fletcher, A.S., Hamilton, S.A., Moores, J.D.: Undersea laser communication with narrow beams. IEEE Commun. Mag. **53**(11), 49–55 (2015)
4. Sendra, S., Lloret, J., Jimenez, J.M., et al.: Underwater communications for video surveillance systems at 2.4 GHz. Sensors **16**(10), 1769 (2016)

5. Arnon, S.: Underwater optical wireless communication network. Opt. Eng. **49**(1), 15001–15005 (2010)
6. Uema, H., Matsumura, T., Saito, S.: Research and development on underwater visible light communication systems. Electron. Commun. Japan **98**(3), 9–13 (2015)
7. Zhao, Y.H., Wang, C., Wang, F.M.: 1.725 Gb/s underwater visible light communication system based on a silicon substrate green LED and equal gain combination receiver. In: International Conference and Exhibition on Visible Light Communications (Invited Talk 3). ICEVLC, Yokohama (2018)
8. Wang, C., Yu, H.Y., Zhu, Y.J.: A long distance underwater visible light communication system with single photon avalanche diode. IEEE Photonics J. **8**(5), 1–11 (2016)
9. Chi, N., Wang, C.F., Li, W.P.: Research progress of underwater visible light communication technology based on blue/green LED. J. Fudan Univ. (Nat. Sci.) **58**(05), 537–548 (2019)
10. Hu, F., Wu, G.F., Wu, G.: Design of two-way underwater communication testing system based on blue light LED. Electron. Des. Eng. **24**(08), 123–126 (2016)
11. Liu, C., Wu, H., Xiong, D.: Nonlinear calibration method of detector in thermal radiation temperature measurement system. Chin. J. Quantum Electron. **34**(2), 227–230 (2017)
12. The Maxim MAX30102 wearable blood oxygen and heart rate biosensor solution. Glob. Electron. China (04), 45–48 (2018)
13. Bai, P.F., Liu, Q., Duan, F.B.: Wearable blood oxygen saturation detection system based on MAX30102. Laser Infrared **47**(10), 1276–1280 (2017)
14. Chaussabel, D., Pulendran, B.: A vision and a prescription for big data-enabled medicine. Nat. Immunol. **16**(5), 435–439 (2018)
15. Zhang, S.M.: Research progress of blood glucose monitoring technology. Cap. Food Med. **24**(10), 23–24 (2017)
16. Deng, S.X., Chen, P.J., Qiao, D.C.: Introduction to Exercise Physiology, vol. 12. Beijing Sport University Press, Beijing, pp. 152–163 (2007)
17. Zhu, S.C., Zhao, L.X., Yang, H.: Device technology of light emitting diodes used in visible light communication. ZTE Technol. J. **20**(06), 29–32 (2014)
18. Zou, P., Liu, Y.F., Wang, F.M., Chi, N.: Mitigating nonlinearity characteristics of gray-coding square 8QAM in underwater VLC system. In: Asia Communications and Photonics Conference, pp. 1–3. IEEE, Hangzhou (2018)
19. Wang, X.D., Cui, Y., Wu, N.: Performance analysis of optical carrierless amplitude and phase modulation for indoor visible light communication system. Acta Photonica Sinica **46**(05), 109–118 (2017)

Study on 3D Reconstruction of Plant Root Phenotype Based on X-CT Technique

Xin Guan[1], Jinpeng Wang[1(✉)], Yang Zhou[1], Kemo Jin[2(✉)], and Nianyu Zou[1]

[1] Information Science and Engineering College,
Dalian Polytechnic University, Dalian 116034, China
wangjp@dlpu.edu.cn
[2] Key Laboratory of Plant-Soil Interactions, Ministry of Education, China Agricultural
University, Beijing 100193, China
kemo.jin@cau.edu.cn

Abstract. The change of the global climate in recent years influences the adaptability of plant roots to the soil environment. Detecting the plant roots in the original soil environment without destroying and moving them and analyzing, processing and reconstructing the root image requires the improvement of technology. Firstly, to conduct the in-situ non-destructive measurement of the root system in soil, this paper investigates and studies the three-dimensional imaging of plant root phenotype using X-CT technology. Secondly, to examine the feasibility of the segmentation methods applied in medical image analysis in root CT images, the experiments of segmentation using fuzzy clustering of CT images were designed. The segmentation results of root features obtained from 10, 30 and 50 times clustering algorithms for the in-situ root CT images are analyzed with the FCM (Fuzzy C-Means) algorithm. The experimental results show that the traditional medical image segmentation method does not produce a good segmentation effect in this particular environment.

Keywords: In situ non-destructive measurement · X-CT tomography · Fuzzy clustering segmentation · FCM algorithm

1 Introduction

The root system is an important organ for plants to obtain nutrients and water. Due to the invisible opacity of the root system in soil, it is difficult to observe and study the underground part of plants [1], this makes the research on the above-ground part of the plant far more than the underground part. In the early days of plant root research, the existing methods of observing plant roots can be divided into two categories [2], one is a destructive method and the other is a non-destructive method. At that time, limited by soil media and imaging technology under plants, researchers often adopted destructive sampling methods, including container method, tube loading method, nail

---

J. Wang and K. Jin—Contributed equally to this work.

© ICST Institute for Computer Sciences, Social Informatics and Telecommunications Engineering 2020
Published by Springer Nature Switzerland AG 2020. All Rights Reserved
X. Jiang and P. Li (Eds.): GreeNets 2020, LNICST 333, pp. 182–192, 2020.
https://doi.org/10.1007/978-3-030-62483-5_20

plate method, air culture method, mesh bag method, etc. Non-destructive methods as opposed to destructive methods, such as underground root chamber method, isotope tracer method, etc. [3]. In 2017, the French Academy of Agricultural Sciences, with the help of the high-throughput plant phenotyping platform, RhizoTube, a high-throughput micro root window measurement system for plant root phenotype. Automatic, high-throughput and noninvasive long-term root growth monitoring of plant root phenotype was realized [4]. While these methods guarantee the complete measurement of plant roots, they are all based on the study of plant root characteristics under an ideal environment, but still cannot reach the original growth environment of plants in reality. The disadvantages lie in the following aspects. Firstly, the cultivation of transplanted seeds cannot completely restore the original growth environment of plants; Secondly, when the root system of a plant is taken out for contact measurement, the roots with a too-small diameter around it will break, which will damage the accuracy of the final measurement; Thirdly, the observation and measurement of the root system are not complete and it can only be measured locally. With the advancement of medical imaging technology in recent years, the imaging methods of root nondestructive measurement are mainly divided into two types: one is based on nuclear magnetic resonance (NMR) imaging [5] and the other is based on X-CT imaging. Because the principle of NMR technology in hydrogen resonance degree impact will be limited by the influence of water in the soil, the energy released by the complex medium in the soil will affect the determination of the position and type of the nucleus, making its imaging resolution lower [6].

This paper investigates and studies a series of principles and methods for 3D imaging of plant root phenotypes based on X-CT [7]. To achieve unified in situ non-destructive measurements of different complex plant root systems. Firstly, it is necessary to conduct high-precision non-destructive imaging of roots in soil media based on a kind of high-resolution X-CT imaging equipment. The principle of fuzzy clustering image segmentation is studied in this paper [8–10] and the FCM algorithm [11–13] was used to perform clustering operations on CT images of the root in situ for 10, 30 and 50 times. It can be found from the segmentation result diagram of the obtained root features that FCM algorithm of fuzzy clustering segmentation can perform an iterative operation on the CT image of the in-situ root system, however, it is difficult to achieve satisfactory results for CT images of the root in situ with more noise and artifacts using traditional medical segmentation methods.

2 Systematic Research Route

The research technology route of three-dimensional imaging of plant root phenotype based on X-CT technology is shown in the following figure (Fig. 1):

The main research contents of this paper include: The X-CT technique was used to conduct three-dimensional imaging of the root phenotype in the soil. The feasibility of applying the fuzzy clustering algorithm commonly used in the medical image to the CT image of the root in situ was also explored.

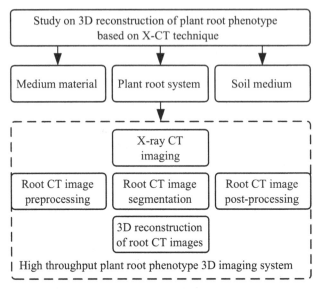

Fig. 1. Research technology roadmap

3 Study on Root Phenotypes with High Throughput

In the past two years, with the rapid development of plant phenotype, high-throughput detection technology has been gradually optimized and applied in the field of plant phenotype detection. For example, the visible light high-throughput optical imaging instrument with relatively high wavelength can be used for non-contact imaging of plant leaf area, seed morphology, spike type, root system and other physical structures. There are also transmission images of subsurface parts of plants using non-visible light with wavelengths between 100 μm and 500 μm [14]. The investigation found that the purpose of high-throughput plant phenotype detection is to better study the genotypes and genetic traits of plants in the next step and build a front-end plant phenotype database. Based on the characteristics of high throughput technology, such as high yield or output per unit time, and combined with the research route in this paper, the research on root phenotype detection of high-throughput plants can be further divided into three parts. Firstly, high-throughput plant root phenotype X-CT imaging system; Secondly, CT image data analysis of plant root phenotype with high throughput; Thirdly, 3D reconstruction of CT images of high-throughput plant root phenotypes.

With the rapid development of deep learning and machine vision research [15], it has brought the possibility of development to the data analysis of high-throughput root phenotypic CT images and the 3D reconstruction of high-throughput CT images. Plant phenotypic omics is an interdisciplinary subject, image processing for the obtained large number of CT images of the root system in situ has also gradually crossed into the field of computer [16]. Based on the analysis of CT image data analysis of high-throughput plant roots, there are few types of research at home and abroad.

Due to the limitations of the research and development of a high-throughput root CT image imaging system, high-throughput root CT image processing will remain the bottleneck of the new development of plant root phenotype for a long time to come. The high-throughput X-CT plant root phenotypic imaging system only completed the collection of root phenotypic data. As an intermediary link, how to analyze the root characteristics from the huge root image database obtained. The phenotypic classification obtained through image analysis further understands the morphological structure and functional process of plants, and it can find out useful information for plant phenotyping. That's all that matters.

4 X-Ray CT Imaging

4.1 Linear Attenuation Coefficient

In recent years, with the development of imaging technology in the medical field, X-CT has established an unshakable position as a high-performance non-destructive diagnostic technology in the medical field. X-CT has not only brought unprecedented revolutionary influence in the field of medicine, but also been successfully applied in many fields such as industrial nondestructive testing, geophysical resource exploration and botanical determination [17]. Preliminary studies at home and abroad show that it is feasible to use X-CT to image plant roots.

X-ray CT imaging is different from traditional X-ray imaging. For the complex spatial structures of plant roots that block and cross each other, traditional X-ray imaging cannot achieve the effect, but CT imaging technology can solve this problem.

As can be seen from Fig. 2, X-CT imaging uses x-rays to create a tomography image of an object's interior, Since some of the X-ray attenuations occurs when the X-ray penetrates the object, X-CT imaging is essentially the imaging of linear attenuation coefficient, the ultimate goal of CT image reconstruction is to solve the μ value of each voxel (Fig. 3).

Known from Lambert's law in physics [18], when the monochromatic wiring harness passes through an object of uniform density, the energy of the wiring harness is weakened by the interaction between the atoms of the material. The weakening degree is related to the thickness and absorption coefficient or composition of the material. It can be expressed by the following formula:

$$I = I_0 e^{-\mu d} \tag{1}$$

Where I_0 is the intensity of incident X-ray; I is the ray intensity transmitted after passing through an object of uniform density; μ is the linear coefficient of matter to this wavelength; d is the path length through the object of uniform density; e is the natural logarithmic base. We can see from the formula, the μ value is related to the X-ray energy, the atomic coefficient and density of the substance, the larger d or μ, the smaller I can be deduced, that is the greater the attenuation of X-rays. From the conditions of lambert's law [19], X-rays penetrate objects of uniform density. For soils with complex structures, the formula still needs to be transformed (Fig. 4).

Suppose the object is divided into equal-length segments, each of length d, and d is small enough, assume that the density and attenuation coefficient of each segment are uniform, and the incident intensity of the first segment of X-ray with length d is I_0,

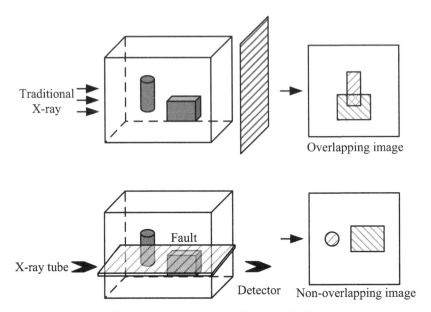

Fig. 2. Comparison of traditional X-ray and X-CT imaging

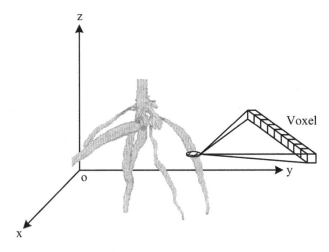

Fig. 3. Root spatial voxel

According to formula (1), the X-ray intensity I_1 of the first transmission can be obtained. According to the principle of calculus, there are:

$$I_n = I_0 e^{-(\mu_1 d + \mu_2 d + \ldots + \mu_n d)} \tag{2}$$

The transformation formula (2) that we can get:

$$\mu_1 + \mu_2 + \ldots + \mu_n = \frac{1}{d} \ln \frac{I_0}{I_n} \tag{3}$$

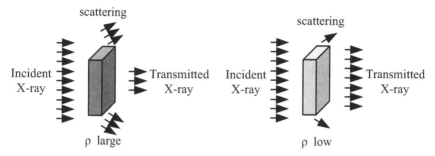

Fig. 4. The effect of density on X-ray attenuation

Therefore, under the premise that d, I_0 and I_n are known, the projection equations in multiple directions must be established to calculate the values of μ.

4.2 CT Number

In CT applications, the linear attenuation coefficient μ is not very descriptive and depends to a large extent on the X-ray spectral energy. Therefore, the CT number needs to be calculated relative to the attenuation of water. The data set generated by CT imaging technology is usually composed of CT values, it is defined as the X-ray attenuation coefficient of an object minus the water X-ray attenuation coefficient divided by the water attenuation coefficient and then multiplied by 1000. The unit of the result is expressed in Hounsfield.

$$CT_{number} = \frac{\mu_{material} - \mu_{water}}{\mu_{water}} \times 1000 \tag{4}$$

In formula (4) μ is the attenuation coefficient of X-ray, by definition, the CT number of air is -1000, the CT number of water is 0. Materials with an X-ray attenuation coefficient greater than water have positive CT value, while those with an X-ray attenuation coefficient less than water have negative CT value. When performing X-ray tomography, X-rays can make the voxels in each tomography scan to generate a CT value. This small

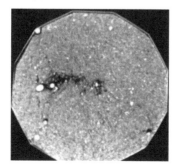

Fig. 5. CT image of in situ root system

space is regarded as a cube and 3D image data after the X-CT scan. The following figure shows the tomographic image data obtained by X-CT imaging (Fig. 5):

5 Root Image Segmentation

5.1 Image Feature

Among the data obtained from the three-dimensional imaging of plant root phenotypes based on X-CT technology, root analysis is a critical and hot CT image processing problem that needs to be accurately solved. Root image segmentation is also regarded as an important pre-processing link before three-dimensional root reconstruction. The central idea is to determine whether a specific voxel in the image space is "root" or "non-root". Unlike the micro-root window technology of transplanting seeds cultivated or visualized in a transparent nutrient base, which can identify the root imaging from the visible image. The root CT image is intended to be segmented from other substances (water, air, small stones, etc.) taking root features out is regarded as a technical problem in CT image processing. Because in the CT image, the root and some non-root matter will have overlapping CT numbers, in this process, the image segmentation methods involved in feature extraction are diverse and varied. Therefore, a root segmentation method suitable for the in-situ root CT image should be found.

5.2 Fuzzy Clustering Image Segmentation

The CT image segmentation of the root system is based on the difference in grayscale, texture, color and spatial characteristics of the root system in the soil medium. It benefits from the mature medical CT technology used in the imaging. After investigation and research, it is found that the clustering segmentation method is suitable for the segmentation and processing of medical images, which can well describe the fuzziness and uncertainty in general CT images.

Blurring is a fundamental feature of most medical imaging. It is difficult to make a short-term breakthrough in the research of imaging equipment in the early stage and there is no set of imaging equipment that can fully apply to the high resolution and high resolution of plant roots at the present stage, so the CT images have an artifact, noise, and fuzzy root boundary. FCM algorithm has been widely used in medical image segmentation [20].

6 FCM Algorithm

In the fuzzy clustering algorithm, the FCM algorithm is also known as the fuzzy c-means algorithm, which is an improved algorithm compared with the traditional algorithm [21]. It divides all the pixels in the image into y fuzzy groups. For each fuzzy group, the clustering center of each group of pixels is solved. Through continuous iteration, the objective function is minimized. The objective function is:

$$J(a, b) = \sum_{i=1}^{m} \sum_{j=1}^{n} \mu_i(x_j)^k \left\| x_j - b_i \right\|^2 \tag{5}$$

In the formula, $\mu_i(x_j)^k$ represents the membership function of the j-th pixel to the i-th category. Satisfy $0 \leq \mu_i(x_j)^k \leq 1$, b_i is used to represent the clustering center of class i, we can get the Euclidean distance $\|x_j - b_i\|^2$. It is also a similar measure of the algorithm.

In the FCM algorithm, $\sum_{i=1}^{m} u_i(x_j)^k = 1$, which is expressed as the membership degree of each cluster center is 1. In this case, the minimum value of Eq. (5) is solved. The iterative and fuzzy operations are used to solve the cluster center value to complete the fuzzy classification of the collected image data.

The specific steps of the FCM algorithm are shown in the figure below (Fig. 6):

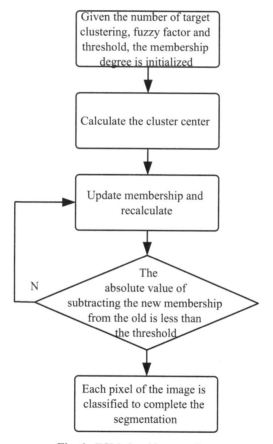

Fig. 6. FCM algorithm step flow

7 Analysis and Discussion

In situ CT tomography of plant roots was performed, as shown in Fig. 7:

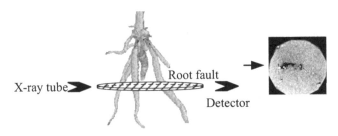

Fig. 7. Root CT imaging

We perform N = 10 times, N = 30 times, and N = 50 fuzzy operations on the obtained in situ root CT images by MATLAB R2017a, the obtained segmentation map is:

In this figure, a, d and g are CT images of the root in situ. b, e and h are the clustering results of the FCM algorithm 10, 30 and 50 times respectively. c, f and i are the results of the FCM algorithm 10, 30 and 50 times respectively.

According to the transverse comparison in Fig. 8, it is found that fuzzy clustering segmentation based on CT images of in-situ roots can segment the characteristics of plant roots within a certain fuzzy threshold, but some problems of the complex structure of soil medium are also exposed. For example, there may be uneven density distribution in the soil, resulting in different imaging grays or grays consistent with plant edge root systems, another example is the effect of a little air and moisture in the soil on the imaging results, however, it is undeniable that fuzzy clustering segmentation can be applied to CT image segmentation of root in situ through the segmentation results of the relatively thick upper root.

From the vertical comparison in Fig. 8, it can be seen that the processing diagram generated by multiple fuzzy clustering segmentation still fails to separate root charac-teristics to a large extent. The gap between the processed segmentation results is getting smaller and smaller, which exposes the problem of basic imaging equipment. There is no imaging equipment exclusive to in-situ CT images based on plant roots and this imaging is still solely dependent on medical imaging equipment. If we don't consider the imaging factor, we can't seem to get the desired segmentation map with root features by using only one segmentation method. At the same time, it is proved that the segmentation method cannot be single and limited in a specific complex environment.

Through the horizontal and vertical comparison of the results, it can be found that the FCM algorithm of fuzzy clustering segmentation can perform an iterative operation on CT images of in-situ roots, but it is difficult to achieve satisfactory results by using the traditional medical segmentation method for CT images of in-situ roots with more noise and artifacts.

In the in-situ CT image processing, after investigation and research, it is found that few segmentation methods are completely applicable to the in-situ root CT image and the traditional image segmentation methods are still used to realize the segmentation of CT image root features, which are specific to the in-situ root CT image segmentation methods that are rarely reported at home and abroad. The present research only improves and applies the mature method of medical CT image segmentation (fuzzy clustering

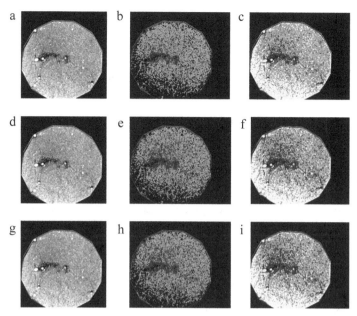

Fig. 8. Comparison of the number of fuzzy clustering segmentation operations for in situ root CT images

image segmentation). A CT image segmentation method based on in situ root system is urgently needed for the future research of root reconstruction and high throughput image processing.

8 Conclusion

This paper introduces the working principle of X-ray computed tomography, X-ray computed tomography technology was employed to reinforce the plant roots in situ nondestructive imaging, to obtain in situ root CT images, using medicine commonly used fuzzy clustering segmentation methods for feature extraction, through 10, 30, 50 times found that the results of the fuzzy arithmetic of single fuzzy clustering FCM segmentation algorithm can be implemented for in situ root feature extraction and segmentation of CT images, but not very good effect. To extract root characteristics accurately and with high quality, the segmentation algorithm of CT images of in-situ roots should be further optimized to make the data analysis of phenotypic images of high-throughput plants more accurate and stable in the future.

Acknowledgements. This research was financially supported via Project of the National Natural Science Foundation of China (61402069), the 2017 Project of the Natural Science Foundation of Liaoning province (20170540059), the General project of Liaoning education department in 2016 (2016J205).

References

1. Luo, X., Zhou, X.: The advancement of research on the methods of observation and measurement for plant root 3D architecture. December CSAE (2005)
2. Luo, X., Zhou, X., Yan, X., Luo, L., Xiang, Z.: Visualization of plant root morphology in situ based on X-ray CT imaging technology. Trans. Chin. Soc. Agric. Mach. (02), 104–106 + 133 (2004)
3. Zhou, X., Luo, X.: 3-D visualization of root system in situ based on XCT technology. Trans. Chin. Soc. Agric. Mach. 40(S1), 202–205 (2009)
4. Jeudy, C., Adrian, M., Baussard, C., et al.: RhizoTubes as a new tool for high throughput imaging of plant root development and architecture: test, comparison with pot grown plants and validation. Plant Methods 12(1), 31 (2016)
5. Wang, N.: Research on Three-Dimensional Reconstruction and Visualization of Maize Roots Based on Magnetic Resonance Imaging and VIK. Zhejiang University, Hangzhou (2013)
6. Zhou, X., Luo, X.: Advances in non-destructive detecting technologies for plant roots. August CSAE (2009)
7. Mairhofer, S., Pridmore, T., Johnson, J., et al.: X-Ray computed tomography of crop plant root systems grown in soil. Curr. Protoc. Plant Biol. 2(4), 270–286 (2017)
8. Li, K., Li, M., Xue, R., Song, W.: CT slice image segmentation of the seedling roots. For. Eng. 30(01), 25–29 (2014)
9. Zhang, J.: Non-Destructive Detection of Plant Roots Based on Magnetic Resonance Imaging Technology. Zhejiang University, Hangzhou (2014)
10. Xu, M., He, S., Zhang, Y.: Segmentation of medical image based on spatial fuzzy clustering and CV model. Intell. Comput. Appl. 9(05), 236–239 + 245 (2019)
11. Jiang, C., Liu, J., Zhong, H., Li, H., Li, D.: Study on CT Image Segmentation of Intracranial Hemorrhage Based on Improved FCM Fuzzy Clustering. Jilin University, Changchun (2018)
12. Zhou, X., Luo, X., Liu, Z.: Advances in CT image segmentation technology for plant roots in situ. Comput. Eng. Des. 2007(17), 4252–4256 (2007)
13. Zhao, Y.: Research on Image Segmentation Algorithms Based on FCM. Hunan Normal University, Changsha (2019)
14. Hu, W., Fu, X., Chen, F., Yang, W.: A path to next generation of plant phenomics. Chin. Bull. Bot. 54(05), 558–568 (2019)
15. Zhou, J., et al.: Plant phenomics: history, present status and challenges. J. Nanjing Agric. Univ. 41(04), 580–588 (2018)
16. Das Choudhury, S., Samal, A., Awada, T.: Leveraging image analysis for high-throughput plant phenotyping. Front. Plant Sci. 10, 508 (2019)
17. Yan, B., Li, L.: CT Image Reconstruction Algorithm. Science Press, Beijing (2014)
18. Yu, X., Gong, J.: CT Principles and Technology. Science Press, Beijing (2014)
19. Zheng, X., Camilo, V., Jennifer, C.: Existing and potential statistical and computational approaches for the analysis of 3D CT images of plant roots. Agronomy 8(5), 71 (2018)
20. Zhao, R.: Multiresolution Analysis and Segmentation of Medical Images. Yunnan University, Kunming (2018)
21. Lei, T., Zhang, X., Jia, X., Liu, S., Zhang, Y.: Research progress on image segmentation based on fuzzy clustering. Acta Electron. Sinica 47(08), 1776–1791 (2019)

Image Matching Algorithm Based on Improved ORB Feature Extraction

Yang Qingjiang and Shan Chuang[✉]

Electronic and Information Engineering Institute, Heilongjiang
University of Science and Technology, Harbin 150027, People's Republic of China
1321134596@qq.com

Abstract. Ambient illumination is an important factor affecting the high mismatching rate of single threshold image matching algorithm. Therefore, this paper proposes a local dynamic threshold extraction algorithm. This algorithm needs to perform light homogenization processing in the image preprocessing stage and calculate the appropriate threshold of each pixel for feature point judgment during feature extraction, which can effectively alleviate the problem of missing extraction and multiple extraction during feature extraction of a single threshold.

Keywords: Illumination · Illumination homogenization · Feature extraction · Dynamic local threshold

1 Introduction

In recent years, image recognition technology in computer vision technology is very popular and widely used in many fields of life. Image matching algorithm based on image feature points is a common technique. With the development of The Times, the requirements for image matching algorithms are becoming increasingly strict. Strong robustness, good real-time performance and fast calculation rate are the important basis for measuring image matching algorithms.

The ORB (Oriented FAST and Rotated BRIEF) algorithm [1] is one of the most popular algorithm in image matching algorithm, it is famous for its calculation speed, and therefore more conspicuous in real time. ORB algorithm consists of two parts: FAST feature extraction algorithm [2] and BRIEF feature descriptor algorithm [3]. ORB algorithm is greatly affected by ambient lighting when extracting feature points. Feature points extracted from the same image under different lighting conditions will vary greatly, resulting in a high mismatching rate for image matching. Many scholars at home and abroad have proposed a variety of improved algorithms to make ORB algorithm have better illumination robustness. In literature [4], the author proposed to combine SURF algorithm [5] with ORB algorithm, so that the feature points extracted by ORB algorithm have the better robustness of SURF algorithm. Literature [6] improves the robustness of the algorithm by combining SIFT [7] algorithm with ORB algorithm. These two improvements have improved robustness, but they are not ideal in terms of computational

X. Jiang and P. Li (Eds.): GreeNets 2020, LNICST 333, pp. 193–204, 2020.
https://doi.org/10.1007/978-3-030-62483-5_21

efficiency and real-time performance. Literature [8] combines the k-means clustering algorithm with the feature detection algorithm [9], classifies the feature points, and then sets the threshold according to the classification results. However, this method greatly increases the computational complexity and is not applicable in the field where the real-time requirement is high.

In this paper, before extracting feature points, light homogenization is carried out to reduce the influence of light on image feature extraction. In the feature extraction stage [10], the first step is to determine the local threshold of gray scale fluctuation to determine whether the pixel point is in sensitive area. The second step is to set the threshold for the pixels in the sensitive area, otherwise the traditional fixed threshold is used.

2 The Original ORB Algorithm

2.1 FAST Feature Extraction Algorithm

FAST feature points are widely used because of their high computational performance, but FAST feature points have no directional feature, so the ORB algorithm adopts the grayscale centroid method to improve this, which is called oFAST algorithm (Oriented FAST). The gray-scale centroid algorithm calculates the centroid within the radius r by calculating a moment. The vector formed by the center of the circle and the center of mass is the directional property of the feature point.

In a 5 × 5 image block, the element expression of the moment of the corresponding image is shown in formula (1):

$$m_{pq} = \sum_{x,y} x^p y^q I(x, y) \tag{1}$$

Where, I(x, y) is the image grayscale expression used to calculate the image grayscale value of this point. X and y are at $[-r, r]$, r is the radius of the image, and here it is $[-3, 3]$.

Then the center of mass of the image window can be expressed as formula (2):

$$C = \left(\frac{m_{10}}{m_{00}}, \frac{m_{01}}{m_{00}} \right) \tag{2}$$

Where, when $p = q = 0$, m_00 is the sum of all the gray values of the image block; When $p = 1, q = 0$, you get m_10; When $p = 0, q = 1$, you get m_01; M_10, m_01 is the first order matrix, m_00 is the zero order matrix.

Then the Angle of the whole image window, namely the direction of the feature point, can be expressed as formula (3):

$$\theta = arc \tan(m_{01}, m_{10}) \tag{3}$$

When the graph is a binary image, m_00 is the area that can be represented as the image. If the gray value of the image is regarded as the weight, then the center of mass is the center of mass of the image, that is, the weighted center of the whole gray value.

2.2 BRIEF Feature Descriptor Subalgorithm

RIEF descriptor is a binary descriptor, which has an incomparable speed advantage in the image matching stage. However, there are also big drawbacks. BRIEF does not have rotation invariance. The ORB algorithm makes use of the feature point orientation feature obtained by the improved FAST algorithm. It first rotates the descriptor obtained by the BRIEF algorithm, and then discriminates the binary code, so that the descriptor has rotation invariance. The improved BRIEF called rBRIEF (Rotated BRIEF). Suppose you have a smooth the image, the size of S × S neighborhood p, tau test:

$$\tau(p; x, y) = \begin{cases} 1 : I(p, x) < I(p, y) \\ 0 : I(p, x) \geq I(p, y) \end{cases} \tag{4}$$

Where, $I(p, x)$ represents the pixel gray value of the smoothed image neighborhood p at point $x = (u, v)^{\wedge}T.N\_d$ (x, y) point pairs are selected to form bitstream binary descriptors of n_d dimension:

$$f_n(p) = \sum_{1 \leq i \leq n} 2^{i-1} \tau(p; x_i, y_i) \tag{5}$$

There are three options for n_d, 128, 256, and 512 (16, 32, and 64 bytes). Considering the generation rate, distribution and accuracy of bitstream binary descriptors, the 256-dimensional descriptors have the best comprehensive performance. To overcome the lack of rotation invariance in the BRIEF descriptor, the ORB algorithm takes advantage of the orientation of oFAST feature points to rotate the BRIEF descriptor in its main direction. The following is the implementation process: For any feature point, the location information in the neighborhood of 31 × 31 is the set of n pairs of points ((x)_i, y_i)), and they are represented as a matrix of order 2 × n:

$$S = \begin{pmatrix} x_1, \dots, x_n \\ y_1, \dots, y_n \end{pmatrix} \tag{6}$$

The corresponding rotation matrix R_ is expressed by using the principal direction theta of the characteristic points of oFAST:

$$R_\theta = \begin{bmatrix} \cos\theta & -\sin\theta \\ \sin\theta & \cos\theta \end{bmatrix} \tag{7}$$

The matrix S is rotated through the matrix R_ to get a new matrix S:

$$S_\theta = R_\theta S \tag{8}$$

RBRIEF description subexpression with rotation invariance:

$$g_n(p, \theta) = f_n(p)|(x_i, y_i) \in S_\theta \tag{9}$$

3 This Article Improves the ORB Algorithm

3.1 Image Preprocessing

The importance of image pre-processing for image feature extraction is self-evident. A high-quality image can greatly reduce the external interference for the subsequent image feature extraction and image feature description. In this paper, adaptive threshold determination is required for feature extraction, and ambient light has a strong influence on threshold selection, which requires light homogenization in the image pre-processing stage. The process of light homogenization is as follows:

The multi-scale gaussian function is used to remove the illumination component of the scene [11],

construct a two-dimensional gamma function to adjust the parameters of the two-dimensional gamma function based on the distribution characteristics of the light component, reduce the brightness value of the over-intense illumination region, and increase the brightness value of the over-dark illumination region, so as to finally realize the adaptive correction processing of the uneven illumination image [00]. The illumination component can be obtained by convolving the gaussian function with the original image [12].

$$G(x, y) = \lambda \exp\left(-\frac{x^2 + y^2}{c^2}\right) \tag{10}$$

In the formula, G is the gaussian function, c is the scale factor, and is the normalization constant, so that the gaussian function $G(x, y)$ satisfies the normalization conditions.

Light component:

$$I(x, y) = F(x, y)G(x, y) \tag{11}$$

Where, $F(x, y)$ is the input image, and $I(x, y)$ is the estimated illumination component.

$$O(x, y) = 255\left(\frac{F(x, y)}{255}\right)^{\gamma}, \gamma = \left(\frac{1}{2}\right)^{\frac{m - I(x, y)}{m}} \tag{12}$$

The two-dimensional gamma function $O(x, y)$ is the output image brightness value after correction, $F(x, y)$ is the source image, where is the index value of brightness enhancement, and m is the brightness mean of the illumination component. In order to avoid interference between RGB channels, the whole processing process needs to be conducted in HSV color space, where the brightness component V is processed, and finally transformed from HSV space to RGB space.

The flow chart of the whole process is as follows:

3.2 FAST Adaptive Threshold Feature Extraction

FAST when making feature point extraction, first traverse each pixel of the image point, is the sole criterion for judging whether pixels feature points in the current pixel as the center of the circle draw three pixels for the radius of a circle (16 pixels circle, see

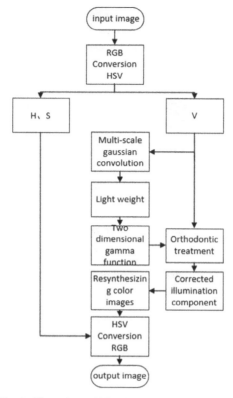

Fig. 1. Flow chart of illumination homogenization

Chart 1), statistical circle on 16 pixels and the center pixel gray value of absolute value of difference, if there are nine consecutive pixels in the circle and the center pixel of the difference between absolute value is greater than the threshold T set beforehand, is considered center pixel is feature points (Fig. 2).

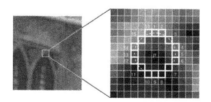

Fig. 2. Pattern diagram of feature extraction

FAST feature extraction USES a single threshold to determine the feature points. In this way, the feature points extracted are densely distributed and the number of redundant feature points is large, which is not conducive to the description of feature points and affects the result of feature matching. In this paper, the location information of the target pixel is determined when the feature points are extracted, and whether there are obvious changes in the pixel gray level in a certain size area of the pixel to be detected. According to this change, the detection area is divided into three categories: ignored area, ignored area and ignored area. The ignored region means that there is no gray level change in the region where the pixel is located, and no feature points are extracted in this region. The unfocused region refers to the region where the pixel is located, and the gray level change range is not obvious. The threshold value of feature point extraction in this region adopts the threshold value of the original ORB algorithm. The attention region refers to the region where the pixel is located where the gray level changes greatly, and the dynamic threshold is adopted to set the threshold to extract the feature points according to the gray level of the region.

Set the current pixel point as P, and select the threshold value in a window with a size of 7 * 7 centered on the pixel point P.

$$T(x_i, y_j) = \begin{cases} t_1, & if\ (MAX - MIN) < S; \\ pass, & if\ MAX = MIN; \\ t_2, & if\ (MAX - MIN) \geq S \end{cases} \tag{13}$$

Where, $T(x\_i, y\_j)$ refers to the threshold of determining whether the pixel point at (I, j) is the feature point. S is the measure to judge the range of gray level change in the region, $S = 15$. T_1 and t_2 are two different thresholds, $t\_1 = 10$, and t_2 is the adaptive threshold.

$$t_2 = c\frac{\sum_{i,j=1}^{i,j=7} I(i, j) - I_{max} - I_{min}}{I_\alpha} \tag{14}$$

Where, I_max is the maximum pixel gray value in this window; I_min is the minimum pixel gray value in this window; $I\_$ is to go out the maximum and minimum of the remaining 47 pixels grayscale average; C is the adaptive parameter and the percentage of the empirical threshold in the absolute pixel gray contrast in the window. The test shows that the feature points selected when c is 0.18 are more stable.

The improved FAST feature extraction process is as follows:

1) traverse the image pixel points and judge the gray level changes in the 7 * 7 neighborhood centered on the pixel points;
2) if the gray level does not change, the feature point extraction will not be carried out in this area;
3) if the change amplitude of gray scale is less than S, the fixed threshold is used for feature point extraction;
4) if the change range of gray level is greater than S, the adaptive threshold is used for feature extraction

4   Experimental Simulation and Analysis

In order to verify the effectiveness of the algorithm in this paper, the algorithm was verified in the hardware test environment of windows 10 operating system, Intel(R) Core(TM) i7-6700hq CPU and memory of 8G, and in the OpenCv2.4 computer vision library and vs2017 development environment.

4.1   Image Feature Point Extraction Experiment

Feature extraction comparison between ORB algorithm and proposed algorithm (in order to visually and clearly compare the extraction effects of the two algorithms, non-maximum suppression is not performed here).

It is obvious from Fig. 3 that the feature points extracted by ORB algorithm are largely clustered and overlapped. It can be seen from Fig. 4 that the feature points extracted by the algorithm in this paper are less overlapped and clustered than those in Fig. 3. Intuition does not mean that the algorithm in this paper is better than the ORB algorithm in extracting feature points. Five experiments have been done here for time-consuming comparison (Table 1).

Fig. 3. Effect diagram of orb algorithm extracting feature points

Fig. 4. Effect diagram of feature points extracted by the algorithm in this paper

To compare the ORB algorithm and the proposed algorithm in extracting feature point differences under different ambient lighting conditions, reduce the illumination of the original image by 10%, 20%, 40% and 80%, and enhance the illumination by 10%, 20%, 40% and 80%, respectively.

Table 1. Comparison of the time taken to extract feature points between the ORB algorithm and the algorithm in this paper

Experimental group	ORB algorithm	Algorithm In this paper	Take down than
1	1.485302	1.111101	25.19%
2	1.248731	1.054183	15.58%
3	1.199899	0.978244	18.47%
4	1.123802	0.919201	18.21%
5	1.151884	0.971725	15.65%

In Fig. 5, (a) the extraction result diagram with the original image reduced by 10%, (b) the result diagram reduced by 20%, (c) the result diagram reduced by 40%, (d) the result diagram reduced by 80%, (e) the result diagram enhanced by 10%, (f) the result diagram enhanced by 20%, (g) the result diagram enhanced by 40%, (h) the result diagram enhanced by 80%. In Fig. 6, (1) the extraction result diagram shows a 10% reduction in brightness of the original image, (2) the result diagram shows a 20% reduction in brightness, (3) the result diagram shows a 40% reduction, (4) the result diagram shows an 80% reduction, (5) the result diagram shows a 10% increase, (6) the result diagram shows a 20% increase, (7) the result diagram shows a 40% increase, and (8) the result diagram shows an 80% increase (Table 2).

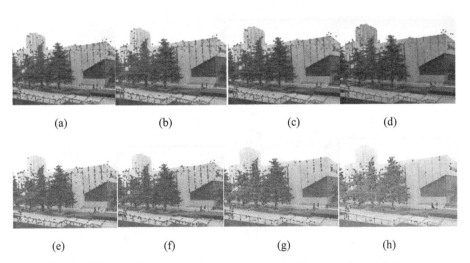

(a)   (b)   (c)   (d)

(e)   (f)   (g)   (h)

Fig. 5. ORB algorithm extracts feature points with different ambient brightness

As you can see, the number of feature points extracted by the ORB algorithm changes dramatically as the illumination changes in the environment. This algorithm is more stable than the ORB algorithm when illumination changes. And extracting feature points is faster than the ORB algorithm (Fig. 7).

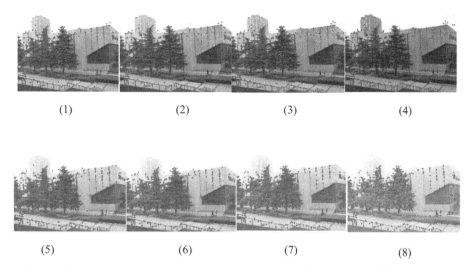

Fig. 6. The algorithm in this paper extracts feature points with different ambient brightness

Table 2. Variation table of ORB algorithm feature points under different ambient lighting conditions

Experimental group	BRIGHTNESS CHANCE	Quantitative ORB characteristics	Feature points in this algorithm	The ORB time-consuming	The algorithm in this paper takes time
1	+10%	597	543	1.76524	1.21317
2	+20%	499	531	1.63414	1.19782
3	+40%	420	549	1.59905	1.18755
4	+80%	334	527	1.52381	1.24416
5	−10%	638	556	1.81021	1.21869
6	−20%	544	546	1.73501	1.27718
7	−40%	403	529	1.65032	1.31011
8	−80%	267	557	1.72591	1.27054

4.2 Image Matching Experiment

Use the ORB algorithm and the algorithm in this article for feature matching. Figure 8 is the ORB algorithm matching result graph, and Fig. 9 is the algorithm matching result graph of this paper (Tables 3 and 4).

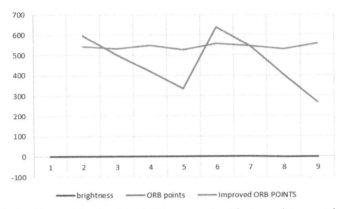

Fig. 7. ORB algorithm and the algorithm in this paper extract feature point comparison line graph

Fig. 8. ORB algorithm matching result diagram

Fig. 9. Results of algorithm matching in this paper

Through the above data comparison, the feature point extraction algorithm proposed in this paper can effectively improve the speed of the algorithm, and at the same time, it can also have better matching accuracy under the changing ambient light.

Table 3. ORB algorithm match table

Experimental group	Successful logarithmic	Match the logarithmic	Precision
1	102	135	75.56%
2	96	129	74.42%
3	109	142	76.76%
4	85	114	74.26%
5	91	111	81.98%

Table 4. Algorithm matching table in this paper

Experimental group	Successful logarithmic	Match the logarithmic	Precision
1	77	91	85.72%
2	74	89	83.15%
3	81	100	81.00%
4	85	105	80.95%
5	71	91	81.98%

5 Conclusion

This paper proposes an improved ORB feature extraction algorithm, which has the following advantages compared with the original ORB algorithm:

1) it combines adaptive threshold and fixed threshold, which can effectively alleviate the problem of too dense and overlapping ORB feature extraction;
2) before extracting the features, the region was determined. If the gray level is consistent and the region has not changed, the feature extraction will not be carried out. Compared with ORB traversing each pixel, the extraction speed can be accelerated;
3) before feature points are extracted, light homogenization is carried out to reduce the difficulty of extraction caused by light changes.

References

1. Rublee, E., Rabaud, V., Konolige, K., Bradski, G.: ORB: an efficient alternative to SIFT or SURF. In: 2011 IEEE International Conference on Computer Vision (ICCV) (2011)
2. Viswanathan, D.G.: Features from accelerated segment test(FAST)[EB/OL]. (2016-04-15)[2017-03-30]
3. Calonder, M., Lepetit, V., Fua, P.: BRIEF: binary robust independent elementary features. In: Daniilidis, K., Maragos, P., Paragios, N. (eds.) Computer Vision – ECCV 2010. Lecture Notes in Computer Science, vol. 6314, pp. 778–792. Springer, Berlin, Heidelberg (2010). https://doi.org/10.1007/978-3-642-15561-1_56

4. Qing-Wei, C., Min-Dong, L., Chuan, L., Jun, Z., Pan-Ling, H., Lei, L.: Research on image feature point extraction and matching algorithm in visual SLAM. Modern Manufact. Eng. **2019**(10), 135–139 + 134

5. Bay, H., Tuytelaars, T., Van Gool, L.: SURF: speeded up robust features. In: Leonardis, A., Bischof, H., Pinz, A. (eds.) Computer Vision – ECCV 2006. Lecture Notes in Computer Science, vol. 3951, pp. 404–417. Springer, Berlin, Heidelberg (2006). https://doi.org/10.1007/11744023_32

6. Wenchao, J: Panoramic image Mosaic technology based on ORB algorithm. Urban Survey **2019**(03), 105–108 + 114 (2019)

7. Lowe, D.G.: Distinctive image features from scale-invariant keypoints. Int. J. Comput. Vis. **60**(2), 91–110 (2004)

8. Dan, M.: Research on image matching algorithm based on local features. Xinjiang University (2018)

9. Lv, H., Huang, X., Yang, L., Liu, T., Wang, P.: A k-means clustering algorithm based on the distribution of SIFT (2013)

10. Tanabe, M., Kinoshita, K.: Absolute irradiance responsivity calibration using diode lasers emitting at three wavelengths for tricolor laser applications. Optik **202**, 163653 (2019)

11. Li, J.: Research on image fuzzy measurement based on multi-scale spatial analysis. Xihua University (2014)

12. Yulei, Huang: An image enhancement algorithm based on L channel illumination estimation and gamma function. Autom. Technol. Appl. **37**(05), 56–60 (2008)

Modeling and Derivation of Small Signal Model for Grid-Connected Inverters

Pengyu An, Xunwen Su$^{(\boxtimes)}$, Xianzhong Xu, and Wenhui Zhu

Heilongjiang University of Science and Technology, Harbin 150022, China
suxunwen@163.com

Abstract. With the development of new energy power generation technology, more and more regions are increasing the construction of new energy power generation facilities. However, due to the unstable factors in the source of new energy, the traditional modeling methods require a lot of work. Modeling the impedance of the system in a d-q frames has become a new option, but there is no reference for the derivation process of the system impedance, control model and mathematical model. In order to reduce the amount of tasks of more complex impedance modeling, this paper correspond the control model and mathematical model of Grid-connected inverters in Matlab/Simulink. Among them, the most important part of the control model is the model building of PLL, which includes power control and current control.

Keywords: Grid-connected inverters · Small signal model · Impedance modeling

1 Introduction

Grid-connected inverters are the key components that deliver renewable energy to the grid [1, 2]. In the process of grid connection, harmonic pollution will occur, which pollutes the electricity. The frequency and phase of the output current of the grid-connected inverter are delayed from the frequency and phase of the system, so the phase-locked loop is needed to achieve synchronization [3]. Compared with state space equation method, impedance-based method has advantages that analysis of high-order equations can be avoided and model doesn't have to be rebuilt when system structure changes [4].

Recently, the impedance model-based analysis [5–7] has been proposed to provide insights into this instability issue.

Because the Grid-connected terminal is a three-phase AC system, it cannot be studied in the static coordinate system. The existing methods need to carry out Park transformation and Clark transformation, so as to realize from three-phase static coordinate system to two-phase static coordinate system or to d-q frames. Impedance model is easier to establish in this paper in d-q frames. The model equivalent input is given in this paper considering rotor current controllers and PLL. And use the small signal method to analyze the stability of the system [8–10].

X. Jiang and P. Li (Eds.): GreeNets 2020, LNICST 333, pp. 205–214, 2020.
https://doi.org/10.1007/978-3-030-62483-5_22

This paper is organized as follows. Section 2 is the topology, mathematical model and control model of Grid-connected inverters. Section 3 is the mathematical model and control model of PLL. Section 4 is the model with current control and power control. Section 5 is based on MATLAB/Simulink, which corresponds the simulation model to the control model and deduces the equivalent impedance. Section 6 concludes this paper.

2 Impedance Modeling of Grid-Connected Inverter

Figure 1 is the structure diagram of three-phase Grid-connected inverter, which can be divided into main power route control circuit. In the main circuit, U_a, U_b and U_c are the three-phase voltage measured with grid entry point; I_a, I_b and I_c are the three-phase current measured with grid entry point; $u_d, u_q, 0$ are the d-axis q-axis grid entry point voltage output by the PLL in the controller coordinate system; i_d, i_{dref}, i_q and i_{qref} are the d-axis q-axis grid entry point current and its reference value; PQ are the active power and reactive power calculated by the measured data in the power control; P_{ref}, Q_{ref} are the active power reference quantity of power and reactive power [11].

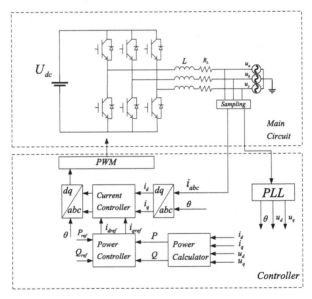

Fig. 1. The structure diagram of three-phase Grid-connected inverter

The small signal model in d-q frames is shown in Fig. 2. In the Fig. 2, the D, U and I respectively represent duty ratio, voltage and current in stable state, and d, u and i are the small signal disturbance values. Among them, the main circuit parameters are indicated by superscript s, and the subscript d and q corresponds to the d-axis q-axis respectively. $u_{dc} = U_{dc} + \tilde{u}_{dc}$; u_d, i_d, u_q, i_q represents the voltage disturbance value of d-axis parallel node, the current disturbance value of parallel node, the voltage disturbance value of q-axis parallel node and the current disturbance value of parallel node in the main circuit coordinate system respectively; D_d, D_q, d_d, d_q represents the duty ratio steady value of d-axis and q-axis and the small signal disturbance value respectively.

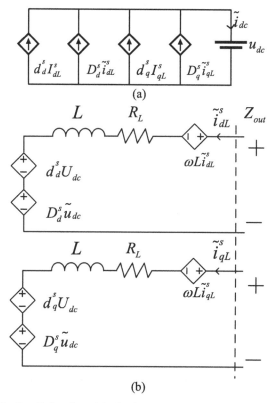

(a)

(b)

Fig. 2. Small signal model of Grid-connected inverter in d-q frames

From the small signal model in Fig. 2 (a),

$$u_{dc} = \tilde{d}_d^s I_{dL}^s + D_d^s \tilde{i}_{dL}^s + \tilde{d}_q^s I_{qL}^s + D_q^s \tilde{i}_{qL}^s \qquad (1)$$

Its matrix form is

$$
\begin{bmatrix} u_{dc} \\ 0 \end{bmatrix} =
\begin{bmatrix} D_d^s & D_q^s \\ 0 & 0 \end{bmatrix}
\begin{bmatrix} \tilde{i}_{dL}^s \\ \tilde{i}_{qL}^s \end{bmatrix} +
\begin{bmatrix} I_{dL}^s & I_{qL}^s \\ 0 & 0 \end{bmatrix}
\begin{bmatrix} \tilde{d}_d^s \\ \tilde{d}_q^s \end{bmatrix}
\qquad (2)
$$

From Fig. 2 (b), we can see the small signal model of AC measurement, and the small signal disturbance of grid entry point voltage is

$$
\begin{bmatrix} \tilde{u}_d^s \\ \tilde{u}_q^s \end{bmatrix} =
\begin{bmatrix} sL + R_L & -\omega L \\ \omega L & sL + R_L \end{bmatrix}
\begin{bmatrix} \tilde{i}_{dL}^s \\ \tilde{i}_{qL}^s \end{bmatrix} +
\begin{bmatrix} D_d^s & 0 \\ D_q^s & 0 \end{bmatrix}
\begin{bmatrix} \tilde{u}_{dc} \\ 0 \end{bmatrix} +
\begin{bmatrix} U_{dc} & 0 \\ 0 & U_{dc} \end{bmatrix}
\begin{bmatrix} \tilde{d}_d^s \\ \tilde{d}_q^s \end{bmatrix}
\qquad (3)
$$

Based on (2) and (3)

$$\begin{bmatrix} \tilde{u}_d^s \\ \tilde{u}_q^s \end{bmatrix} = \begin{bmatrix} sL + R_L + D_d^{s2} & -\omega L + D_d^s D_q^s \\ \omega L + D_d^s D_q^s & sL + R_L + D_d^{s2} \end{bmatrix} \begin{bmatrix} \tilde{i}_{dL}^s \\ \tilde{i}_{qL}^s \end{bmatrix} + \begin{bmatrix} U_{dc} + D_d^s I_{dL}^s & I_{qL}^s D_d^s \\ I_{dL}^s D_d^s & U_{dc} + D_q^s I_{qL}^s \end{bmatrix} \begin{bmatrix} \tilde{d}_d^s \\ \tilde{d}_q^s \end{bmatrix}$$

(4)

The transfer function matrix of the voltage to the current at the grid entry point is

$$G_{iu} = \begin{bmatrix} sL + R_L + D_d^{s2} & -\omega L + D_d^s D_q^s \\ \omega L + D_d^s D_q^s & sL + R_L + D_d^{s2} \end{bmatrix}$$

(5)

The matrix of current transfer function from duty ratio to grid entry point is

$$G_{id} = \begin{bmatrix} sL + R_L + D_d^{s2} & -\omega L + D_d^s D_q^s \\ \omega L + D_d^s D_q^s & sL + R_L + D_d^{s2} \end{bmatrix}^{-1} \begin{bmatrix} U_{dc} - D_d^s I_{dL}^s & I_{qL}^s D_d^s \\ I_{dL}^s D_q^s & U_{dc} - D_q^s I_{qL}^s \end{bmatrix}$$

(6)

Figure 3 depicts the open-loop input admittance model.

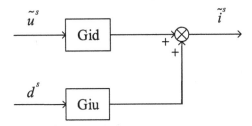

Fig. 3. Control program block diagram of main circuit

3   Control Model of PLL

From [12], the function of phase-locked loop is synchronize the control system with the main system and the control system, the PI control method is used in the phase-locked loop. A grid-connected converter usually needs PLL to synchronize with the grid.

The transfer function between two systems can be expressed as

$$T_\theta = \begin{bmatrix} \cos\theta & \sin\theta \\ -\sin\theta & \cos\theta \end{bmatrix}$$

(7)

When the output small disturbance of the PLL between the main system and control system is small enough, $\cos\theta = 1$, $\sin\theta = \theta$, we can get

$$\vec{U}^c = \begin{bmatrix} 1 & \theta \\ -\theta & 1 \end{bmatrix} \vec{U}^s$$

(8)

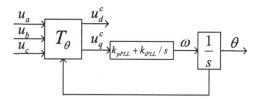

Fig. 4. Control block diagram of PLL in d-q frames

It can also be written as

$$\begin{bmatrix} U_d^c + \tilde{u}_d^c \\ U_q^c + \tilde{u}_q^c \end{bmatrix} = \begin{bmatrix} 1 & \theta \\ -\theta & 1 \end{bmatrix} \begin{bmatrix} U_d^s + \tilde{u}_d^s \\ U_q^s + \tilde{u}_q^s \end{bmatrix} \tag{9}$$

$$\begin{bmatrix} \tilde{u}_d^c \\ \tilde{u}_q^c \end{bmatrix} \approx \begin{bmatrix} U_d^s\theta + \tilde{u}_d^s \\ -U_q^s\theta + \tilde{u}_q^s \end{bmatrix} \tag{10}$$

From Fig. 4

$$\theta = \tilde{u}_q^c K_{PLL} \frac{1}{s} \tag{11}$$

Where $K_{PLL} = k_{pPLL} + \frac{k_{iPLL}}{s}$, so (11) can also be written as

$$\theta = \frac{K_{PLL}}{s + U_d^s K_{PLL}} \tilde{u}_q^c \tag{12}$$

Similarly,

$$G_{PLL}^i = \begin{bmatrix} 0 & I_q^s G_{PLL} \\ 0 & -I_d^s G_{PLL} \end{bmatrix} \tag{13}$$

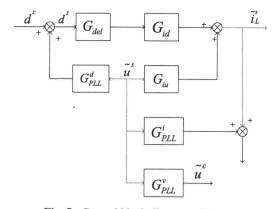

Fig. 5. Control block diagram with PLL

$$G_{PLL}^d = \begin{bmatrix} 0 & -D_q^s G_{PLL} \\ 0 & D_d^s G_{PLL} \end{bmatrix} \tag{14}$$

The control block diagram with PLL is as shown in the Fig. 5

4 Model with Current Control and Power Control

According to the main circuit structure, the PI current loop control mode is adopted and the current loop control matrix is

$$G_{ci} = \begin{bmatrix} k_{pi} + \frac{k_{ii}}{s} & 0 \\ 0 & k_{pi} + \frac{k_{ii}}{s} \end{bmatrix} \tag{15}$$

Since the current loop control contains coupling, the d-axis and q-axis are decoupled, the matrix form is

$$G_{dei} = \begin{bmatrix} 0 & -\frac{3\omega L}{U_{dc}} \\ \frac{3\omega L}{U_{dc}} & 0 \end{bmatrix} \tag{16}$$

After the current controller is added, the power controller is add to achieve more accurate control effect. The transfer function matrix of power control is

$$G_{cPQ} = \begin{bmatrix} k_{pPQ} + \frac{k_{iPQ}}{s} & 0 \\ 0 & k_{pPQ} + \frac{k_{iPQ}}{s} \end{bmatrix} \tag{17}$$

Through linearization, the transfer function of active power and reactive power in d-q frames can be obtained

$$G_{PQ}^i = \begin{bmatrix} U_d^s & U_q^s \\ -U_q^s & U_d^s \end{bmatrix} \tag{18}$$

$$G_{PQ}^v = \begin{bmatrix} I_d^s & I_q^s \\ I_q^s & -I_d^s \end{bmatrix} \tag{19}$$

The control block diagram including current control, power control and PLL is shown in Fig. 6.

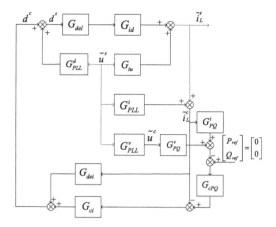

Fig. 6. Control block diagram including current control, power control and PLL

5 Matlab/Simulink Partial Simulation Model and Equivalent Impedance Derivation

From the control block diagram in Fig. 6, we can get

$$\begin{bmatrix} P \\ Q \end{bmatrix} = G^v_{PQ} \begin{bmatrix} u^c_d \\ u^c_q \end{bmatrix} + G^i_{PQ} \begin{bmatrix} i^c_{dL} \\ i^c_{qL} \end{bmatrix} \tag{20}$$

$$\tilde{i}^c_{Lref} = G_{cPQ}(PQ_{ref} - PQ) \tag{21}$$

$$\tilde{d}^c = G_{dei}\tilde{i}^c_L + G_{ci}(\tilde{i}^c_{Lref} - \tilde{i}^c_L) \tag{22}$$

Based on (15) to (19), the simulation model of duty ratio of control system can be obtained as follows

$$\tilde{d}^c = \left[(P_{ref} - P)\left(k_{pPQ} + \frac{k_{iPQ}}{s}\right) - \tilde{i}^c_{dL} \right]\left(k_{pi} + \frac{k_{ii}}{s}\right) - \frac{3\omega_0 L}{U_{dc}}\tilde{i}^c_{qL}$$
$$+ \left[(Q_{ref} - Q)\left(k_{pPQ} + \frac{k_{iPQ}}{s}\right) - \tilde{i}^c_{qL} \right]\left(k_{pi} + \frac{k_{ii}}{s}\right) + \frac{3\omega_0 L}{U_{dc}}\tilde{i}^c_{dL} \tag{23}$$

The simulation model in Matlab /Simulink is (Fig. 7).
Continue to derive the control model based on (21), (22)

$$\tilde{i}^c_L = \tilde{i}^s_L + G^i_{PLL}\tilde{u}^s \tag{24}$$

$$\tilde{i}^s_L = G_{del}G_{id}\tilde{d}^s + G_{iu}\tilde{u}^s \tag{25}$$

$$\tilde{d}^s = \tilde{d}^c + G^d_{PLL}\tilde{u}^s \tag{26}$$

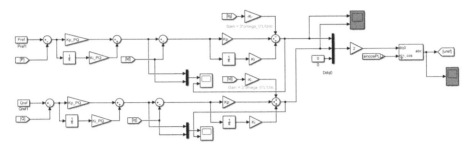

Fig. 7. Simulation model of duty ratio $\tilde{d}^c$ of control system in Matlab/Simulink

If the reference value of active power and reactive power is 0, then

$$\tilde{d}^c = \left(G_{dei} - G_{ci} - G_{ci}G_{cPQ}G^i_{PQ}\right)\tilde{i}^c_L - G_{ci}G_{cPQ}G^v_{PLL}G^v_{PQ}\tilde{u}^s \tag{27}$$

Substituting Eq. (24) into Eq. (27)

$$\tilde{d}^c = \left(G_{dei} - G_{ci} - G_{ci}G_{cPQ}G^i_{PQ}\right)\tilde{i}^s_L + \left(G_{dei}G^i_{PLL} - G_{ci}G^i_{PLL} - G_{cPQ}G^i_{PLL}G^i_{PQ} - G_{cPQ}G^v_{PLL}G^v_{PQ}\right)\tilde{u}^s \tag{28}$$

$$\tilde{d}^s = \left(G_{dei} - G_{ci} - G_{ci}G_{cPQ}G^i_{PQ}\right)\tilde{i}^s_L$$
$$+ \left(G_{dei}G^i_{PLL} - G_{ci}G^i_{PLL} - G_{ci}G_{cPQ}G^i_{PLL}G^i_{PQ} - G_{ci}G_{cPQ}G^v_{PLL}G^v_{PQ} + G^d_{PLL}\right)\tilde{u}^s \tag{29}$$

Based on (29) and (25)

$$\tilde{i}^s_L = G_{del}G_{id}\left(G_{dei} - G_{ci} - G_{ci}G_{cPQ}G^i_{PQ}\right)\tilde{i}^s_L$$
$$+ \left[G_{iu} + G_{del}G_{id}\left(G_{dei}G^i_{PLL} - G_{ci}G^i_{PLL} - G_{ci}G_{cPQ}G^i_{PLL}G^i_{PQ} - G_{ci}G_{cPQ}G^v_{PLL}G^v_{PQ} + G^d_{PLL}\right)\right]\tilde{u}^s \tag{30}$$

So the equivalent impedance is

$$Z_{out} = \frac{\left[E + G_{del}G_{id}\left(G_{ci} + G_{ci}G_{cPQ}G^i_{PQ} - G_{dei}\right)\right]}{\left[G_{iu} + G_{del}G_{id}\left(G_{dei}G^i_{PLL} - G_{ci}G^i_{PLL} - G_{ci}G_{cPQ}G^i_{PLL}G^i_{PQ} - G_{ci}G_{cPQ}G^v_{PLL}G^v_{PQ} + G^d_{PLL}\right)\right]} \tag{31}$$

Simulation verification is carried out in MATLAB/Simulink, the results are as follow Figure.

From the Fig. 8, the theoretical results of the model are in good agreement with the actual results of frequency sweep the reliability of the impedance model has been deduced.

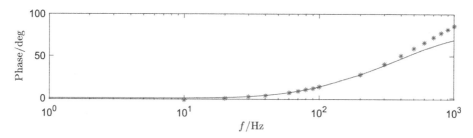

Fig. 8. Simulation result

6   Conclusion

In this paper, the solution process of the impedance model of the Grid-connected inverters in the d-q frames is deduced. According to the small signal model in the d-q frames, the transfer matrix of the main circuit with disturbance is derived firstly, then the influence of the phase-locked loop in the control system is considered, same as the influence of the current control and the power control. Finally, the output equivalent impedance is derived. It provides a bridge for the later establishment and research of impedance model, reduces the workload of model understanding, and lays foundation for the subsequent stability analysis in a longer term.

Acknowledgements. This work was financially supported by the National Science Foundation of China under Grant(51677057), Local University Support plan for R&D, Cultivation and Transformation of Scientific and Technological Achievements by Heilongjiang Educational Commission(TSTAU-R2018005) and Key Laboratory of Modern Power System Simulation and Control & Renewable Energy Technology, Ministry of Education(MPSS2019-05).

References

1. Gu X-H., Zhang, J.: Research on harmonic suppression of grid-connected inverter based on active damping method. J. Liaoning Univ. Technol. Natural Science Edition., 39(6) (2019)
2. Fu, S., Zhang, X., Xu, D.: Improved control method of grid-connected converter based on voltage perturbation compensation under weak grid conditions. In: Conference 2019, ECCE, Korea (2019)
3. Se-Kyo, C.: A phase tracking system for three phase utility interface inverters. IEEE Trans. Power Electron. **15**(3), 431–438 (2000)
4. Wen, B., Dushan, B., Rolando, B., et al.: Small-signal stability analysis of three-phase AC system in the presence of constant power loads based on measured dq frame impedances. IEEE Trans. Power Electron. **30**(10), 5952–5963 (2015)
5. Sun, J.: Impedance-based stability criterion for grid-connected inverters. IEEE Trans. Power Electron. **26**(11), 3075–3078 (2011)
6. Piyasinghe, L., Miao, Z., Khazaei, J., Fan, L.: Impedance model-based SSR analysis for TCSC compensated type-3 wind energy delivery systems. IEEE Trans. Sustain. Energy **6**(1), 179–187 (2015)
7. Cespedes, M., Lei, X., Jian, S.: Constant-power load system stabilization by passive damping. IEEE Trans. Power Electron. **26**(7), 1832–1836 (2011)

8. Liu, H., Xie, X., Liu, W.: An oscillatory stability criterion based on the unified dq-frame impedance network model for power systems with high-penetration renewables. IEEE Trans. Power Syst. **13**(3), 511–521 (2018)
9. Du, E., Zhang, N., Kang, C., Kroposki, B., Miao, M.: Managing wind power uncertainty through strategic reserve purchasing. IEEE Trans. Power Syst. **8**(2), 2547–2559 (2018)
10. Bottrell, N., Prodanovic, M., Green, T.C.: Dynamic stability of a microgrid with an active load. IEEE Trans. Power Electron. **28**(11), 5107–5119 (2013)
11. Wen, B., Boroyevich, D., Burgos, R.: Analysis of DQ small-signal impedance of grid-connected inverters. IEEE Trans. Power Electron. **31**(1), 675–676 (2016)
12. Blaabjerg, F., Teodorescu, R., Liserre, M., Timbus, A.V.: Overview of control and grid synchronization for distributed power generation systems. IEEE Trans. Ind. Electron. **53**(5), 1398–1409 (2006)

Performance Analysis of Signal Detection Algorithm in Data Link System

Jiang Xiaolin[1,2], Qu Susu[1(✉)], and Tang Zhengyu[1]

[1] Heilongjiang University of Science and Technology, Harbin 150000, China
3216739483@qq.com
[2] Harbin Institute of Technology, Harbin 150000, China

Abstract. Data link is a defined message format and communication protocol. It is a real-time transmission system between sensors, control systems and weapon platforms. Data links connect geographically dispersed forces, sensors, and weapon systems to create seamless connectivity, information sharing, and increased command speed and coordination.In this paper, MIMO technology is applied to data link system to improve the information rate. The performance of data link system is closely related to MIMO technology. The comparative analysis of the (Zero Forcing)ZF, (Minimum Mean-Squared Error)MMSE, (Zero Forcing-Ordered successive interference cancellation)ZF-OSIC and (Minimum Mean-Squared Error-Ordered successive interference cancellation)MMSE-OSIC algorithms in MIMO technology was carried out. The results are as follows: MMSE-OSIC algorithm is the best among the four algorithms, MMSE, ZF algorithm is the worst, and ZF-OSIC is between them.

Keywords: Data link system · MIMO. MMSE - OSIC · MMSE. ZF - OSIC · ZF

1 Introduction

Data link communication system, full name controller pilot data link communications (CPDLC), mainly uses data instead of voice to provide traffic management for controllers and pilots. Data link communication system can provide communication services in air traffic service facilities, including release, application, report and so on in standard format [1, 2]. Meanwhile, it can also compensate channel congestion, signal mishearing and signal distortion in voice communication. The data link communication system provides the pilot with control information in text form. With the development of international SATCOM technology, CPDLC has been widely used by international airlines and traffic control systems as a means of communication through data link.In a single antenna system, there are two ways to improve the communication rate, which are to increase the transmission power and bandwidth. MIMO technology is introduced into data link system to further improve its performance [3].

MIMO technology transceiver antenna number is more than one, by a plurality of components. The signal from the sending end goes through multiple paths to the receiving

© ICST Institute for Computer Sciences, Social Informatics and Telecommunications Engineering 2020
Published by Springer Nature Switzerland AG 2020. All Rights Reserved
X. Jiang and P. Li (Eds.): GreeNets 2020, LNICST 333, pp. 215–222, 2020.
https://doi.org/10.1007/978-3-030-62483-5_23

end, maximizing the utilization of spectrum and increasing the capacity [4, 5]. In the same channel, space diversity technology is used to transmit multiple antennas. Compared with other technologies, MIMO technology has obvious advantages in increasing system capacity.

2   Data Link System MIMO System

2.1   V - BLAST Design

Beli-laboratories Layered spacetime (BLAST) was originally proposed by Foschini. Because of its simple structure, BLAST is of great significance to the improvement of frequency band utilization. BLAST simply means sending data in parallel using multiple transmitting antennas, and then data on each receiving antenna are separated [6]. It can be seen from its structure that the spectrum utilization of the system can be very high, As the number of antennas changes, so does the volume.

2.2   MIMO Model of Data Link System

Multiple antennas are installed on the data link end machine, and a MIMO system is formed between the two data link end machines. In this system, each antenna at the transmitting end sends signals at the same time, and the signals received at the receiving end are the superposition of signals sent by the transmitting end [7, 8]. The model is shown in Fig. 1:

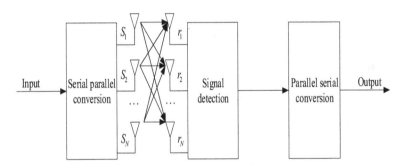

Fig. 1. MIMO signal model of data link system

Suppose at a certain time, the signal vector sent by the sender is $x = [x_1\ x_2\ \dots\ x_N]^T$, the signal vector received by the receiver is $r = [r_1\ r_2\ \dots\ r_M]^T$, and the additive noise of the signal is $z = [z_1\ z_2\ \dots\ z_M]^T$. Assuming that the transmission channel of the signal is A flat channel, that is, the channel H remains unchanged during transmission time of one frame of data, then the relationship between the signal received by the receiving end and the input end is:

$$r = Hx + z \tag{1}$$

In the type, H represents the transmission matrix of $N \times M$, as follows:

$$H = \begin{Bmatrix} h_{11} & h_{12} & \cdots & h_{1N} \\ h_{21} & h_{22} & \cdots & h_{2N} \\ \vdots & \vdots & \ddots & \vdots \\ h_{M1} & h_{M2} & \cdots & h_{MN} \end{Bmatrix} \tag{2}$$

Where, h_{MN} is the transmission channel coefficient, N is the N antenna at the transmitting end, and M is the M antenna at the receiving end.

The data link terminal converts the high-speed data stream through a series and forms a low-speed parallel data stream, which is simultaneously transmitted through N antennas on the same frequency band.In nature, there is a multipath effect in the channel, and the signal reaches the receiving end through various fading, and the receiving end recovers the original data stream from the obtained signals. Under the condition of constant signal bandwidth and transmitting power, the system capacity is significantly improved [9, 10].

3 Signal Detection Algorithm

In the data link system, a reasonable signal detection algorithm can be designed and the spatial separation gain can be used to eliminate the interference and noise between the sending end and the receiving end, so as to maximize the transmission rate and increase the capacity of the system.

3.1 ZF Detection Algorithm

ZF algorithm is the simplest detection, and it is effective. Its idea is to use the breaking matrix to enhance the received signal vector linearly, and finally detect the received signal to obtain the sent signal vector [11]. Then the receiving vector is:

$$r = Hx + z \tag{3}$$

ZF technology uses a weighted matrix to eliminate interference:

$$H^+ = (H^H H)^{-1} H^H \tag{4}$$

Where $(\bullet)^H$ is emmett transpose.

$$\tilde{x}_{ZF} = H^+ r = x + (H^H H)^{-1} H^H z = x + \tilde{x}_{ZF} \tag{5}$$

Among them, $\tilde{x}_{ZF} = H^+ z = (H^H H)^{-1} H^H z$. The resulting

$$r = x + H^{-1} z \tag{6}$$

It can be seen that the ZF detection algorithm is simple in implementation and low in computational complexity. The multi-stream interference between signals is eliminated, but the noise is amplified.

3.2 MMSE Detection Algorithm

ZF amplifies the noise during detection, MMSE takes advantage of the statistical characteristics of noise to optimize the ZF algorithm, that is, to minimize the minimum mean square error.

$$
\begin{aligned}
W_{MMSE} &= \arg \min_w E\left[\left\|x - W^H r\right\|^2\right] \\
&= \left(E\left(rr^H\right)\right)^{-1} E\left(rx^H\right) \\
&= \left(H^H H + \sigma_Z^2 I\right)^{-1} H^H
\end{aligned}
\tag{7}
$$

In formula (7), σ_Z^2 is the variance of noise, It can be seen that MMSE detection algorithm needs statistical information σ_Z^2 of noise. From the weighted matrix W_{MMSE} of MMSE, the estimated value of the transmitter can be expressed as:

$$
\tilde{x}_{MMSE} = W_{MMSE} r = \left(H^H H + \sigma_Z^2 I\right)^{-1} H^H r
\tag{8}
$$

Thus, the corresponding mean square error can be obtained as follows:

$$
\begin{aligned}
MMSE &= E\left[(x - W_{MMSE} r)(x - W_{MMSE} r)^H\right] \\
&= I - H\left(HH^H + \sigma_Z^2 I\right)^{-1} H \\
&= \left(I + \sigma_Z^2 H^H H\right)^{-1}
\end{aligned}
\tag{9}
$$

3.3 ZF-Based Sorting Serial Interference Elimination (ZF-OSIC) Detection Algorithm

ZFdetection algorithm is a one-time detection process by multiplying the inverse matrix of the channel matrix by the left, without considering the influence of the detection order of different layers on the detection [12].The sorted serial interference elimination (OSIC) method does not calculate all the solution vectors at one time, but achieves the solution of each layer by sorting and layer by layer detection. Theoretically, the best detection order is that the large signal-to-noise(SNR) ratio, the more sub vector is detected first. Therefore, the main idea of ZF-OSIC detection algorithm is to firstly detect the sub-vector in the maximum sub-layer of SNR, then detect the interference caused by this vector from the received signal, and then sort iteratively to complete the detection of the whole signal. The signal detection algorithm of each layer adopts ZF detection algorithm. The detailed steps of ZF-OSIC detection are shown below:

First, the initialization process, that is, when i = 1, let

$$
G_1 = H^+
\tag{10}
$$

$$
k_1 = \arg \min_j \left\|(G_1)_j\right\|^2
\tag{11}
$$

And then recursion, every time $i = i+1$, until i is equal to the number of antennas, let

$$W_{ki} = (G_1)_{ki} \tag{12}$$

$$y_{ki} = w_{ki}^T r_i \tag{13}$$

$$\hat{a}_{ki} = Q(y_{ki}) \tag{14}$$

$$r_{i+1} = r_i - \hat{a}_{ki}(H)_{ki} \tag{15}$$

$$G_{i+1} = H_{\underline{ki}}^+ \tag{16}$$

$$k_{i+1} = \underset{j \notin \{k_1 \dots k_i\}}{\arg \min} \left\| (G_{i+1})_j \right\|^2 \tag{17}$$

This algorithm uses the ZF algorithm, Where, $H_{\underline{ki}}^+$ represents the pseudo-inverse matrix of H after deleting column ki,$(G_1)_{ki}$ is the ki-th row vector of G_i,Q stands for quantitative decision function.In the whole calculation process, the SNR of each layer is not directly calculated, but the layer with the minimum SNR is indirectly found by finding the minimum row of G.This is because the average noise power of each layer is the same. When H of the layer is larger, G is smaller. That is, the larger the channel gain, the smaller the proportion of noise power in the receiving vector, and the higher the SNR.In addition, MMSE detection algorithm can also be used in the detection process of each layer, which is the only difference between MMSE-OSIC and ZF-OSIC algorithms [13–15].

4   Simulation of Detection Algorithm

This section simulates the MIMO detection algorithm of serial interference cancellation in multi-antenna data link system. The analog channel is Rayleigh channel, The noise on the receiving antenna is random and follows the independent distribution of zero mean.

As shown in Fig. 2 and Fig. 3, There are two antennas at the transmitter and two at the receiver, the modulation modes are QPSK and BPSK respectively, and the detection performance of the two modulation modes is very close. Compared with ZF detection algorithm, MMSE detection algorithm has better performance, According to theoretical analysis,ZF algorithm is offset by the interference between different antenna, separates the different algorithms of data flow, but has amplified noise, while MMSE algorithm considered the effect of noise and interference between antennas, makes the detection results of the receiver and transmitting between the minimum error in statistical sense. Therefore, the performance of MMSE detection is generally better than ZF detection on the whole, but the calculation of MMSE algorithm is large and the complexity is high. After sorting these two algorithms respectively, namely ZF-OSIC and MMSE-OSIC, it

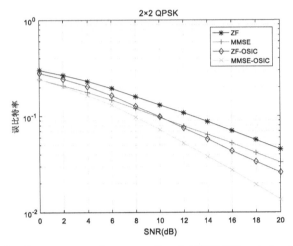

Fig. 2. Bit error rate performance of 2 × 2 QPSK different algorithms

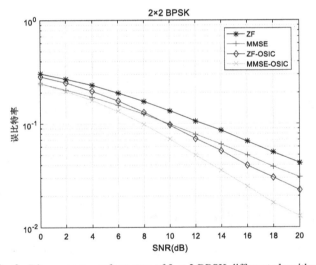

Fig. 3. Bit error rate performance of 2 × 2 BPSK different algorithms

can be seen that their detection performance is superior to ZF algorithm and MMSE algorithm, and MMSE-OSIC algorithm is the best.

According to the simulation comparison in Fig. 4 and Fig. 5, it is observed that when the number of transmitting and receiving antennas is 8 × 8 respectively and the modulation mode is QPSK and BPSK, the bit error rate in QPSK modulation mode is better than that in BPSK modulation mode. The bit error rate of ZF-OSIC and MMSE-OSIC is much better than that of algorithm ZF and MMSE respectively. When the SNR is greater than 10 dB, the effect of ZF-OSIC and MMSE-OSIC is more obvious. The higher the SNR is, the lower the bit error rate of the detection of both is, In a word, MMSE-OSIC has the best detection performance.

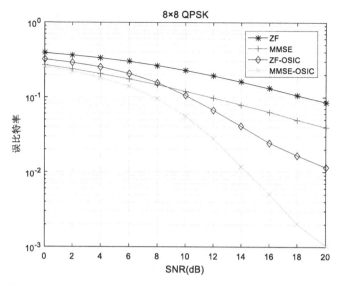

Fig. 4. Bit error rate performance of 8 × 8 QPSK different algorithms

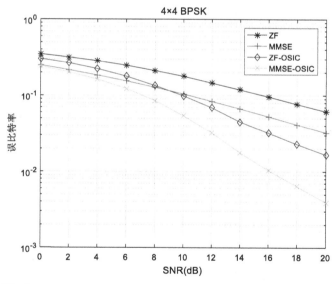

Fig. 5. Bit error rate performance of 8 × 8 BPSK different algorithms

5   Conclusion

In the design of data link end, MIMO technology is introduced into it, and the system model of multi-antenna data link is given. This model uses MIMO spatial multiplexing technology to send data stratified and sent out through different antennas. After independent subchannel fading, through theoretical analysis, the system capacity of the data link end is improved. The algorithm is verified by several simulations, it shows that MMSE-OSIC has the best performance, followed by ZF-OSIC, MMSE and ZF, which have the worst performance. When the number of transmitting antennas increases, the corresponding interference is bound to increase, so it is an inevitable trend to improve the detection of the system.

References

1. Ngo, H., Larsson, E., Marzetta, T.L.: Energy and spectral efficiency of very large multiuser MIMO systems. IEEE Trans. Wirel. Commun. **61**(4), 1436–1449 (2012)
2. Lee, H.J., Kim, D.: A hybrid zero-forcing and sphere-decoding method for MIMO systems. In: International Conference on Wireless Communications, Networking and Mobile Computing, pp. 1–4. IEEE (2006)
3. Hughes-Hartogs, D.: Ensemble Modem Structure for Imperfect Transmission Media: USA4731816[P] (1988)
4. Sun, D., Zheng, B.: A novel multi-user low complexity bit allocation algorithm in cognitive OFDM networks based on AM-GM inequality. In: Proceedings of the 3rd International Conference on Wireless, Mobile and Multimedia Networks, pp. 217–220, Beijing, China (2010)
5. Jian, Z., Qi. Z.: A novel adaptive resource allocation algorithm for multiuser OFDM-based cognitive radiosystems. In: Proceedings of International Conference on Network Computing and Information Security, pp. 442–445, Guilin, China (2011)
6. Ding, Z., Yang, Z., Fan, P., Poor, H.V.: On the performance of non-orthogonal multiple access in 5G systems with randomly deployed users. IEEE Sig. Process. Lett. **21**(12), 1501–1505 (2014)
7. Basar, E., Aygölü, U., Panayirc, E., Poor, H.V.: Orthogonal frequency division-multiplexing with index modulation. IEEE Trans. Sig. Process. **61**(22), 5536–5549 (2013)
8. Dai, L., Wang, B., Yuan, S., Chih-Lin, I., Wang, Z.: Non-orthogonal multiple access for 5G: solutions, challenges, opportunities, and future research trends. IEEE Commun. Mag. **53**(9), 74–81 (2015)
9. Mishra, P., Singh, G., Vij, R., et al.: BER analysis of Alamouti space time block coded 2x2 MIMO systems using Rayleigh dent mobile radio channel. In: 2013 IEEE 3rd International Advance Computing Conference (IACC), pp. 154–158. IEEE (2013)
10. Zhou, S., Zhao, M., Xu, X., et al.: Distributed wireless communication system: a new architecture for future public wireless access. IEEE Commun. Mag. **41**(3), 108–113 (2003)
11. Windpassinger, C., Fischer, R., Huber, J.B.: Lattice-reduction -aided broadcast precoding. IEEE Trans. Commun. **52**(12), 2057–2060 (2004)
12. Dohler, M., Aghvami, H.: On the approximation of MIMO capacity. IEEE Trans. Wirel. Commun. **4**(1), 30–34 (2005)
13. Loyka, S.: Multi- -antenna capacities of waveguide and cavity channels. IEEE Trans. Veh. Technol. **54**(3), 863–872 (2005)
14. Beach, M.A., McNamara, D.P., Fletcher, P.N., et al.: MIMO-a solution for advanced wireless access. In: Proceedings of IEEE 11th International Conference on Antennas and Propagation, pp. 231–235 (2001)
15. Tarokh, V., Seshadri, N., Calderbanb, A.R.: Space time codes for high data rate wireless communication: performance criterion and code construction. IEEE Trans. Inf. Theor. **44**(2), 744–765 (1998)

Detection Algorithm of Compressed Sensing Signal in GSM-MIMO System

Jiang Xiaolin, Tang Zhengyu$^{(\boxtimes)}$, and Qu Susu

Heilongjiang University of Science and Technology, Harbin 150000, China
18317859938@139.com

Abstract. For the generalized spatial modulation in the underdetermined system with the number of transmitting antennas larger than the number of receiving antennas, the activation of the antenna is small and inaccurate. The traditional MMSE algorithm and the ZF algorithm still perform the pseudo-inverse operation on the entire channel matrix, which results in a large number of redundancy. Although the ML algorithm has the best detection performance, the complexity is difficult to meet the actual requirements. In this paper, for the sparse characteristics of GSM signals, a detection algorithm is improved based on the compressed sensing recovery algorithm SWOMP. The algorithm first selects multiple or one active antenna sequences conforming to the spatial modulation to form an index set according to the situation, and then uses the backtracking principle to select the atomic column according to the threshold, rejects the unreliable sequence, and finally uses the minimum mean square error algorithm to detect the modulation symbol according to the activated antenna index. The pseudo-inverse operation of the entire channel matrix is avoided, and the bit error rate is lower than the ZF algorithm, the MMSE algorithm and the OMP algorithm, and the performance of the proposed algorithm is closer to the ML algorithm, which is a better way that balance between complexity and detection performance.

Keywords: GSM-MIMO · Compressed sensing · Signal detection

1 Introduction

The MIMO system not only makes the system's throughput increase in proportion, but also makes the system's reliability stronger by the diversity technology. However, while high-speed and large-scale multi-dimensional data brings high-efficiency transmission, the detection method and performance of the receiving end face higher challenges. For a traditional multi-antenna spatial multiplexing system, how many independent RF links are there in the system. With the surge in data demand, in order to further increase the data transmission rate, the MIMO system needs to increase the number of transmitting antennas, and the number of corresponding RF links becomes enormous, which not only brings a small challenge to the miniaturization of the transmitter. It also makes the hardware implementation cost high. Unlike spatial multiplexing, SM-MIMO maps information bit blocks into two information bearing units: spatial constellation symbols

© ICST Institute for Computer Sciences, Social Informatics and Telecommunications Engineering 2020
Published by Springer Nature Switzerland AG 2020. All Rights Reserved
X. Jiang and P. Li (Eds.): GreeNets 2020, LNICST 333, pp. 223–232, 2020.
https://doi.org/10.1007/978-3-030-62483-5_24

and signal constellation symbols. Only one or a few antennas are activated when transmitting, which reduces the number of links and opens up the possibility of hardware cost reduction at the receiving end.

Spatial modulation technology is a relatively new multi-antenna transmission technology, except for the amplitude phase in the general real and imaginary fields. Amplitude and Phase Modulation (APM), which also introduces the spatial dimension as the third dimension, mines the serial number of the transmitting antenna for additional mapping, and establishes the mapping relationship between the antenna number and the input bits to complete the spatial modulation target. Since the number of receiving antennas of the user is small and the number of antennas of the BS is small, signal detection is a challenging large-scale uncertain problem. When the number of transmitting antennas becomes large, the optimal maximum likelihood (ML) signal detector suffers from excessive complexity [3], which is unacceptable in practical engineering. Since the number of active antennas is smaller than the total number of transmit antennas, the SM signal has inherent sparsity, and the sparsity of the signal can be improved by utilizing the compression sensing (CS) theory [1, 2]. That is, the signal detection scheme based on the compressed sensing theory can be used for detection at the receiving end. And CS-based signal detectors have been proposed for uncertain small SM-MIMO [4, 5]. However, compared with the best ML detectors, their bit error rate (BER) performance still has a large gap, and Gao et al. [6] proposed a packet transmission scheme suitable for large-scale SM-MIMO systems. A structured Subspace Pursuit (SSP) detection algorithm is presented. This algorithm is an improvement of the OMP algorithm. Each iteration selects multiple atoms and expands the signal search space. Therefore, its bit error rate is lower than that of the OMP algorithm, but the complexity is also increased accordingly. A CoSaMP-based Spatial Matching Pursuit (SMMP) detector is proposed in [7], which is applied in a multiple access channel with a large-scale antenna base station. These compression-sensing detectors have a large performance penalty. Therefore, the balance between detection performance and complexity of detection algorithms needs further discussion.

2 System Model

In this paper, it is assumed that the number of input signal antennas and the number of output signal antennas in the MIMO system are respectively N_T and N_R, noting $N_T \leq N_R$. If the number of active antennas on the sender is N_A, there are $C_{N_T}^{N_A}$ a total possible combination. There are a combination of spatial symbols $N = 2^{\log C_{N_T}^{N_A}}$. The system model for spatial modulation is (Fig. 1):

Assume that there is perfect channel information. The receiving model of the system is

$$y = Hx + \sqrt{N_T E_x / \rho}\, n \tag{1}$$

Where y is a column vector of $N_R \times 1$. S is a constellation symbol set, x is a column vector of $N_T \times 1$, and the sparsity is N_A. Which is $x \in S^{N_T \times 1}$. H is the channel matrix

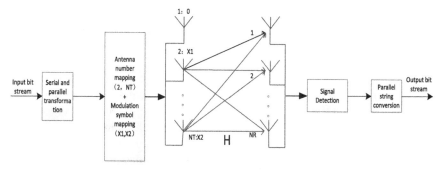

Fig. 1. Block diagram of GSM system

of $N_T \times N_R$. n is Gaussian white noise, ρ is signal to noise ratio, $E_x = \dfrac{\sum\limits_{i=1}^{N_T} \|x_i\|_2^2}{N_T}$ is the mean of the transmitted symbol energy.

3 Detection Algorithm Based on Compressed Sensing Reconstruction

According to the above GSM system model, the transmitted signal vector x is an N_A sparse signal, that is, there are only N_A non-zero elements in x, which is much smaller than the number of transmitting antennas N_T. For MIMO channels, if the channel matrix H satisfies the N-order RIP, then for any sparse signal x, the following formula is satisfied.

$$(1 - \delta_k)\|x\|_2^2 \leq \|Hx\|_2^2 \leq (1 + \delta_k)\|x\|_2^2 \tag{2}$$

It can be accurately recovered from the received signal, where is the K-order RIP parameter [5]. According to the research literature [6], it can be seen that when $N_R \geq cN_A \cdot \log(N_T/N_A)$, $\delta_{N_A} \leq 0.1$ satisfies the requirement of formula (2). Therefore, a compression-aware recovery algorithm can be used to solve the GSM detection problem.

3.1 GSM Detection Based on SWOMP Algorithm

In theory, the OMP algorithm can reconstruct the sparse signal x after iteration. SWOMP selects multiple atomic columns at a time to effectively reduce the number of iterations. However, in the wireless channel, since the received signal is affected by additive noise, only one inner product maximum or multiple is selected for each iteration, and it is very likely that the wrong activated antenna index is selected, so that the detection performance of the OMP algorithm is large. discount. Therefore, we use an improved flexible and optional atomic selection strategy and backtracking ideas to make use of Eq. (3) to make an compromise between algorithm complexity and detection performance.

$$|\theta_i| \geq T_g = g * \max\left|\Phi^T r\right| \tag{3}$$

Where, $\forall i \in J$, set J is the set of serial numbers of Φ in the first selection sensing matrix, with coefficient $g \in (0, 0.5]$. The threshold standard of pruning is related to the residual error of this iteration, which can determine the correct atomic scolumn better. In order to adapt the recovery algorithm to the system, we will use the sparse signal to take the characteristics of the constellation symbol in the modulation constellation. In the recovery algorithm, we will select the point in the constellation as the estimated value of the signal instead of the original algorithm. The value obtained by direct calculation. Finally, the MMSE is used to detect the antenna number that retains the closest and satisfies the spatial symbol as the spatial symbol detection value (Fig. 2).

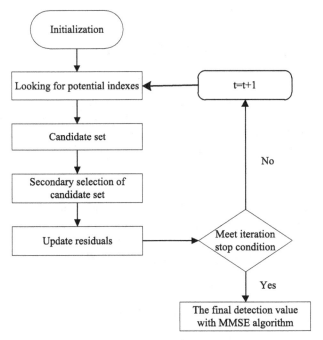

Fig. 2. Flow chart of improved swomp detection algorithm

The algorithm steps are as follows:

Input: Received signal y ,Channel matrix A=H;

Output: Input signal x estimate θ_t

Initialization: $\theta_t = 0$, Residual r_0 = y ,Support set sequence $\Lambda_0 = \phi$,

Support set $A_0 = \phi$, t=1, F=0.

While t < max_iter or $\left\| r_t \right\|_2 \leq \varepsilon_1$ **do**

$$r_t = y - A_{\Lambda_t} \theta_t \ , \ t=t+1 \qquad \text{\{Update residual\}}$$

$$u = \left| A^T r_{-1} \right| \qquad \text{\{Correlation test\}}$$

If F=0 do

$j_a = find \ (u \ \geq T_h))$

else do $j_a = find(u = \max \left| A^T r \right|)$ { First selection}

end   if

$\Lambda_t = \Lambda_{t-1} \cup j_a$ {Merge supports}

$$\theta_t = \left(A_{\Lambda_t}^T A_{\Lambda_t} \right)^{-1} A_{\Lambda_t}^T y \qquad \text{\{Caculate a new estimite\}}$$

$$\Lambda_t = find \left(\left| \theta_i \right| \geq T_g = g^* \max \left| A^T r \right| \right) \qquad \text{\{Second selection\}}$$

end while

$$\theta_t = \theta_{A_{\Lambda_t}} = Q\left(\left((H_{A_{\Lambda_t}})^H H_{A_{\Lambda_t}} + \sigma^2 I \right)^{-1} (H_{A_{\Lambda_t}})^H y \right), \ \{\text{Final estimite}\} \quad (4)$$

Where Q is the quantization function. Each vector is quantized to an integer. σ^2 is the statistical information of noise. The above iterative process can ensure that the final antenna combination is a spatial symbol.

Transform selection criteria: $\|rt\|_2 \geq \|rt - 1\|_2$; According to the comparison of the residual energy values of the adjacent stages, if the current stage is larger than the previous stage, it means entering the small step long stage, then switch the selection criteria. The maximum number of iterations is the sampling value M. if the iteration stop condition is met, the output is direct. The stop iteration condition is set to the following two cases: in the first case $\|rt\|_2 \leq \varepsilon_1$, ε_1 is the noise measurement value under ideal condition; in the second case, there is no qualified atomic column in step (5), when there is no suitable selection column for the first time, it also means that there is no atom matching with the Acolumn of the sensing matrix in the residual error, and the result can be directly output; the above two conditions are full Any one of them will exit the cycle and output the result.

3.2 Complexity Analysis

As a measure, the number of floating-point operations (only considering multiplication, one complex number multiplication equals to four floating-point operations) needed to realize a GSM signal detection is used to compare and analyze the computational complexity of ML algorithm, ZF algorithm, OMP algorithm and the algorithm proposed in this paper.

For MMSE algorithm, floating point operand required [8]:

$$C_{MMSE} = 4N_T^3 + 12N_T^2 N_R + 7N_T^2 + 6N_T N_R \tag{5}$$

For ZF algorithm, floating point operand required [9]:

$$C_{ZF} = 4N_T^2 N_R + \frac{4}{3}N_T^3 + 8N_R N_T + 11N_T^2 \tag{6}$$

For ML algorithm, floating point operand required [10]:

$$C_{ML} = (6N_A + 3)N_R M^{N_A} N \tag{7}$$

For the OMP algorithm, the required floating-point operands are:

$$C_{OMP} = 4N_T N_R N_A + 4N_R \sum_{t=1}^{N_A}(t)^2 + \frac{4}{3}\sum_{t=1}^{N_A}(t)^3 + 8N_R \sum_{t=1}^{N_A}t + 11\sum_{t=1}^{N_A}(t)^2 \tag{8}$$

The algorithm proposed in this paper: the number of iterations is less than or equal to, and the operation amount of each iteration is mainly concentrated in two parts: inner product operation part and generalized inverse operation part.

The inner product operation part is similar to OMP, t represents the total number of iterations, $t \leq N_A$ in the proposed algorithm.

$$C_a = 4N_T N_R t \tag{9}$$

The generalized inverse operation part can be expressed as:

$$C_b = 4(G_{avg}^k)^2 N_R + \frac{4}{3}(G_{avg}^k)^3 + 8G_{avg}^k N_R + 11(G_{avg}^k)^2 \tag{10}$$

Finally, MMSE algorithm is used to estimate symbols:

$$C_c = 4(G_{avg}^k)^3 + 12G_{avg}^k N_R + 7(G_{avg}^k)^2 + 6G_{avg}^k N_R \tag{11}$$

Where G_{avg}^k represents the number of indexes contained in the selected set in iteration K.

From the above analysis, the complexity of the algorithm can be expressed as follows:

$$C_{proposed} = 4N_T N_R t + 4(G_{avg}^k)^2 N_R + \frac{4}{3}(G_{avg}^k)^3 + 8G_{avg}^k N_R + 11(G_{avg}^k)^2 +$$
$$4(G_{avg}^k)^3 + 12G_{avg}^k N_R + 7(G_{avg}^k)^2 + 6G_{avg}^k N_R \tag{12}$$

From the above analysis, it can be seen that ML increases exponentially with the number of active antennas and modulation order, and the complexity is the highest. Compared with MMSE and ZF algorithm, because of the inverse of the whole channel matrix, the improved compressed sensing detection algorithm in this paper firstly uses the uncertain number of active antennas to obtain the sequence, which is equivalent to reducing the number of transmitting antennas, and then uses MMSE algorithm to obtain the signal modulation symbols. Although the complexity of the algorithm is higher than that of the OMP algorithm, the other advantages of compressed sensing are guaranteed and the bit error rate is reduced. See Table 1 for details. When the number of transmitting antennas NT = 10 and the number of receiving antennas NR = 16, the specific complexity is:

Table 1. Complexity comparison

Algorithm	Average complexity	
	NA = 2	NA = 3
ML	122880	172032
MMSE	24860	24860
ZF	10113	10113
OMP	2052	2692
proposed	4562	4577

4 Simulation Analysis

In this paper, the bit error rate (BER) is used as the detection performance index to evaluate the BER performance of the above algorithms. The simulation parameters are configured as follows: in the single symbol GSM system, the number of transmitting antennas $NT = 10$, the number of receiving antennas $NR = 16$, the observed SNR range

[0,10], and the step size is 2. The channel adopts quasi flat Rayleigh fading channel, assuming that the receiver knows the exact channel state information.

As for the threshold coefficient g of the middle backtracking of the algorithm proposed in this paper, when the signal-to-noise ratio is 0, the influence of the parameters on the algorithm is as follows:

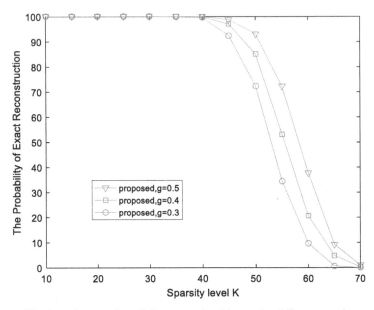

Fig. 3. Influence of coefficient g on algorithm under different sparsity

It can be seen from Fig. 3 that under the same sparsity K, with the increase of coefficient g, the recovery probability of the algorithm is improved, and the attenuation of the increase of sparsity is also slow. This is because the increase of coefficient g will make the backtracking filtering more accurate. However, when the coefficient g is greater than 0.5, the overall recovery probability will be affected due to the small number of selected atomic column sets. Therefore, the coefficient g is taken as 0.5 in the later simulation.

Figure 4 shows the trend of the BER of ML, ZF, MMSE, OMP algorithm and the improved algorithm with the signal-to-noise ratio when the number of active antennas is unknown and QPSK modulation is adopted. The simulation results show that the BER of the proposed algorithm is lower than that of ZF algorithm, MMSE algorithm and OMP algorithm, and more close to the performance of ML algorithm. Among them, the method of selecting atomic columns in OMP algorithm is fixed and cannot eliminate the wrong sequence. Although the initial error rate is lower than ZF, MMSE algorithm, but with the improvement of SNR, the error rate has not improved significantly. Floor effect exists in the detection algorithms of compressed sensing [11]. The algorithm proposed in this paper can reduce the error rate and the convergence floor at the same time.

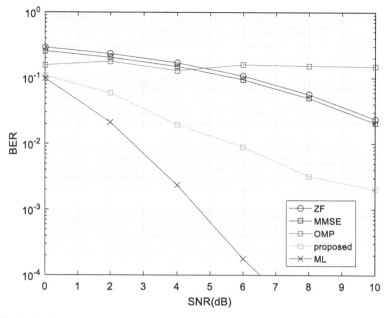

Fig. 4. Performance comparison of ML, ZF, MMSE, OMP and improved algorithm

5   Conclusion

In this paper, a low complexity signal detection algorithm based on sparse reconstruction theory is proposed by using the sparsity of GSM signal. Simulation results show that compared with the traditional OMP algorithm, the threshold value is used to select multiple atoms for fast approximation, and the switch selection strategy is used to accurately approximate the sparse when the sparse degree of the original signal may be overestimated. At last, we trace back the support set, prune the error columns which may be caused by multiple atom selection, and effectively reduce the system error rate at the expense of a small amount of complexity. Although there is some performance loss compared with ML algorithm, it effectively solves the problem of high complexity of ML algorithm and to some extent, the balance between complexity and detection performance is achieved, so it has certain application significance.

References

1. Duarte, M., Eldar, Y.: Structured compressed sensing: from theory to applications. IEEE Trans. Signal Process. **59**(9), 4053–4085 (2009)
2. Dai, W., Milenkovic, O.: Subspace pursuit for compressive sensing signal reconstruction. IEEE Trans. Inf. Theory, **55**(5), 22302249 (2009)
3. Zheng, J.: Signal vector based list detection for spatial modulation. IEEE Wireless Commun. Lett. **1**(4), 265–267 (2012)
4. Baraniuk, R.G.: Compressive sensing. IEEE Signal Processing Mag. **24**(4), 118–121 (2007)

5. Candes, E.J., Wakin, M.B.: An introduction to compressive sampling. IEEE Signal Process. Magazine **25**(2), 21–30 (2008)

6. Gao, Z., Dai, L., Qi, C., et al.: Near – optimal signal detector based on structured compressive sensing for massive SM – MIMO. IEEE Trans. Vehicular Technol. **66**(2), 1860–1865 (2017)

7. Garcia-Rodriguez, A., Masouros, C.: Low-complexity compressive sensing detection for spatial modulation in large-scale multiple access channels. IEEE Trans. Commun. **63**(7), 2565–2579 (2015)

8. Xiao, Y., Yang, Z., Dan, L., et al.: Low-complexity signal detection for generalized spatial modulation. IEEE Commun. Lett. **18**(3), 403–406 (2014)

9. Bhat, S., Chockalingam, A.: Sparsity - exploiting detection of large - scale multiuser GSM - MIMO signals using FOCUSS. In: Proceedings of 2016 IEEE 83rd Vehicular Technology Conference, Nanjing, IEEE, 1–5 (2016)

10. Wang, C., Cheng, P., Chen, Z., et al.: Near – ML low – complexity detection for generalized spatial modulation. IEEE Commun. Letters **20**(3), 618–621 (2016)

11. Wu, C.H., Chung, W.H., Liang, H.W.: OMP - based detector design for space shift keying in large MIMO systems. In: Proceedings of 2014 IEEE Global Communications Conference, Austin, TX, USA: IEEE, 4072–4076 (2014)

Communication Model and Performance Analysis of Frequency Modulation-Correlation Delay-Orthogonal Chaotic Phase Shift Keying

Juan Wang[1(✉)], Wei Li[1], Zhiming Qi[2], and Shu Wang[3]

[1] Electronic and Information Engineering Institute, Heilongjiang
University of Science and Technology, Harbin 150022, China
76115347@qq.com
[2] State Grid Liaoning Maintenance Company, Shenyang, Liaoning 110006, China
[3] State Grid Anshan Power Supply Company, Anshan, Liaoning 114001, China

Abstract. In the Multi-ary chaos shift keying communication represented by CD-QCSK, each symbol carries multiple information bits, which can effectively improve the transmission rate. Since the bit energy of chaotic carrier is not constant, which may easily lead to misjudgment when receiving. Based on the feasibility study of combination in FM modulation and CD-QCSK communication, this article proposes a multi-ary FM-CD-QCSK communication model of frequency modulation-correlation delay-orthogonal chaos shift keying. By the FM modulator, The chaotic signal is changed into the carrier signal with constant envelope and random change of frequency within a certain range, so that each symbol has equal bit energy for transmission to improve the anti-interference and anti-noise ability of the system effectively. The simulation results show that, whatever the spreading spectrum factor changed, compared with existing Multi-ary chaos shift keying communication QCSK and CD-QCSK, the FM-CD-QCSK has better error performance and can better meet the requirements of high rate and communication quality under the same signal to noise ratio.

Keywords: Multi-ary · Chaos keying · FM modulation

1 Introduction

The chaotic signal takes advantage of the characteristics of aperiodic, high bandwidth, random-like and numerous, and is applied to spread spectrum and other wireless communication systems to act as carriers to achieve spectrum expansion while completing digital modulation. It has strong anti-noise, anti-interference, anti-multi-path fading and anti-parameter sensitivity [1, 2]. As the mainstream of binary chaotic keying communication technology, differential chaotic shift keying based on incoherent demodulation (Differential Chaos Shift Keying, DCSK) [3] does not need channel estimation and spread spectrum code synchronization, but it takes about half of the time to transmit power, use chaotic carriers that do not contain any data information as reference signals,

X. Jiang and P. Li (Eds.): GreeNets 2020, LNICST 333, pp. 233–242, 2020.
https://doi.org/10.1007/978-3-030-62483-5_25

which is difficult to meet the application requirements of high-speed, high-quality data transmission [4, 5]. Therefore, the Multi-ary chaotic keying communication with multiple information bits carried by each symbol has gradually become the focus of scholars at home and abroad [6].

In the article [7], a Multi-ary quadrature chaotic shift keying(QCSK) communication model is proposed based on DCSK. The model adopts four-phase orthogonal modulation and makes use of the zero correlation between the two orthogonal signals to increase its data transmission rate to twice as much as DCSK. In order to further improve the information transmission rate, the article [8], a Multi-ary correlation delay-quadrature chaotic shift keying (CD-QCSK) communication model is proposed by combining correlation delay shift keying (CDSK) [9] with QCSK. The model can have the advantages of high transmission rate of CDSK and QCSK at the same time, and its transmission rate can be 4 times of that of DCSK [10] and 2 times of that of QCSK. However, due to the aperiodic characteristics of chaotic signal, the bit energy will change with time after CD-QCSK modulation, which leads to the increase of bit error rate (BER) caused by decision problem.

Based on the above analysis, this paper proposes a combination of frequency modulation (FM) [11] and CD-QCSK communication, and proposes a Multi-ary frequency modulation-correlation delay-quadrature chaos shift keying (FM-CD-QCSK) communication model. At the transmitter, the chaotic signal is transformed into a carrier whose envelope is constant and the frequency varies randomly in a certain range by FM modulation, so that each symbol has the same bit energy and reduces the bit error rate (BER). At the receiving end, the reference signal and its orthogonal signal are divided into two ways to correlate with the carrier signal. As well as the information transmitted by each symbol can be doubled without FM resolution, and the system can improve the information transmission rate while reducing the bit error rate. The communication performance of FM-CD-QCSK system in Gaussian white noise channel and Riley channel is simulated and analyzed, and compared with the Multi-ary chaotic keying communication system QCSK and CD-QCSK. The results show that the FM-CD-QCSK system can effectively combine the error performance with the transmission rate, and will have a certain application prospect in the field of high-speed wireless communication.

2   FM-CD-QCSK Communication Model

2.1   Emission Model

Figure 1 shows emission model of FM-CD-QCSK. If high-frequency carrier $c(t) = A_c \cos \omega_c t$ is assumed, the chaotic signal $x(t)$ with random amplitude variation can be output after frequency modulation as follows:

$$x'(t) = A_c \cos(\omega_c t + k_f \int_0^t x(\tau) d\tau) = A_c \cos \varphi(t) \tag{1}$$

In the formula (1), A_c is the carrier amplitude, k_f is the frequency modulation coefficient, and $x'(t)$ is the chaotic frequency modulation signal with constant amplitude and random frequency variation.

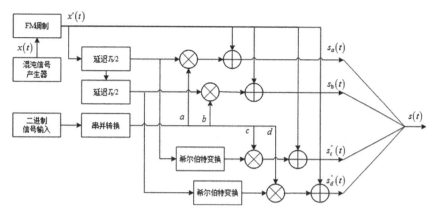

Fig. 1. Emission model of FM-CD-QCSK

In view of the shortage of transmitting reference signal in half of each symbol period of QCSK, FM-CD-QCSK uses the communication principle of CDSK chaotic keying for reference, converts binary information sequence into information signal as a, b, c, d, modulates four information signals at the same time by using chaotic signal of delay $T_b/2$, T_b and its orthogonal signal, and then transmits it after superposition with chaotic frequency modulation signal $x'(t)$.

In the l symbol period, the output signal of the FM-CD-QCSK transmitter, as follows.

$$s(t) = \begin{cases} s_a(t) = x'(t) + ax'(t - T_b/2), & (l-1)T_b \leq t < (l-3/4)T_b \\ s_b(t) = x'(t) + bx'(t - T_b), & (l-3/4)T_b \leq t < (l-2/4)T_b \\ s_{\hat{c}}(t) = x'(t) + c\hat{x}'(t - T_b/2), & (l-2/4)T_b \leq t < (l-1/4)T_b \\ s_{\hat{d}}(t) = x'(t) + d\hat{x}'(t - T_b), & (l-1/4)T_b \leq t < lT_b \end{cases} \quad (2)$$

In the formula (2), $\hat{x}'(t)$ is a chaotic signal with which $x'(t)$ is orthogonal by Herbert transform. After the information sequence is transformed by bipolar transformation, a, c is modulated with the chaotic frequency modulation signal $x'(t - T_b/2)$ of delay $T_b/2$ and its orthogonal chaotic frequency modulation signal $\hat{x}'(t - T_b/2)$, and b, d is modulated with the chaotic frequency modulation signal $x'(t - T_b)$ and its orthogonal chaotic frequency modulation signal $\hat{x}'(t - T_b)$ with a delay of T_b, respectively. If the information signal is "+1", the information signal is the same as the reference signal; if the information signal is "−1", the information signal is opposite to the reference signal. Therefore, the transmitted information signal is included in the correlation value of symbol sample values of two adjacent symbol.

2.2 Receiving Model

Figure 2 shows receiving model of FM-CD-QCSK. It is assumed that the transmitted signal is interfered by additive white Gaussian noise $n(t)$ in the transmission process, the received signal is correlated with its delay $T_b/2$, T_b signal respectively. For example $r_a(t)$, $r_{\hat{c}}(t)$, the output $y_a(t)$, $y_c(t)$ through the correlator are

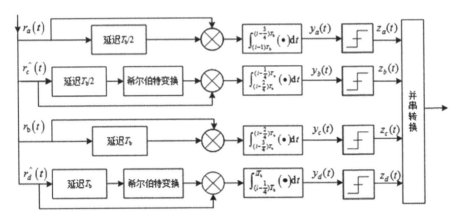

Fig. 2. Receiving model of FM-CD-QCSK

$$y_a(t) = \int_{(l-1)T_b}^{(l-\frac{3}{4})T_b} r_a(t)r_a(t - T_b/2)dt$$

$$= \int_{(l-1)T_b}^{(l-\frac{3}{4})T_b} [sa(t) + n(t)][sa(t - T_b/2) + n(t - T_b/2)]dt$$

$$= \int_{(l-1)T_b}^{(l-\frac{3}{4})T_b} [sa(t)sa(t - T_b/2)]dt + \int_{(l-1)T_b}^{(l-\frac{3}{4})T_b} [sa(t)n(t-T_b/2)]dt$$

$$+ \int_{(l-1)T_b}^{(l-\frac{3}{4})T_b} [sa(t - T_b/2)n(t)]dt + \int_{(l-1)T_b}^{(l-\frac{3}{4})T_b} [n(t)n(t - T_b/2)]dt$$

$$= \int_{(l-1)T_b}^{(l-\frac{3}{4})T_b} [sa(t)sa(t - T_b/2)]dt \tag{3}$$

$$y_c(t) = \int_{(l-\frac{2}{4})T_b}^{(l-\frac{1}{4})T_b} r_c^\wedge(t)r_c^\wedge(t - T_b/2)dt$$

$$= \int_{(l-\frac{2}{4})T_b}^{(l-\frac{1}{4})T_b} [s_c^\wedge(t) + n(t)][s_c^\wedge(t - T_b/2) + n(t - T_b/2)]dt$$

$$= \int_{(l-\frac{2}{4})T_b}^{(l-\frac{1}{4})T_b} [s_c^\wedge(t)s_c^\wedge(t - T_b/2)]dt + \int_{(l-\frac{2}{4})Tb}^{(l-\frac{1}{4})Tb} [s_c^\wedge(t)n(t-T_b/2)]dt$$

$$+ \int_{(l-\frac{2}{4})T_b}^{(l-\frac{1}{4})T_b} [s_c^\wedge(t - T_b/2)n(t)]dt + \int_{(l-\frac{2}{4})Tb}^{(l-\frac{1}{4})Tb} [n(t)n(t - T_b/2)]dt$$

$$= \int_{(l-\frac{2}{4})T_b}^{(l-\frac{1}{4})T_b} [s_c^\wedge(t)s_c^\wedge(t - T_b/2)]dt \tag{4}$$

In the formula (3) and (4), due to the good auto-correlation of chaotic signals, the correlation values of the other three terms are all zero except that the first term is a useful term carrying information. When the threshold value of the decision circuit is set to zero, the FM modulation will make the signal energy tend to be constant. When $y_a(t)$, $y_c(t) > 0$, the value of the decision output is "+1", and when $y_a(t)$, $y_c(t) < 0$ the value of the decision output is "-1".

3 FM-CD-QCSK Simulation Analysis

In order to study the communication performance of FM-CD-QCSK, the FM-CD-QCSK communication model based on Gaussian (AWNG) channel and Riley (Rayleigh) channel is simulated and analyzed, and the influence of the system parameters on the communication performance is obtained, and compared with other similar chaotic keying communication models.

Figure 3 shows the variation of error performance of FM-CD-QCSK with spread spectrum factor in AWNG channel. The simulation time is T = 2000 s and the spread spectrum factor is m = 8, 16, 32. It can be seen from the simulation diagram that when $E_b/N_0 < 4$ dB, the error performance of the FM-CD-QCSK system at different m values is comparable, which indicates that the influence of the spreading factor on the system performance is not obvious in this range. When the signal-to-noise ratio (SNR) is larger than 4 dB, the spread spectrum factor increases, which leads to the deterioration of the error performance of the system. This is because the noise interference term of the receiver correlator demodulation is proportional to the spread spectrum factor. With the increase of the spread spectrum factor, the introduced noise component will increase, which will lead to the deterioration of the error code performance of the system. Therefore, there is an optimal m value parameter setting.

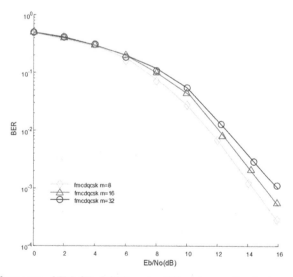

Fig. 3. BER performance of FM-CD-QCSK under different spreading factors of AWNG channel

Figure 4 shows the variation of error performance of FM-CD-QCSK system with signal-to-noise ratio (SNR) in AWNG channel. The simulation time is T = 2000 s, and the signal-to-noise ratio is $E_b/N_0 = 4, 10, 16$ respectively. The simulation results show that the bit error rate (BER) decreases with the increase of SNR. At the same time, when $E_b/N_0 = 4$ dB, the BER curve is almost a straight line, which shows that the influence of spread spectrum factor on BER is not obvious at low SNR. When $E_b/N_0 = 10, 16$ dB, it can be found that when $0 < m < 8$, as the spreading factor increases, the bit error rate gradually decreases, but when the spread spectrum factor increases to the critical value $m = 8$, the bit error rate increases with the increase of spread spectrum factor. This is due to the increase of the noise component introduced in the receiver correlator with the increase of the spread spectrum factor, which leads to the increase of the bit error rate (BER) of the system. Therefore, under the condition of certain signal-to-noise ratio (SNR), the optimal error performance of FM-CD-QCSK system will be obtained by properly selecting the spread spectrum factor $m = 8$.

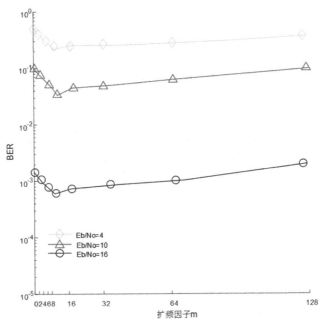

Fig. 4. BER performance of FM-CD-QCSK under different signal-to-noise ratios in AWNG channel

Figure 5 shows the simulation results for all systems (including FM-CD-QCSK 、CD-QCSK and QCSK) with different values of M for BER performance comparison. The simulation time is T = 2000 s, and the spread spectrum factor is $m = 8, 16$ respectively. By comparing diagram (a), diagram(b), it can be found that when $0 < E_b/N_0 < 16$, the bit error rate (BER) of FM-CD-QCSK system is better than that of QCSK and CD-QCSK at different m values. When the bit error rate is the same, the signal-to-noise ratio (SNR) of FM-CD-QCSK is about 1 dB higher than that of CD-QCSK and 2 dB higher than that of QCSK. Thus it can be seen that by introducing FM modulation into CD-QCSK, FM-CD-QCSK can achieve better communication performance.

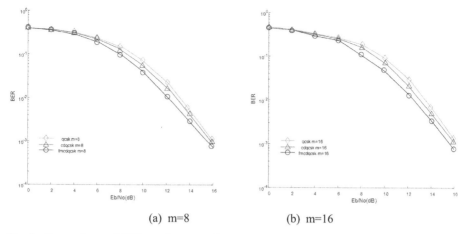

(a) m=8   (b) m=16

Fig. 5. Comparison of BER performance between FM-CD-QCSK,CD-QCSK and QCSK in AWNG channel

Figure 6 shows the variation of BER performance of FM-CD-QCSK with spread spectrum factor in Rayleigh channel. The simulation time is T = 2000 s, and the spread spectrum factor is m = 8, 16, 32 respectively. The maximum Doppler translation of channel parameters is 1 Hz, and the M-path parameter is L = 3. The simulation results show that when $E_b/N_0 < 4$ dB, FM-CD-QCSK has similar bit error performance under different spread spectrum factors, which shows that the spread spectrum factor has little effect on the bit error rate at low signal-to-noise ratio (SNR) of Riley channel. When the signal-to-noise ratio (SNR) is larger than 4 dB, the error performance of the system is improved with the increase of spread spectrum factor. By comparing with Fig. 5, it can be found that the optimal value of spread spectrum factor of FM-CD-QCSK system in Rayleigh channel is larger than that in Gaussian channel, which is due to the fact that multipath delay will increase the decision energy, thus offsetting the partial correlation noise interference caused by the increase of spread spectrum factor.

Figure 7 shows the BER performance simulation curve of FM-CD-QCSK in AWNG channel and Rayleigh. channel. The simulation time is T = 2000 s, the spread spectrum factor is m = 8, the maximum Doppler translation of channel parameters is 1 Hz, and the multipath channel parameters are L = 2 and L = 3 respectively. When $0 < E_b/N_0 < 18$, the bit error rate (BER) of FM-CD-QCSK in Gaussian channel is close to 0, and the bit error rate (BER) of FM-CD-QCSK in Rayleigh fading channel is higher than that in Gaussian channel, and the attenuation is slower. It is shown that the effect of multipath delay effect on signal transmission in Riley channel model is more serious than that in AWGN channel, so the error performance of the system in Gaussian channel is obviously better. In Riley channel, when $E_b/N_0 < 10$ dB, the bit error rate (BER) of FM-CD-QCSK under different multipath channel parameters is approximately the same, which indicates that the multipath channel parameters have little effect on the system performance in this range. When the signal-to-noise ratio (SNR) is larger than 10 dB, the difference between L = 3 and L = 2 is gradually obvious. When BER = 10^{-2}, the signal-to-noise ratio (SNR) of the system with multipath parameter L = 3 is about 2 dB higher than that

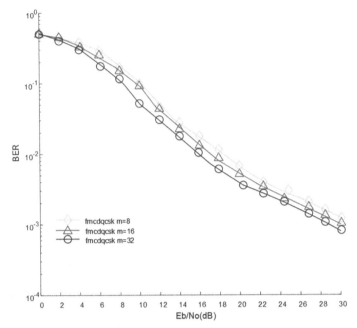

Fig. 6. BER performance of FM-CD-QCSK under different spread spectrum factors in rayleigh channel

of L = 2, which is due to the increase of signal energy in the decision variable due to the superposition of signal energy in each path channel, which makes it easier for the decision valve to judge the correct value.

Figure 8 shows the BER performance simulation curves of FM-CD-QCSK, CD-QCSK and QCSK over Rayleigh channel. The simulation time is T = 2000 s, the spreading factor is m = 16, the maximum Doppler shift of the channel parameters is 1 Hz, and the multipath channel parameters are L = 2 and L = 3, respectively. Compared with diagram (a), diagram (b), the BER values of the three systems are very similar when $0 < E_b/N_0 < 10$. It is shown that the error performance of the three systems is approximately the same when the signal-to-noise ratio (SNR) of the three systems is low. When $10 < E_b/N_0 < 30$, with the increase of signal-to-noise ratio (SNR), the bit error rate (BER) of FM-CD-QCSK decays faster, and the advantages of BER are more and more obvious. In the case of $E_b/N_0 = 30$, the bit error rate (BER) of FM-CD-QCSK approaches zero. It can be concluded that the error performance of FM-CD-QCSK system in Riley channel is better than that of CD-QCSK and QCSK.

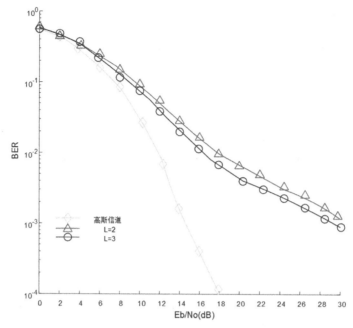

Fig. 7. Comparison of BER performance of FM-CD-QCSK in AWNG channel and rayleigh channel

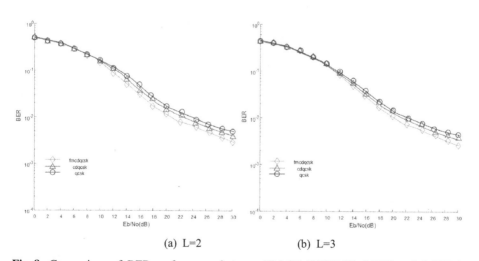

(a) L=2   (b) L=3

Fig. 8. Comparison of BER performance between FM-CD-QCSK,CD-QCSK and QCSK in rayleigh channel

4 Conclusion

In order to further improve the anti-interference and anti-noise ability of Multi-ary chaotic keying communication represented by CD-QCSK, a new type of FM-CD-QCSK Multi-ary chaotic keying communication model is proposed by combining it with FM modulation. The system makes each symbol have equal bit energy by FM modulation, and can ensure high-speed transmission of information while reducing the bit error rate. The communication performance of FM-CD-QCSK in AWNG and Rayleigh channels are simulated and compared with CD-QCSK and QCSK. It is concluded that FM-CD-QCSK has better communication performance and can better meet the high quality and high speed application requirements of wireless communication.

Funding. This work was supported by Heilongjiang Fundamental Research Foundation for the Local Universities in 2018 (2018KYYWF1189) and Science and Technology Innovation Foundation of Harbin (2017RAQXJ082).

References

1. Zhang, G., Meng, W., Zhang, T.: Multiple ary multiple user DCSK based on FDM. Syst. Eng. Electron. Technol. **3901**, 183–187 (2017)
2. Wang, J., Zhang, S., Yan, C.Y.: Communication model and performance of frequency modulation dependent delay difference chaotic keying. J. Heilongjiang Univ. Sci. Technol. **26**(05), 581 (2016)
3. Wang, J., Li, W., Zhang, Y., Qi, Z.M.: Delay-frequency-difference chaotic keying communication model and performance of orthogonal WALSH code. Journal of heilongjiang university of science and technology **29**(04), 478–484 (2019)
4. Pan J.: Research on Multi-ary chaotic keying Modulation method [D]. Jilin University (2005)
5. Meng W.: Research on multi-user chaos keying communication system [D]. Chongqing university of posts and telecommunications (2017)
6. Wang, J., Li, W., Qin, Y.Y.: Analysis of MIMO-CD-FM-DCSK communication performance in coal mine. Commun. Technol. **51**(07), 1505–1510 (2018)
7. Wu, Y.Q., Wang, M.: Performance analysis and simulation of QCSK in Rayleigh slow fading channel. J. Guilin Inst. Electron. Technol. **01**, 33–35 (2004)
8. Ke, J.X.: Design of a new CD-QCSK chaotic secure communication system. Radio Commun. Technol. **39**(06), 8–9+33 (2013)
9. Duan, J., Jiang, G., Yang, H.: Multiple input multiple output CDSK chaotic communication system and its performance analysis. Wuhan Univ. J. Nat. Sci. **21**(3), 221–228 (2016). https://doi.org/10.1007/s11859-016-1163-8
10. Kolumban, G., Kenney, M., Kis, G., et al.: FM-DCSK: a novel method for chaotic communications. IEEE Int. Symp. Circ. Syst. **4**, 477–480 (1998)
11. Wang, J., Li, J., Na, Y.: Communication model and performance of ultra-wideband chaotic keying in underground coal mine. J. Heilongjiang Univ. Sci. Technol. **27**(05), 555–559 (2017)

Green Networking

Short Term Wind Power Prediction Based on Wavelet Transform and BP Neural Network

Shuang Zheng, Zhaoju Jia, Ziwei Zhang, Fugang Liu[✉], and Long Han

Heilongjiang University of Science and Technology, Harbin 150022, China
zs1980225@126.com, liufugang@mail.usth.edu.cn

Abstract. Wind power generation has great randomness because of its randomness and uncontrollability. Due to the instability of wind energy, the power system access to large-scale wind power will pose a serious threat to the system. The accuracy of wind power prediction is very important to the security and stability. In this paper, a prediction model of electric power based on wavelet and BP neural network is proposed. The wavelet can further refine the periodic and nonlinear characteristics of electric power, and it solves many uncontrollable features when testing with BP neural network alone. The simulation shows that the prediction results of this method is better than that of BP neural network.

Keywords: BP neural network · Wavelet transform · Wind power prediction

1 Introduction

Wind energy is a conversion of solar energy, which is an important strategic choice for many countries to develop new energy and achieve sustainable development. According to the 2018 Global Wind Report by Global Wind Energy Council, the share of wind power is steadily increasing. In 2018, the new installed capacity of the global wind energy industry was 51.3 GW, and the global offshore market grew by 0.5%. During the utilization of wind energy, the randomness and fluctuation of wind speed will result in the unstable output of turbine. The power quality cannot be guaranteed [1]. In addition, due to the instability of wind energy, the power system access to large-scale wind power will pose a serious threat to its safe and stable operation. Accurate wind power prediction can solve the grid connection problem well, which is conducive to reducing the operating cost of wind energy plants.

The influence factors of wind power forecast include wind turbine arrangement, terrain, roughness, air pressure, temperature, speed, direction and other environmental conditions. It is also affected by the actual conditions such as unit operation status, wake effect, turbulence and other factors. Currently, the common statistical model methods for wind power forecast are Kalman filter method [2], time series method [3, 4], artificial neural network method [5–8], wavelet analysis method [9, 10], SVM regression method [11] and combined model method [12, 13]. They are based on the statistics of historical wind patterns, the physical data from the site and the physical properties of wind

X. Jiang and P. Li (Eds.): GreeNets 2020, LNICST 333, pp. 245–254, 2020.
https://doi.org/10.1007/978-3-030-62483-5_26

turbines. Kalman filter method needs to be closely combined with the estimated target in application. Accurate mathematical modeling of the estimated target is helpful to the design and implementation of Kalman filter algorithm. In [2], the authors explored a new Kalman filter method, which can get the curves of wind speed and predict it. The prediction models of random time series include several models. In [4], the authors proposed an improved Markov chain model. They utilized a three-dimensional state transition probability matrix to generate wind power time series. However, the forecast error of high wind speed is still large, especially when the wind speed changes violently in a short time, the power will climb rapidly. To further improve the forecast accuracy of high wind speed period, some scholars choose the deep convolution neural network with more layers of network structure. In [11], the authors presented a Support Vector Machine method using Dutch Hill Wind Farm data and got a better result than linear Support Vector Machine model. In this paper, we study the recent researches and establish a wind power forecast model based on wavelet transform and neural network. In Sect. 2, we introduce the basic theory of BP neural network and the specific process of wind power forecast by BP neural network. In Sect. 3, we explore wind power forecast method based on wavelet neural network. In Sect. 4, we took a wind power plant in North China as an example and compared the application effect of wavelet neural network and BP neural network model in wind power forecast based on one month's local weather data. Finally, we present the conclusion and discuss the issues involved in wind power forecast.

2 Wind Power Forecast by BP Neural Network Algorithm

BP (Back Propagation) neural network algorithm is a reverse transmission algorithm simulating human brain neurons. The structure of BP neural network is shown in Fig. 1. In this algorithm, the two processes of "signal forward transmission" and "error back propagation" are carried out in a reciprocating manner, to meet the conditions of the error between the output and the expected value.

The forward transmission of signal is to calculate the output of network with given signals. Set the input vector of input layer as $U_k(u_1, u_2, \cdots, u_n), k = 1, 2, \cdots, n$. The corresponding input information actual vector is $d_k = (d_1, d_2, \cdots, d_n)$, so the hidden layer input s is:

$$s_j = \sum_{i=1}^{n} w_{ij}u_i - \theta_j \, j = 1, 2, \cdots, p \tag{1}$$

Where w_{ij} is the connection weight, θ_j is the threshold value, p is the number of neurons in the hidden layer.

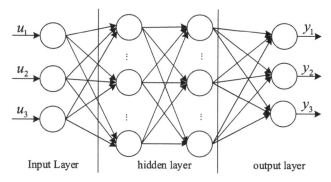

Fig. 1. Topological structure of BP neural network

2.1 Back Propagation of Error

Calculate the error and change the weight accordingly. The error E of the output layer is:

$$E = \frac{1}{2}(d - O)^2 = \frac{1}{2}\sum_{k=1}^{q}(d_k - O_K)^2 \tag{2}$$

By expanding the above error definitions, we can see that the error is a function of the weight of each layer, and then we can change the error by adjusting the weight. The change of the weight of each neuron is directly proportional to the decrease of the error gradient. The weight adjustment amount is:

$$\Delta v_{jk} = -\eta \frac{\partial E}{\partial v_{jk}}, \quad j = 0, 1, 2, \cdots, p, \ k = 1, 2, \cdots q \tag{3}$$

$$\Delta w_{jk} = -\eta \frac{\partial E}{\partial v_{ij}}, \quad i = 0, 1, 2 \cdots, m, \ k = 1, 2, \cdots p \tag{4}$$

Where η is the proportional coefficient.

2.2 Wind Power Forecast Based on BP Neural Network

The historical weather forecast data (including wind speed and numerical weather forecast, such as humidity, wind direction, atmospheric pressure and temperature) are used as input data for training, and the weight coefficients of different layers of neural network are obtained. The power forecast is realized according to the future weather forecast data by applying this model.

The specific learning process of BP neural network is as follows:

1. Initializing the weights w_{ij} and v_{jt}, the thresholds θ_j and γ_t;
2. Input initial learning sample $U_k = (u_1, u_2, \cdots, u_n)^\mathrm{T}$ and the final target sample $Y_k = (y_1, y_2, \cdots, y_m)^\mathrm{T}$;

3. Calculate hidden layer net input s_j and output b_j:

$$s_j = \sum_{i=1}^{n} w_{ij} u_i - \theta_j, \ j = 1, 2, \cdots, p \tag{5}$$

$$b_j = f(s_j), \ j = 1, 2, \cdots, p \tag{6}$$

4. Calculate output layer net input l_t and net output c_t:

$$l_t = \sum_{j=1}^{p} v_{jt} b_j - \gamma_t, \ t = 1, 2, \cdots, m \tag{7}$$

$$c_t = f(l_t) \quad t = 1, 2, \cdots, m \tag{8}$$

5. According to $Y_k = (y_1, y_2, \cdots, y_m)^{\mathrm{T}}$ and the output, the correction error d_t is calculated as:

$$d_t = (y_t - c_t) f(l_t) \ t = 1, 2, \cdots, m \tag{9}$$

6. According to the known v_{jt}, d_t, b_j and e_j, the correction error of hidden layer is calculated as

$$e_j = \left[\sum_{t=1}^{m} v_{jt} d_t \right] f(s_j), \ j = 1, 2, \cdots, p \tag{10}$$

7. Modify the implicit connection weight v_{jt} and threshold γ_t of the output layer according to the known d_t and b_j

$$\Delta v_{jt} = \eta d_1 b_j, \ j = 1, 2, \cdots, p; \ t = 1, 2, \cdots, m \tag{11}$$

$$\Delta \gamma_t = \eta d_1, \ t = 1, 2, \cdots, m \tag{12}$$

8. According to the known $e_j \ U_k = (u_1, u_2, \cdots, u_n)^{\mathrm{T}}$, correct the connection weight input of the hidden layer

$$\Delta w_{ij} = \beta e_j u_i, \ i = 1, 2, \cdots, n; \ j = 1, 2, \cdots, p \tag{13}$$

$$\Delta \theta_j = \beta e_j, \ j = 1, 2, \cdots, p \tag{14}$$

In the above process, steps 3. and 4. are the forward propagation, and steps 5. to 8. are the error back propagation, to correct the weight and threshold error. The global error satisfying the condition is obtained by using all the samples to repeat the cycle.

In the process of solving BP algorithm, the local optimal solution is easily got by gradient descent operation along the error value. According to the experimental simulation in Sect. 4 of this paper, the forecast accuracy cannot meet the corresponding accuracy requirements only by applying BP neural network algorithm.

3 Wavelet Neural Network Algorithm for Forecast Wind Power

3.1 Wavelet Neural Network Algorithm

Wavelet analysis method is developed for the deficiency of Fourier transform. It can obtain local time interval information. A basic function $\varphi(t)$ is given:

$$\varphi_{a,b}(t) = \frac{1}{\sqrt{a}} \varphi(\frac{t-b}{a}) \tag{15}$$

where $\varphi_{a,b}(t)$ is the base wavelet, a, b are the scale and displacement parameter respectively. Given that the square integrable signal $x(t)$, then the continuous wavelet transforms of $x(t)$ is defined as:

$$WT_x(a, b) = \frac{1}{\sqrt{a}} \int x(t) \varphi^* (\frac{t-b}{a}) dt = \int x(t) \varphi_{a,b}^*(t) dt = \langle x(t) | \varphi_{a,b}(t) \rangle \tag{16}$$

If a and b are discretized, $a = a_0^j$, $b = a_0^j b_0$,

$$WT_x(j, k) = \int_{-\infty}^{+\infty} x(t) \varphi_{j,k} dt, j, k \in Z \tag{17}$$

Combining the multi-resolution feature of wavelet with the tower method, the pyramid decomposition and reconstruction algorithm of discrete signal based on wavelet is expressed as:

$$C_n^k = \frac{1}{\sqrt{2}} \sum_{j \in Z} C_j^{k-1} h_{j-2a} \tag{18}$$

$$b_n^k = \frac{1}{\sqrt{2}} \sum_{j \in Z} C_j^{k-1} \overline{g}_{j-2a} \tag{19}$$

After the decomposition operation, the formula is reorganized as:

$$\left(\overline{F}_n\right)_n = \frac{1}{\sqrt{2}} \sum_{j \in Z} f_{n-2ja_j} \tag{20}$$

Where c_n^k, b_n^k denote the decomposition coefficient of wavelet; h_j, g_j denote the discrete filter of wavelet. The decomposition and reconstruction of wavelet is to divide the wind power time series into n layers at different frequencies, and then get n high frequency signal components and one approaching signal component.

The BP neural network algorithm mentioned above is easy to get local optimal solution. In addition, the output power of turbine unit fluctuates greatly, and forecast accuracy of the traditional neural network model is not satisfactory. In this paper, we studied the method of using wavelet neural network to forecast wind power. The actual power of a single unit in each wind farm has a quasi-periodic characteristic within 24 h. Wavelet decomposition can bring forward some periodic and nonlinear characteristics of electric power, and further refine them.

According to the rules of each sub sequence, the method of matching prediction is used. BP neural network can play the advantages of dealing with the problems of nonlinearity and no clear rules. Combining the models of the two algorithms can effectively improve the accuracy of wind power forecast. Wavelet neural network is based on wavelet basis function. The input parameters are weighted by the weight of wavelet neural network in hidden layer. The topological mechanism is shown in Fig. 2.

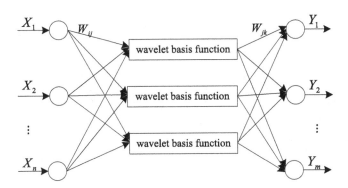

Fig. 2. The topological mechanism of wavelet neural network

When the signal sequence is $X_i(i = 1, 2, \cdots, k)$, the hidden layer calculation formula is

$$h(j) = \left(\sum_1^k w_{ij} x_i - b_j \middle/ a_j \right) \tag{21}$$

The output layer calculation formula of wavelet neural network is as follows:

$$y(k) = \sum_{k=1}^{m} w_{ik} h(i) \quad k = 1, 2, \cdots m \tag{22}$$

Where W_{ik} is the weight from hidden layer to output layer;

In the process of autonomous learning, wavelet neural network needs to modify the parameters of neural network. The forecast error can be got from the formula (23):

$$e = \sum_{k=1}^{m} yn(k) - y(k) \tag{23}$$

The weights and coefficients of wavelet neural network are modified according to the above formula. They are shown in formula (24)–(26).

$$W_{n,k}^{(i,j)} = W_{n,k}^{j} + +\Delta W_{n,k}^{(i,j)} \tag{24}$$

$$a_k^{(i,j)} = a_k^j + +\Delta a_{(n,k)}^{(i,j)} \tag{25}$$

$$b_k^{(i,j)} = b_k^j + +\Delta b_{(n,k)}^{(i,j)} \tag{26}$$

According to the calculation of network prediction error, the results of the above formula are as follows (γ is the learning rate):

$$\Delta W_{(n,k)}^{(i,j)} = -\gamma \frac{\partial e}{\partial W_{n,k}^{(i)}} \tag{27}$$

$$\Delta a_k^{(i,j)} = -\gamma \frac{\partial e}{\partial a_k^{(i)}} \tag{28}$$

$$\Delta b_k^{(i,j)} = -\gamma \frac{\partial e}{\partial b_k^{(i)}} \tag{29}$$

3.2 Wind Power Forecast Based on Wavelet Neural Network

Because the single sequence of wavelet decomposition on different scales has different characteristics, the prediction method matching with itself is determined according to the change of wind power. The large scale of the subsequence corresponds to the low frequency part of the wavelet space, which reflects the curve characteristics of the wind power time series. The small scale of the sequence reflects the relationship between the nonlinear characteristics of the system. Therefore, in order to accurately depict the power subsequence on different scales, the matching BP neural network should be determined respectively for modeling and forecast. In this paper, the wavelet transform divides the wind power time series into n layers at different frequencies, and obtains n high frequency signal components and an approaching signal component. Then, the decomposed single sequence is respectively forecasted by BP neural network matching with this. Finally, combined with the forecast results of each sequence, the forecast value of wind power series is reconstructed. The specific training process is as follows:

(1) Network initialization such as expansion factor a_k, translate factor b_k, network connection weight W_{jk}, and network learning rate γ;

(2) Classifying the samples into training samples and test samples.
(3) Prediction output. Input the training samples, calculate the network output and calculate the error between network output and actual output.
(4) Correct the network weight based on wavelet function parameters and the error. Make network prediction approach real value.

4 Experimental Simulation Analysis

Taking a wind power plant in North China as an example, the weather data of a month are used, including wind speed and numerical weather forecast. The numerical weather forecast includes humidity, wind direction, atmospheric pressure and temperature. These data were measured every 15 min from 10:00 a.m. to 11:45 p.m. every day, and a total of 1753 samples are collected. In this experiment, 1533 data groups were used for training and 120 data groups for prediction. Figure 3 shows the forecast results of the two algorithms.

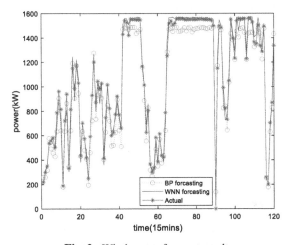

Fig. 3. Wind power forecast result

In Fig. 3, there is a big difference between the actual power result and the wind power prediction by only using BP neural network. When the actual power is high, even if the actual power data fluctuates little, the prediction error of the model cannot reach the ideal effect. The reason is that the model does not grasp the overall law in depth and fails to consider the data variation interval of each time period. The prediction effect of wavelet neural network model is better than the former, but the prediction effect of the algorithm is not good when the actual power is small, and the fluctuation is large.

Figure 4 and Fig. 5 show the forecast errors of the two algorithms. Figure 4 (a) and Fig. 5 (a) are directly forecasted by using BP neural network, and Fig. 4 (b) and Fig. 5 (b) are the forecast results by using wavelet neural network. The maximum error by using BP neural network forecast method is 23%, and the overall error change is relatively large. Generally, it cannot meet the prediction standard. The prediction error of wavelet

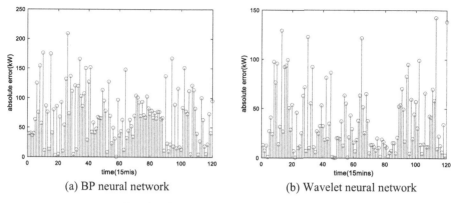

(a) BP neural network   (b) Wavelet neural network

Fig. 4. The absolute error of two algorithms

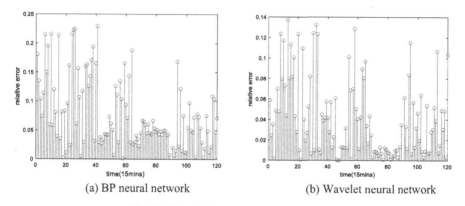

(a) BP neural network   (b) Wavelet neural network

Fig. 5. The relative error of two algorithms

neural network is about 10%, the mean square deviation is 84.16 kW, and the overall error is relatively small.

5 Conclusion

In this paper, the application effect of wavelet neural network and BP neural network model in wind power prediction is compared. It can be concluded from the theoretical analysis that the feature information of wind power can be extracted quickly by decomposing the wind power sequence in different frequencies through wavelet transform, which can be used as the input data of neural network prediction model. The simulation results show that the accuracy of forecast model can be improved by using the good time-frequency analysis ability of wavelet to non-stationary signals and the nonlinear mapping ability of BP neural network. In addition, for some limitations of the Morlet function, we can change the prediction effect of the model by changing the wavelet basis function in future.

Acknowledgements. This work has been partially supported by the National Natural Science Foundation of China project (51674109) and 2017 scientific research project of basic scientific research business expenses of provincial colleges and universities in Heilongjiang Province.

References

1. Deshmukh, S., Bhattacharya, S., Jain, A., Paul, A.R.: Wind turbine noise and its mitigation techniques: a review. Energy Procedia **160**, 633–640 (2019)
2. Salgado, P., Igrejas, G., Afonso, P.: Multi-Kalman filter to wind power forecast. In: 2018 13th APCA International Conference on Automatic Control and Soft Computing (CONTROLO), pp. 110–114. IEEE, Ponta Delgada, Azores (2018)
3. Sun, G., Jiang, C., Cheng, P.: Short-term wind power forecasts by a synthetical similar time series data mining method. Renew. Energy **115**, 575–584 (2018)
4. Li, J., Li, J., Wen, JinYu., Cheng, S., Xie, H., Yue, C.: Generating wind power time series based on its persistence and variation characteristics. Sci. China Technol. Sci. **57**(12), 2475–2486 (2014). https://doi.org/10.1007/s11431-014-5720-0
5. Sharifian, A., Ghadi, M.J., Ghavidel, S.: A new method based on Type-2 fuzzy neural network for accurate wind power forecast under uncertain data. Renew. Energy **120**(MAY), 220–230 (2018)
6. Heydari, A., Garcia, D.A., Keynia, F., Bisegna, F., De Santoli, L.: A novel composite neural network based method for wind and solar power forecast in microgrids. Appl. Energy **251**(1), 113–123 (2019)
7. Naik, J., Dash, S., Dash, P.K., Bisoi, R.: Short term wind power forecast using hybrid variational mode decomposition and multi-kernel regularized pseudo inverse neural network. Renew. Energy **118**, 180–212 (2018)
8. Su, Y., Wang, S., Xiao, Z., Tan, M., Wang, M.: An ultra-short-term wind power forecast approach based on wind speed decomposition, wind direction and Elman neural networks. In: 2018 2nd IEEE Conference on Energy Internet and Energy System Integration (EI2), pp. 1–9. IEEE. Beijing, China (2018)
9. Liu, Z., Hajiali, M., Torabi, A.: Novel forecast model based on improved wavelet transform, informative feature selection, and hybrid support vector machine on wind power forecast. J. Ambient Intell. Hum. Comput. **9**(6), 1919–1931 (2018)
10. Afshari-Igder, M., Niknam, T., Khooban, M.-H.: Probabilistic wind power forecasting using a novel hybrid intelligent method. Neural Comput. Appl. **30**(2), 473–485 (2016). https://doi.org/10.1007/s00521-016-2703-z
11. Pawar, A., Jape, V.S., Mathew, S.: Wind power forecast using support vector machine model in RStudio. In: Mallick, P., Balas, V., Bhoi, A., Zobaa, A. (eds.) Cognitive Informatics and Soft Computing. Advances in Intelligent Systems and Computing, vol. 768, pp. 289–298. Springer, Singapore (2019)
12. Du, P., Wang, J., Yang, W., Niu, T.: A novel hybrid model for short-term wind power forecast. Appl. Soft Comput. **80**, 93–106 (2019)
13. Liu, F., Xu, J., Hu, F., Wang, C., Wu, J.: Lightweight trusted security for emergency communication networks of small groups. Tsinghua Sci. Technol. **23**(2), 195–202 (2018)

Transformer Fault Diagnosis Based on BP Neural Network by Improved Apriori Algorithm

Chang Guoxiang[1] , Gao Qiaoli[1](✉) , Gao Xinming[2] , and Cheng Junting[1]

[1] Heilongjiang University of Science and Technology, Heilongjiang, Harbin, China
270080877@qq.com, 763790270@qq.com, yj74615@qq.com
[2] Shandong Transport Vocational College, Weifang, Shandong, China
287743080@qq.com

Abstract. With the continuous expansion of power system, it is increasing that the fault rate of transformer equipment in power system. Through the fault diagnosis technology, it can be found that the transformer fault in advance, the accident rate is can taken to reduce by measures in time, so the high accuracy of transformer fault diagnosis is required. In this paper, the BP neural network based on the optimized Apriori algorithm which is used to diagnose the transformer fault. It is found that the Apriori algorithm reveals association rules by mining the high frequency term of feature data. The data is directly analyzed and infered to achieve the purpose of simplifying data association by rough set algorithm. Apriori algorithm combines the rough set is used to accurately mine the confidence of association rules, which is used as the weight of BP neural network link, it is simplifying the complexity of data training. Through the simulation experiment, this new method is compared with the traditional BP neural network method, it has the advantages of high accuracy and fast speed of fault diagnosis, and has certain practical value.

Keywords: Apriori algorithm · BP neural network · Transformer · Fault diagnosis

1 Introduction

Power transformer is very important for power system. In recent years, with the continuous improvement of science and technology, the power system is also developing towards the direction of high voltage and ultra-high voltage. Once the transformer breaks down, it often causes serious accidents, so it is very import that the safe and stable work of power transformer to the production and people's life. The fault of power transformer can be extracted and diagnosed, and effective measures can avoid accidents, and save the time and life, which requires a more accurate and rapid transformer fault diagnosis technology [1]. Now, the three ratio method is widely used in all transformer fault diagnosis methods, which has the characteristics of simple principle and simple calculation. But in practical application, there are some problems such as incomplete coding, too absolute coding boundary and limited multi fault judgment ability. In order to improve

© ICST Institute for Computer Sciences, Social Informatics and Telecommunications Engineering 2020
Published by Springer Nature Switzerland AG 2020. All Rights Reserved
X. Jiang and P. Li (Eds.): GreeNets 2020, LNICST 333, pp. 255–266, 2020.
https://doi.org/10.1007/978-3-030-62483-5_27

the accuracy of fault diagnosis, neural network technology is applied to the research of power transformer fault diagnosis, and good results are achieved. BP neural network can store and train many input-output mapping relationships without knowing how to describe them [2]. It has the ability of nonlinear mapping, self-learning and good fault tolerance. In this paper, the improved association rules and BP neural network are used to diagnose the transformer fault. The improved association rules method is to combine the rough algorithm with the association rules. First, the rough set of data is reduced, and then the data is associated. BP neural network is used to train and lean the processed data, predict the operation state of transformer, simplify the operation of transformer, and the complexity of data calculation improves the accuracy and speed of fault diagnosis [3].

2 BP Neural Network

BP neural network is a multilayer feedforward neural network which is trained by error back propagation algorithm. It is an effective classification and recognition tool, which can classify the fault types of transformer. The rule of BP neural network algorithm is to modify the weight and threshold of neural network by error back propagation, and use the method of the fastest descent, so as to reduce the square sum of the gap between the ideal value and the actual value [4]. The topological structure of BP neural network model is shown in Fig. 1, including input layer, hidden layer and output layer.

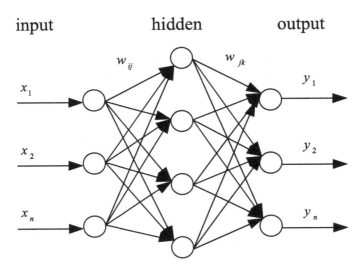

Fig. 1. BP neural network model

BP algorithm is a kind of back propagation algorithm according to the error, and its learning process is the process of constantly adjusting the weight.

The relationship between input layer vector and hidden layer vector is as follows:

$$e_j = f(\sum_{i=0}^{n} w_{ij} x_i) \tag{1}$$

1) The value of the output layer is:

$$e_k = f_x \left(\sum_{j=0}^{n} w_{jk} o_j \right) \tag{2}$$

2) The actual output value is:

$$o_k = f_x(e_k) \tag{3}$$

3) Mapping relation:

$$f(x) = 1/(1 - e^{-x}) \tag{4}$$

There is an error:

$$E_P = \frac{1}{2} \sum_k \left(e_{pk} - o_{pk} \right)^2 \tag{5}$$

The square difference formula is:

$$E = \frac{1}{2P} \sum_P \sum_K \left(e_{pk} - o_{pk} \right)^2 \tag{6}$$

3 Association Rules and Apriori Algorithm

3.1 Association Rules

Association rule is an important data mining method and an unsupervised machine learning method. It is used to establish the association between projects, study the correlation between projects, and give opinions according to the correlation analysis. Apriori algorithm is the most widely used one.

Association rule definition: set sets $I = \{i_1, i_2, i_3, \ldots., i_n\}$, the data set is: $D = \{T_1, T_2, \ldots., T_m\}$.

The subset of itemset I is the corresponding transaction set t, and A is a itemset, if $A \in T$, then $A \Rightarrow B, A \in I, B \in I, A \cap B = \Phi$, Indicates that if item set A appears in a transaction, y will also appear with a certain probability.

N is the number of transactions. In association rules, support and confidence are the two most commonly used standards. If count $(A \subseteq T)$ is the number of transactions containing x in transaction set D, then the support of item set A is:

$$Support(A) = \frac{count(A \in T)}{N} \tag{7}$$

$$Support(A \Rightarrow B) = \frac{count(A \cup B)}{N} \tag{8}$$

The confidence formula is:

$$Confidence(A \Rightarrow B) = \frac{Support(A \Rightarrow B)}{Support(A)} \tag{9}$$

3.2 Apriori Algorithm

If an itemset is frequent, then all subsets of it are frequent. Conversely, if an item set is infrequent and all of its supersets are infrequent, then the entire set of supersets containing infrequent sets can be immediately cut off. This strategy is called support-based pruning.

The basic working idea of the prior algorithm is to retrieve the entire database, calculate the number of each item, and generate the item with the minimum support, which is represented by L1. On this basis, the second frequent item is distributed L2, and continue to execute until the K frequent item set cannot be found. Prior algorithm flow:

Step 1: It is got all the frequent itemsets in the transaction database are implemented through iterative operations, that is, the itemsets whose support is no less than the threshold set by the user [5];

Step 2:It is to use frequent item sets to construct rules that satisfy the minimum trust of users.

The specific method is: first of all, it is been found out the frequent 1-term set, and record it as L_1; then use L_1 to generate candidate item set C_2, determine the items in C_2, and mine L_2, i.e. frequent 2-term set; keep cycling until no more frequent k-term set can be found [6]. That is to say, when generating a k-i candidate, if a subset of the candidate is not in $(k-1) - i$ (which has been determined to be frequent), then the candidate will be deleted without taking it out and judging the support [7]. However, the traditional Apriori algorithm has two defects:

1) A great number of candidate sets are generated in the process of algorithm, which takes up a large amount of memory;
2) Every calculation needs to scan the whole database, so the algorithm has a high time and complexity.

4 Rough Set

4.1 Definition of Rough Set

1) Information system: In general, a knowledge expression system or the information system can be expressed as, in which, it is a domain, which is a collection of all samples; it is an attribute collection, in which the subset is a conditional attribute set, which reflects the characteristics of the object, D is a decision attribute set, which reflects the class of the object; it is an attribute set, which represents the value range of attribute R; it is an information function, which is used to determine each one in U Property values of objects, that is, any.
2) Indiscernible relation: when two objects are described by the same attribute, they are classified into the same category in the system. Their relation is called indiscernible relation, that is, for any attribute subset, if the object, if and only then, and is indiscernible relation, the indiscernible relation is called equivalent relation for short.

3) Upper approximation set and lower approximation set: the lower approximation set is defined as the set composed of all objects that must belong to A in U according to the existing knowledge R, i.e. in the formula, the equivalence relation R contains the equivalence class of relation A; the upper approximation set is defined as the set composed of objects that must belong to and may belong to X in U according to the existing knowledge R.

The main characteristics of rough set are: it does not need to provide subjective evaluation of knowledge or data, only based on the observation data, it can delete redundant information, compare incomplete knowledge roughness, defining dependencies and importance between attributes.

Main research fields: attribute reduction, rule acquisition, and intelligent algorithm based on rough set.

4.2 The Principle of Rough Sets

The function of rough set theory is very powerful, especially for the processing of uncertain information, it has its own congenital advantages. There are two differences between rough set theory and other uncertain and imprecise theories: first, it is not been needed to provide additional hints beyond the data set required for the problem; second, it has strong applicability and is easy to apply to any field. Because the purpose and starting point of rough set theory is to analyze and reason data directly. The fuzzy theory and evidence theory Compared with the methods of dealing with uncertainty problems, tThe most obvious advantage of this algorithm is that it does not need any information other than data as supplement and support, and it has strong complementarity with the theories dealing with other uncertainty problems (especially fuzzy theory) [8]. In this paper, the optimized Apriori algorithm is applied to transformer fault diagnosis.

4.3 Advantages of Rough Set

The main feature of rough set: the rough set is that it does not need to provide subjective evaluation of knowledge or data, but can remove redundant information, compare the incomplete knowledge roughness and define the relevance and importance of the attributes according to the observed data.

1) It can handle all the data which is included incomplete data and multivariable data;
2) It can deal with the uncertainty and ambiguity of data, including the situation of certainty and uncertainty;
3) It can get the minimum expression of knowledge and different particle levels of knowledge;
4) It can reveal the simple concept and easy operation mode from the data;
5) It can generate rules with high accuracy and easy to verify. It is especially suitable for intelligent control and can generate rules automatically.

4.4 The Function of the Rough Set

Rough set theory is greatly used in data mining, involving medical research, market analysis, business risk prediction, meteorology, speech recognition, engineering design, etc. in many data mining systems, the role of rough set theory mainly focuses on the following aspects:

1) Data reduction

Rough set algorithm has a strong ability in data reduction. In the preprocessing stage of data mining system, redundant information (object, attribute value, etc.) is deleted by rough set theory. in the data mining system, which can greatly improve the running speed of the system. [2].

2) Rule extraction

The feature of rough set theory algorithm is simple and direct, and its association rules are one-to-one correspondence. The general steps of generating rules in rough set method are:

(1) It can be got a reduction of condition attributes and deleting redundant attributes;
(2) It can be deleted redundant attribute values of each rule;
(3) It can merged the remaining rules.

 3) Incremental algorithm

In the face of large-scale and high-dimensional data in data mining, it is a research hotspot to find an effective incremental algorithm.

4) Integration with other methods

The combination of rough set theory and neural network method can show their advantages to the greatest extent and improve the speed and accuracy of data operation.

5 Improved Association Rule Algorithm

5.1 Combination of Rough Set and Association Rules

According to the research on the characteristics of traditional Apriori algorithm, it is found that the traditional Apriori algorithm has its own defects, this paper combines the rough set algorithm with the Apriori algorithm to get an improved Apriori algorithm, in order to obtain a better speed effect of transformer fault detection. The basic idea of the improved Apriori algorithm is: firstly, the rough set algorithm is used to process the original data to achieve the purpose of data reduction, and then the Apriori algorithm is used to analyze the reduced data and generate frequent item sets and association rules. The transformer fault classification is realized. The specific process is as follows

1) Select attribute to generate decision table

Select the appropriate commonality as the commonality and decision attribute of the project, and generate the decision table.

2) Attribute reduction

The method based on discernibility matrix is applied to attribute reduction, conditional attribute reduction and kernel calculation. When there are multiple kernel values, the minimum reduction set containing kernel values is selected. After reduction, we get the reduction set of decision table, and use the reduction set to transform the decision table into a new decision table with Boolean generality suitable for Apriori analysis.

3) Use set operation instead of scanning attribute set

Apriori algorithm is used to calculate and generate the frequent set of one item on the new decision table. Delete the items whose support is less than the threshold value, expressed as C_1. The frequent item set is generated on the basis of C_1', $C_i' = C_{i-1} * C_i$, i is greater than 1, less than or equal to the length of the reduction set, and the items with support less than the threshold in C_i' are deleted.

4) For decision attribute set D_j', $D_j' = D_{j-1} * D_j$ j is greater than 1 and less than or equal to the length of decision attribute set calculated according to D_1;
5) C_i' and D_j' generate frequent itemsets with length of $i + j$ through ptwojoin operation with attribute 1, expressed as L_{i+j};
6) in L_{i+j}, frequent item sets with support and trust greater than or equal to threshold can be put into rule set R.

In the above calculation steps, C_i', D_j' and L_{i+j} can be used for each calculation according to attribute 1 to avoid rescanning the decision table (Fig. 2).

6 Example Analysis

The gas composition in the transformer oil is captured, and the gas detection data are obtained. After the gas composition analysis, five characteristic gases CH_4, H_2, C_2H_2, C_2H_4 and C_2H_6, which can be obtained. The content percentage of these five gases is taken as the input quantity, and the output results are the five states of normal, low energy discharge, high energy discharge, medium and low temperature overheating and high temperature overheating. The collected data is shown in Table 2. The improved Apriori algorithm is used to analyze and simplify the association rules between features, and the BP neural network method is selected for training and experiment. The simulation results in this paper are compared with the traditional algorithm diagnosis results to verify the effectiveness of the proposed method. The input of the neural network takes the percentage of the volume of petroleum dissolved in the five gases during synthesis, so as to determine that the input layer node is a $= 5$. The operation states of the above

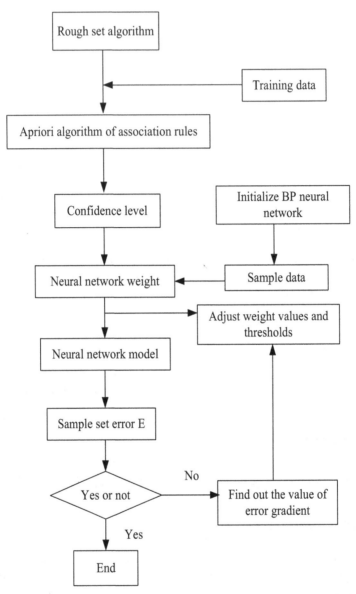

Fig. 2. Fault diagnosis flow chart of BP neural network based on improved Apriori algorithm

five transformers are taken as the output of neural network, the output layer node number $b = 5$ is determined. In general, the number of hidden layer neural nodes is selected according to the empirical formula:

$$h = \sqrt{(a + b)} + \alpha \tag{10}$$

Where α is any number between 1 and 10. According to experience, $\alpha = 2$, so t h = 5.

In order to solve the problem of complex calculation and slow speed, this paper uses a BP neural network based on optimal association rules to diagnose the transformer fault. Let the input value be expressed as A, and the transformer state as B, that is, since the confidence value represents the association rule $A \Rightarrow B$, It contains the conditional probability that B exists when A exists, and this condition reflects the mapping relationship between the content of dissolved gas in transformer oil and the state of transformer. According to the comparison of the original weights obtained by the traditional association rules, the weight matrix obtained by the optimized association rules is more suitable for the initial weights of BP neural network. Therefore, in this paper, the improved Apriori algorithm is used to process the data, and the initial weights between the input layer and the hidden layer of BP neural network, and between the hidden layer and the output layer, ω_n is obtained.

$$w_n = \frac{Confidence_n}{Confidence_1 + Confidence_2 + \ldots + Confidence_5} \tag{11}$$

There: w_n is the weight of the transformer when the content of n-dissolved gas exceeds the standard; Confidence is the confidence of the transformer when the content of n-dissolved gas exceeds the standard.

To verify the effectiveness of this method for sample data processing and classification, 150 groups of DGA data that have been identified fault types are collected for analysis, and divided into training samples and test samples in a ratio of 2:1. Among them, 100 groups of training sample data are used to optimize the Apriori algorithm to optimize the weight and threshold parameters of BP neural network, and 50 groups of test sample data are used to verify the diagnosis accuracy of the optimized parameter model. See Table 2 for the specific distribution of transformer fault sample data. Table 3 shows the DGA data of 10 groups of test samples. In this paper, The oil content and operation state data of 150 transformers in a certain area of Southwest China are selected as the test data set: 5 fault characteristic gases were marked respectively. H_2 exceeds the standard, CH_4 exceeds the standard, C_2H_2 exceeds the standard, C_2H_4 exceeds the standard, C_2H_6 exceeds the standard, five operation states of transformer are marked respectively, the statistical conditions of various gases exceeding the standard and transformer operation are shown in Table 1. The gas frequency for standard or transformer operating conditions is shown in brackets:

It is used to test the transformer fault diagnosis performance of BP neural network based on the improved Apriori algorithm, Matlab is used to write the fault diagnosis program, and the improved Analysis of gas content in transformer oil by Apriori algorithm, so as to obtain the confidence degree of the input quantity and the output state. As the weight coefficient of BP neural network, the average detection accuracy of each gas is counted. The initial weights are trained by BP neural network. The diagnosis speed and accuracy results are compared and shown in Table 4.

Table 1. Fault status table

No.	Fault Type	Fault output status
1	Normal	Y_1
2	Low temperature overheating	Y_2
3	High temperature overheating	Y_3
4	Low energy discharge	Y_4
5	High-energy discharge	Y_5

Table 2. Sample data of transformer fault

Fault Type	Training Sample	Test Sample
Normal	20	10
Low temperature overheating	18	9
High temperature overheating	22	11
Low energy discharge	18	9
High-energy discharge	22	11
All	100	50

It can be seen from Table 4 that the average detection accuracy of the improved Apriori algorithm for transformer fault diagnosis is 86.3%, the operation speed is 1.365 s, the average detection accuracy of BP neural network is 77.6%, and the operation speed is 0.892 s. The improved Apriori algorithm greatly reduces the detection error rate, improves the detection speed, overcomes some problems existing in the current transformer fault diagnosis process, and reduces the harm caused by the transformer fault. This method has certain advantages.

Table 3. Sample data

No.	H_2	CH_4	C_2H_2	C_2H_4	C_2H_6	Fault condition
1	90.7	23	9.2	9.0	5.1	Y_5
2	172	26	14.5	11.6	6.0	Y_5
3	159	25.3	15.6	12.4	5.4	Y_5
4	59.0	45.0	23.4	20.7	11.2	Y_4
5	86.0	110	86.0	45.9	66.8	Y_4
6	93.0	58.2	0.1	37.2	45	Y_2
7	68.5	58.9	1.9	59.7	70.3	Y_2
8	53.6	70.2	4.5	17.0	158.0	Y_2
9	183	291.0	0.33	341.9	109.8	Y_3
10	242.8	76.2	12.5	173.2	76.5	Y_3

Table 4. Analysis of diagnosis results

Methods	Classification accuracy	Running time
BP neural network	77.6%	1.365
BP neural network with improved association rules	86.3%	0.892

7 Conclusion

Traditional BP neural network algorithm has a wide range of applications, but there are also some defects and deficiencies. After the simulation experiment analysis, the traditional BP neural network and the BP neural network optimization association rule algorithm diagnosis results are compared, it is found that the optimization association rule BP neural network algorithm can effectively improve the accuracy of transformer fault diagnosis, shorten the running time, this optimization algorithm has a certain guiding value for the research of transformer fault diagnosis technology.

References

1. Long, J., et al.: Research on transformer fault diagnosis based on BP neural network improved by association rules. In: 2019 2nd International Conference on Electrical Materials and Power Equipment (ICEMPE) (2019)
2. Lv, C.T.: Implementation and application of clustering algorithm based on k-frequent association rules in embedded system. Yunnan university (2013)
3. Liu, Y., Lipo, W., Zhao, L., Yu, Z., (eds.) Advances in Natural Computation, Fuzzy Systems and Knowledge Discovery. Springer Science and Business Media LLC, Heidelberg (2020). https://doi.org/10.1007/978-3-030-32456-8
4. Zheng, W., Shiguan, G.: Research on group purchase recommendation method based on apriori-bp algorithm. Wirel. Internet Technol. **15**(23), 82-84 + 92 (2008)
5. Xiang, J.: Association rule mining of glacier catalog data based on rough set and Apriori algorithm. Mod. Comput. (professional edition) **2009**(11), 34–37 (2009)
6. Kai, S., Peng, C.: Association rule mining and screening of experimental data on glucose tolerance based on APRIORI algorithm and interes. Med. Inf. **22**(08), 1397–1399 (2009)
7. Wang, M., Jiang, Y., Wang, Y., Minlett, Z.T., Zhou, Z.: Association rule mining algorithm based on rough set and single transaction item combination. Comput. Sci. **38**(11), 234–238 (2011)
8. Li, D.: Application of association rule mining in cross-selling of Banks. Jiangsu University of Science and Technology (2016)

Modeling and Simulation of Photovoltaic Grid-Connected System

Dongni Zhang, Xunwen Su$^{(\boxtimes)}$, Xianzhong Xu, Dawei Wang, and Siyu Chen

Heilongjiang University of Science and Technology, Harbin 150022, China
suxunwen@163.com

Abstract. Based on the mathematical model of the photovoltaic array, we can construct a model of a three-phase photovoltaic grid-connected system consisted of a Photovoltaic Array, boost circuit, Maximum Power Point Tracking and photovoltaic inverter. Through the model of PSCAD/EMTDC simulation software, we can understand the principle of Maximum Power Point Tracking, comprehend the working principle of the photovoltaic inverter controller, analysis the influence of harmonics on power quality of power grid, and verify the correctness of the three-phase photovoltaic grid-connected system model.

Keywords: Photovoltaic power generation · Photovoltaic system modeling · Grid-connected control strategy · Compensation measures · Power quality

1 Introduction

With global warming and energy depletion, solar energy, as a clean, renewable energy source, has become the focus of many countries. The solar photovoltaic industry is one of the fastest growing and most stable fields in the world. Due to the subsidy policies of various countries, the cost of photovoltaic power generation has gradually decreased. In China, the number of grid-connected photovoltaic power stations is increasing, so the integral modeling and grid-connected characteristic analysis of photovoltaic system are particularly important. The analysis of the power grid characteristics caused by the connection of solar energy needs to be further improved. This paper is divided into four parts, based on PSCAD/EMTDC to study the three-phase photovoltaic grid-connected system.

2 Main Equipment of Photovoltaic Power Station

Primary equipment of photovoltaic power station is mainly composed of photovoltaic array, bus box, photovoltaic inverter, and other common electrical primary equipment. Photovoltaic cells are the most important components in photovoltaic power stations. Their role is to convert solar energy into electric energy. The output voltage and energy of a single photovoltaic cell is very low, so dozens of photovoltaic cells are usually

© ICST Institute for Computer Sciences, Social Informatics and Telecommunications Engineering 2020
Published by Springer Nature Switzerland AG 2020. All Rights Reserved
X. Jiang and P. Li (Eds.): GreeNets 2020, LNICST 333, pp. 267–275, 2020.
https://doi.org/10.1007/978-3-030-62483-5_28

encapsulated to form a photovoltaic module [1]. Then the photovoltaic modules connected in series and parallel to a photovoltaic array through the junction box and dc
distribution cabinet. Inverter is the most important component of photovoltaic power
station. Its function is to convert the direct current generated by the photovoltaic power
station into the alternating current needed by the grid [2]. Photovoltaic inverters are
usually three - phase full - bridge structures. The grid-connected control of the inverter
and the self-protection function of the inverter are all included in the controller of the
inverter. We can construct a model of a three-phase photovoltaic grid-connected system
consisted of a Photovoltaic Array, boost circuit, Maximum Power Point Tracking and
photovoltaic inverter [3, 4].

3   Maximum Power Point Tracking Technology

MPPT can keep the photovoltaic cell in the best working state constantly, that is, the
maximum output power. The goal of MPPT is to control the output voltage of the
photovoltaic array to track the MPP voltage, so that the photovoltaic array has the
maximum photoelectric conversion efficiency [5]. The current Maximum Power Point
Tracking technology includes constant voltage tracking method, short circuit current
method, disturbance observation method, incremental conductance method, based on
fuzzy control theory and artificial neural network intelligent theory and so on [6]. This
paper will be based on the incremental conductance method of Maximum Power Point
Tracking theory.

The output power of the photovoltaic cell is P = UI
Derivative U on both sides of the equation:

$$\frac{\mathrm{d}P}{\mathrm{d}U} = I + U\frac{\mathrm{d}I}{\mathrm{d}U} = 0$$

$$\frac{\mathrm{d}I}{\mathrm{d}U} = -\frac{I}{U}$$

$$\mathrm{d}U = U(\mathrm{k}) - U(\mathrm{k}-1)$$

When:

$$\frac{\mathrm{d}I}{\mathrm{d}U} < -\frac{I}{U} \quad U > \text{maximum power point voltage}$$

$$\frac{\mathrm{d}I}{\mathrm{d}U} > -\frac{I}{U} \quad U < \text{maximum power point voltage}$$

$$\frac{\mathrm{d}I}{\mathrm{d}U} = -\frac{I}{U} \quad U = \text{maximum power point voltage}$$

In this way, we can obtain the voltage of maximum power point, and realize the power
tracking. Generally, the photovoltaic cell's output voltage in the DC side is relatively
low. Therefore, it is necessary to use the Boost circuit to increase the photovoltaic cell's

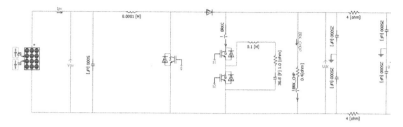

Fig. 1. Boost module of MPPT controlled

output voltage in the DC side and obtain the voltage needed by the power grid. The schematic diagram of Boost circuit in PSCAD is shown in Fig. 1.

First the instantaneous voltage V_{PV} and current I_{PV} of the photovoltaic cell are fed into the MPPT control module, then calculated the working voltage V_{MPPT} at the maximum power point. Compared with V_{MPPT}, the instantaneous voltage V_{PV} is controlled by PI control, V_{PV} outputs PWM drives signal "g" to control the switching of IGBT. When the duty cycle increases, appears in $V_{PV} > V_{MPPT}$; By adjusting duty cycle can obtain the expected PV output voltage.

The function of BRK-CHP is to prevent the damage caused by excessive current on the left side when the right side is empty. The part controlled by S1 and S2 is the crowbar protecting, which can play a role in accelerating and attenuating the rotor winding current, and the value of resistance seriously affects the effectiveness of the crowbar protection circuit (Fig. 2).

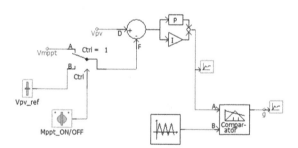

Fig. 2. PI closed loop control module

4   Grid-Connected Control Strategy Based on Rotating Coordinate System

Since the grid-connected terminal is a three-phase AC system, the research cannot be carried out in the static coordinate system [7]. All the existing methods require Park transformation and Clark transformation to realize the transformation from three-phase stationary coordinate system to two-phase rotating coordinate system. The following is the derivation of the formula: firstly, through the Park transformation, converted the coordinate quantity of ABC into the coordinate quantity of dq0:

$$\begin{bmatrix} f_d \\ f_q \\ f_0 \end{bmatrix} = \frac{2}{3} \begin{bmatrix} \cos\theta & \cos(\theta - \frac{2\pi}{3}) & \cos(\theta + \frac{2\pi}{3}) \\ \sin\theta & \sin(\theta - \frac{2\pi}{3}) & \sin(\theta + \frac{2\pi}{3}) \\ \frac{1}{2} & \frac{1}{2} & \frac{1}{2} \end{bmatrix} \begin{bmatrix} f_a \\ f_b \\ f_c \end{bmatrix} \qquad (1)$$

Make:

$$T_{abc-dq0} = \begin{bmatrix} \cos\theta & \cos(\theta - \frac{2\pi}{3}) & \cos(\theta + \frac{2\pi}{3}) \\ \sin\theta & \sin(\theta - \frac{2\pi}{3}) & \sin(\theta + \frac{2\pi}{3}) \\ \frac{1}{2} & \frac{1}{2} & \frac{1}{2} \end{bmatrix} \qquad (2)$$

$$\begin{bmatrix} \mathbf{u}_d \\ \mathbf{u}_q \end{bmatrix} = T_{abc-dq} \begin{bmatrix} \mathbf{u}_a \\ \mathbf{u}_b \\ \mathbf{u}_c \end{bmatrix} = \begin{bmatrix} \mathbf{u}_M \\ 0 \end{bmatrix} \qquad (3)$$

Under the rotating coordinate system, $U_d = U_M$, $U_q = 0$

Generally speaking, if the grid does not have special requirements for reactive power, the inverter does not output reactive power, so make $Q = 0$, to achieve three-phase photovoltaic grid-connected system [8].

Similarly, the current of the photovoltaic system is transformed by parker, namely:

$$\begin{bmatrix} i_d \\ i_q \end{bmatrix} = T_{abc-dq} \begin{bmatrix} i_a \\ i_b \\ i_c \end{bmatrix} \qquad (4)$$

Have:

$$\begin{cases} P = \mathbf{u}_d i_d \\ Q = -\mathbf{u}_d i_q \end{cases} \qquad (5)$$

According to (5), power control can be converted into current control, where i_d controls active power and i_q controls reactive power (Fig. 3).

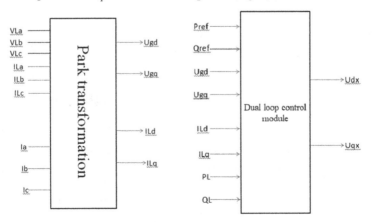

Fig. 3. Photovoltaic system inverter control part

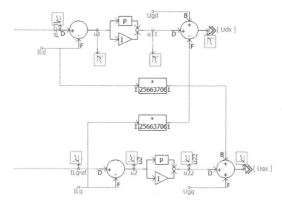

Fig. 4. Current loop control principle

The three-phase static state is changed into two-phase rotation [9]. The calculated DQ axis is used as the input of the current loop, and then the current loop is controlled, as shown in Fig. 4

To achieve the independence of the current decoupling control, we can introduce the current state feedback and grid voltage feed-forward compensation $i_{L\omega L}$, it eliminated the coupling voltage power grid voltage u of d and q axis current, the influence of the formula is as follows:

$$Udref = (idref - iLd)(K3 + \frac{K4}{S}) - iLq\omega L + Ud$$

$$Uqref = (iqref - iLq)(K3 + \frac{K4}{S}) + iLd\omega L + Uq$$

The PI controller can adjust the current according to the difference to ensure that the current output voltage can follow the grid voltage. The d, q axis voltage is converted into the sinusoidal modulation signal after inverse transformation, and the PWM controlled the switching of the three-phase inverter, so that the direct current of the photovoltaic power source is changed into three-phase alternating current with the same frequency and phase as the grid, and finally merged into the grid (Fig. 5).

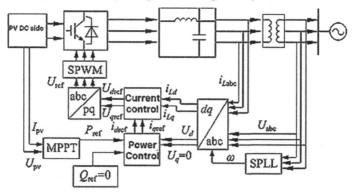

Fig. 5. Photovoltaic system PQ double loop control structure

I_{PV} and U_{PV} on the DC side of the photovoltaic array are sent to MPPT, and obtained the stable maximum power P_{ref} by using Maximum Power Point Tracking, at the same time the Q_{ref} is set to 0. The current and voltage in the three-phase power grid are converted into two-phase rotating d and q coordinates by a rotating coordinate system, and they are sent to PI power control together. I_{dref} is the reference value of the DC capacitor voltage U_{dcref} subtracts the actual value of the DC capacitor voltage U_{dc}, and then obtained by PI control. The reference value of I_{qref} is directly set to zero. The goal of outer loop power control is to make the active power sent by the inverter to the grid equal to the power emitted by the photovoltaic array, while the reactive power sent to the grid is equal to zero. U_{dref} and U_{qref} are obtained by current loop control, and finally they are applied to control the switch of the inverter through SPWM [10].

5   Simulation Analysis

5.1   Simulation Model

The photovoltaic array, combiner box, three-phase inverter, step-up transformer components, and inverter control module are used to build a grid-connected PSCAD simulation model of a photovoltaic power station, as shown in Fig. 6.

The photovoltaic modules connected in series and parallel to a photovoltaic array through the junction box and dc distribution cabinet. Inverter is the most important component of photovoltaic power station. Its function is to convert the direct current generated by the photovoltaic power station into the alternating current needed by the grid.

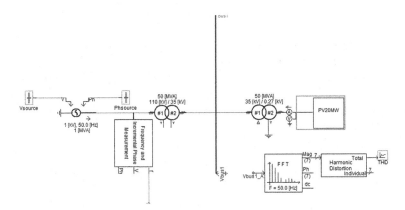

Fig. 6. Photovoltaic power station grid connected simulation module

5.2   Analysis of Simulation Results

Figure 7 is the power diagram of the photovoltaic power station after grid connection. It can be seen from the figure that the maximum power output after grid connection can reach 2.5 MW, meeting the power transmission requirements.

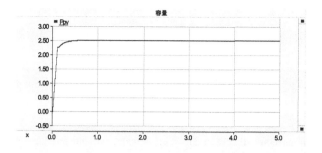

Fig. 7. Power diagram of 2.5 MW PV power grid after grid connection

According to the requirements, we adopted the 0.33 kV/110 kV boost technical scheme,and finally realize block power generation and centralized grid connection. It can be seen that the grid-connected side voltage can reach about 0.32 kV in a short time (Fig. 8), and the voltage deviation is $\frac{0.33-0.32}{0.33} \times 100\% = 3.03\%$, it can maintain stability, but the voltage has jitter.

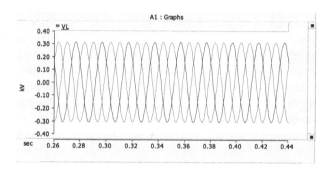

Fig. 8. Voltage waveform diagram of grid-connected slid after grid connection

In order to eliminate harmonics in the power grid as much as possible to obtain the best filtering effect, it is necessary to combine the parameters of the capacitor and the inductor. The LCL filter can be used as the first choice of grid-connected photovoltaic filter because of its compensation effect and low cost.

Add LCL filter in DC-AC inverter unit, the simulation is conducted again, and the total harmonic distortion rate after compensation is shown in Fig. 9.

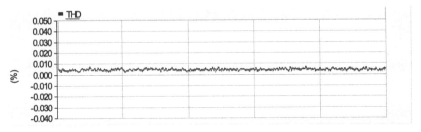

Fig. 9. Total harmonic distortion rate of grid-connected voltage after compensation

After the LCL filter is added to compensate, the total harmonic distortion rate of the a-phase voltage on the power grid side is reduced to about 0.08%. It can be concluded that the LCL filter has achieved a relatively good effect.

6   Conclusion

This paper introduces the photovoltaic array model based on engineering calculation, the Boost circuit with maximum power tracking function, and the inverter control with PQ decoupling, and realizes the overall modeling of the grid connection system of three-phase solar photovoltaic power generation. Finally, we analyzed the simulation model. Based on PSCAD/EMTDC, this paper studied the characteristics of three-phase grid connection of photovoltaic power generation. We built the model of photovoltaic power station and verified by simulation, which the result is correct. This paper analyzes the factors that affect the power quality of power grid, puts forward a feasible solution, and carries out simulation and verification, which provides a convenient method for the subsequent construction and research of photovoltaic power generation models.

Acknowledgment. This work was financially supported by the National Science Foundation of China under Grant (51677057), Local University Support plan for R & D, Cultivation and Transformation of Scientific and Technological Achievements by Heilongjiang Educational Commission (TSTAU-R2018005) and Key Laboratory of Modern Power System Simulation and Control & Renewable Energy Technology, Ministry of Education (MPSS2019-05).

References

1. Xing, G., Chen, F.: Modeling and simulation of solar photovoltaic cells. Autom. Electr. Power Syst. **39**(2), 27–29 (2019)
2. Lu, C.: A simulation study of an improved conductance incremental MPPT control strategy. Inf. Technol. **3**, 111–115 (2019)
3. Ding, M., Wang, W., Wang, X., et al.: A review on the effect of large-scale PV generation on power systems. Proc. CSEE **34**(1), 1–14 (2014)
4. Huang, X., Wang, H., Wang, Y., et al.: Principle and strategies of voltage rise regulation for grid-connected photovoltaic generation system at point of common coupling. Autom. Electr. Power Syst. **38**(3), 112–117 (2014)

5. Pan, H.: Research on PQ control strategy of photovoltaic power generation. Electron. World **21**, 52–54 (2018)

6. Chen, Q., Li, L., Wang, Q., et al.: Simulation model of photovoltaic generation grid-connected system and its impacts on voltage stability in distribution grid. Trans. China Electrotech. Soc. **28**(9), 241–247 (2013)

7. Omran, W.A., Kazerani, M., Salama, M.M.A.: Investigation of methods for reduction of power fluctuations generated from large-scale grid connected photovoltaic systems. IEEE Trans. Energy Convers. **26**(1), 318–327 (2011)

8. Shah, R., Mithulananthan, N., Bansal, R.C.: Oscillatory stability analysis with high penetrations of large-scale photovoltaic generation. Energy Convers. Manag. **65**, 420–429 (2013)

9. Han, D., Li, P., An, S., Shi, P.: Multi-frequency weak signal detection based on wavelet transform and parameter compensation band-pass multi-stable stochastic resonance. Mech. Syst. Signal Process. **71**, 995–1010 (2016)

10. Clark, K., Miller, N.W., Walling, R.: Modeling of GE Solar Photovoltaic Plants for Grid Studies. General Electrical International Inc., Schenectady (2010)

Study on Winding Force Distribution of Huge Nuclear Power Turbo-Generators

Pin Lv[1,2(✉)], Xiaojie Wu[1], Xunwen Su[2], Xianhui Zhu[2], Yin Yang[2], Weiguang Zhao[2], and Pen Xin[3]

[1] China University of Mining and Technology, Beijing, China
lvpinhkj@163.com
[2] Heilongjiang University of Science and Technology, Harbin, China
[3] Jilin Institute of Chemical Technology, Jilin City, China

Abstract. Unlike other polluted energies, the nuclear power is a novel energy which has the characteristics of clean, green and saving material. Due to the unbalanced power flow in the power system or on the load, the winding force appears large amount of harmonics and they are linked to the sound operation of the nuclear plant. In this paper, firstly the technology related nuclear power is discussed and the present central technology is reported. Then based on the experimental data, the finite element model of a third generation huge nuclear power turbo-generator is established. The arrangement of the field winding is given and the law of the field winding force distribution is discussed. And under unbalanced load, the constant forces of each field winding are found. This could bring new technology for the nuclear power turbo-generators safe running and fault diagnosis. The easiest damage part in field winding is shown and the conclusions in this paper give very useful information for future design of nuclear power turbo-generators.

Keywords: Nuclear power · Huge turbo-generators · Field winding force distribution · Easiest damage part

1 Introduction

1.1 Development of Nuclear Power Turbo-Generators

Since the nuclear power has been found, the replacement of nuclear material for traditional fuels becomes very practical and has great application value [1, 2]. Because compared with the traditional fuels, the nuclear material has the characteristics of saving space, clean and green. In the process of generating electricity, the nuclear power does not produce no air pollutants such as sulfur dioxide, nitrogen oxides, soot and so on. By the year of 2018, the nuclear power electricity took 4% of all electricity power in China. However, compared with the developed country especially France whose nuclear power takes 73% electricity generation rate and nuclear power has completely replaced

X. Jiang and P. Li (Eds.): GreeNets 2020, LNICST 333, pp. 276–284, 2020.
https://doi.org/10.1007/978-3-030-62483-5_29

the base load power plant of thermal power, in China in the field of nuclear power utilization rate there is large space to be developed.

So far, the nuclear power technology can be viewed as **four generations**. **First generation nuclear power plant:** They started from 1950s and belonged to the prototype nuclear power plant. **Second generation nuclear power plant:** They started from 1970s and the rising price of fuels pushed their fast development. In this stage, the economic factor is the first priority. **Third generation nuclear power plant:** The concept of safety and reliability had been brought out and took a very important role. **Fourth generation nuclear power plant:** The safety and economy of this kind of nuclear power plant would be more superior. The amount of waste would be very small, and there was no need for emergency measure outside the plant and an inherent ability to prevent nuclear proliferation.

Nowadays, the main technology used in China for nuclear power is the third generation nuclear power plant. In this paper, a AP1000 short for Advanced Passive PWR nuclear power turbo-generator is researched. Figure 1 is the first AP1000 nuclear power turbo-generator in the world. Usually, the nuclear power plants are created near the water.

Fig. 1. First AP1000 nuclear power turbo-generator in the world

1.2 Central Core Technology for Nuclear Power Turbo-Generators

As everyone knows, the generator is a core part of electric power system. Whether electric power system can operate safely or not is directly related to the stability and quality of power system, so the research concerning the turbo-generators is very important. Turbo-generators are operating under damp weather and rotating all the time. Therefore, the insulation fault and mechanical fault are the common reasons that the turbo-generators endure. When the turbo-generators experience failure, the whole electric power system is in danger. Typically the faults of turbo-generators can be classified into the following faults:

1) Stator core failure
In the process of manufacturing and in small turbo-generators due to self-vibration, if unluckily the stator core is damaged and short circuit between pieces, the circulating

current flowing through the short circuit increases gradually with time and silicon steel sheets melt. The melting silicon steel sheets flow into the stator slot and burn the winding insulation [3].

2) Winding main insulation failure
These kinds of failures have two reasons. Firstly, it is insulation aging. Secondly, it is congenital defect of insulation [4].

3) Stator winding strand failure
Stator winding strand failure mostly consists of the strand short circuit failure. The stator bar is usually composed of multiple strands, moreover possessing insulation between strands and needing transposition. If vibrations in turbo-generators happen, it is most likely to damage insulation between strands. Then ground failure or phase to phase short circuit failure maybe occur [5].

4) Stator end winding failure
When the stator end winding endures large impact force, partial discharge occurs due to cracks in insulating materials [6].

5) Rotor winding failure
This failure happens when turn to turn short circuit in rotor happens. However, under external asymmetry, the field winding can make harmonic forces. As a result, it may burn the insulation of the field winding. Then, sometimes, turn to turn short circuit of rotor winding takes places. Therefore, the investigations on the harmonic and constant electromagnetic torque of turbo-generators are very important and deserves to be paid attentions [7].

6) Rotor body failure
In nuclear power turbo-generators, the rotor body is often made of mixed metal. Because of its material characteristics and turbo-generators running characteristics, the rotor body can suffer heat and torsional vibration [8].

7) Cooling water system failure
Mostly, it is because of unclean cooling water. As seen in Fig. 2, the cooling water is directly emitting to the large river or ocean systems [9].

2 Model Establishment

In this paper, a huge capacity turbo-generator is served as a researching object. The huge turbo-generators are quite long and own a relative thin shape. For the physical characteristic of turbo-generators and also for the simplest way to achieve the simulation result, the two-dimensional finite element model of the third generations of nuclear power turbo-generators AP1000 is built up. Its physical model after division in finite element model is shown in Fig. 3.

Fig. 2. Cooling water directly emits to waters

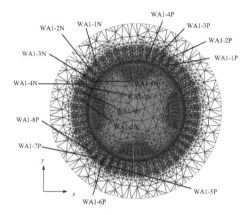

Fig. 3. Physical model after subdivision

In the division part in Fig. 3, the manual division method is adopted. And P presents the current flow direction which is paper inward and N presents the current flow direction which is paper outward. Clearly as seen in Fig. 3, the stator winding WA1-1, WA1-2, WA1-3, WA1-4, WA1-5, WA1-6, WA1-7 and WA1-8 is displayed and WA2-1, WA2-2, WA2-3, WA2-4, WA2-5, WA2-6, WA2-7 and WA1-8 are mechanical 180° away from WA1-1, WA1-2, WA1-3, WA1-4, WA1-5, WA1-6, WA1-7 and WA1-8. The rotor is moving in the anti-clockwise direction. The phasor B and phasor C differ 120° and 240° anti-clockwise electrical degree from the phasor A.

Further, the distributions of field windings are detailed displayed in Fig. 4. In each pole, the field winding has very good symmetry. The distribution and excitation of the field winding guarantees to keep the air-gap waveform at sinusoidal wave. So the field winding current in field winding slot is not the same to give the armature winding sinusoidal voltage. For instance, in field winding 3 and 4, the field current is bigger than field winding 1 and 8.

The huge nuclear power turbo-generator is 4 poles and the pole distribution in the rotor is shown vividly in Fig. 5. According to the China power grid frequency and the 1400 MV nuclear power turbo-generator's pole numbers, the rotor is rotating at the speed of 1500 r/min. In this paper, the field winding tangential electromagnetic force

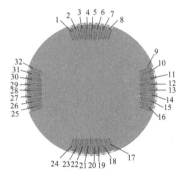

Fig. 4. Field winding number sequence of each slot

is calculated by finite element method, when the damper winding and damper wedge winding is present or not at different current asymmetrical degree.

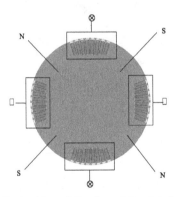

Fig. 5. Position of N-pole and S-pole of rotor

3   Results

When the positive sequence network is 1/2 rated load and 4% current asymmetry, in Fig. 6 the finite element simulation results show the constant component of tangential electromagnetic force of field winding that there is no damper winding and damper wedge, there is no damper wedge but damper winding, and there is both damper wedge winding and damper wedge at the same time. When the positive sequence network is rated load and 6% current asymmetry, in Fig. 7 the finite element simulation results show the constant component of tangential electromagnetic force of field winding that there is no damper winding and damper wedge coded 1, there is no damper wedge but damper winding coded 2, and there is both damper wedge winding and damper wedge at the same time coded 3.

From Fig. 6 and Fig. 7, due to the symmetry of the magnetic field in the nuclear power turbo-generator, under the condition of asymmetric external load the constant component

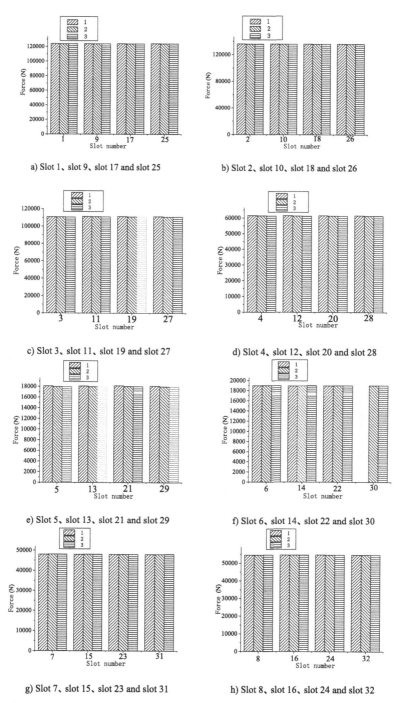

a) Slot 1, slot 9, slot 17 and slot 25

b) Slot 2, slot 10, slot 18 and slot 26

c) Slot 3, slot 11, slot 19 and slot 27

d) Slot 4, slot 12, slot 20 and slot 28

e) Slot 5, slot 13, slot 21 and slot 29

f) Slot 6, slot 14, slot 22 and slot 30

g) Slot 7, slot 15, slot 23 and slot 31

h) Slot 8, slot 16, slot 24 and slot 32

Fig. 6. Constant component of tangential electromagnetic force of field winding when positive sequence network is 1/2 rated condition and asymmetrical degree is 4%

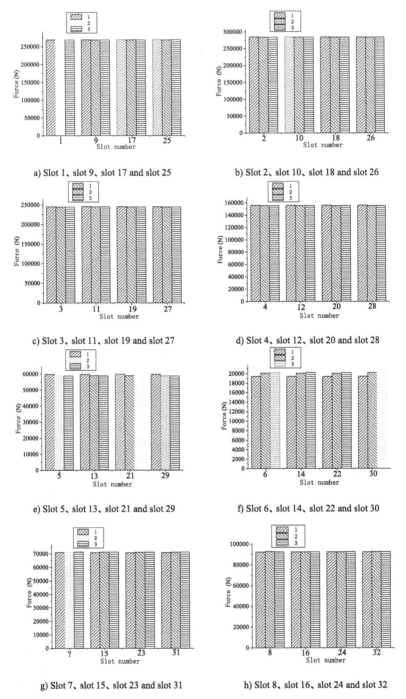

a) Slot 1、slot 9、slot 17 and slot 25

b) Slot 2、slot 10、slot 18 and slot 26

c) Slot 3、slot 11、slot 19 and slot 27

d) Slot 4、slot 12、slot 20 and slot 28

e) Slot 5、slot 13、slot 21 and slot 29

f) Slot 6、slot 14、slot 22 and slot 30

g) Slot 7、slot 15、slot 23 and slot 31

h) Slot 8、slot 16、slot 24 and slot 32

Fig. 7. Constant component of tangential electromagnetic force of field winding when positive sequence network is rated condition and asymmetrical degree is 6%

of the tangential electromagnetic force in the composite magnetic field is approximately equal when the slot number difference is the total slot number of each pole field winding pole. In this case, field winding slot 1, field winding slot 9, field winding slot 17 and field winding slot 25 take the same constant tangential electromagnetic force. Field winding slot 2, field winding slot 10, field winding slot 18 and field winding slot 26 take the same constant tangential electromagnetic force. Field winding slot 3, field winding slot 11, field winding slot 19 and field winding slot 27 take the same constant tangential electromagnetic force. Field winding slot 4, field winding slot 12, field winding slot 20 and field winding slot 28 take the same constant tangential electromagnetic force. Field winding slot 5, field winding slot 13, field winding slot 21 and field winding slot 29 take the same constant tangential electromagnetic force. Field winding slot 6, field winding slot 14, field winding slot 22 and field winding slot 30 take the same constant tangential electromagnetic force. Field winding slot 7, field winding slot 15, field winding slot 23 and field winding slot 31 take the same constant tangential electromagnetic force. Field winding slot 8, field winding slot 16, field winding slot 24 and field winding slot 32 take the same constant tangential electromagnetic force.

The damper winding and damper wedge show no influence on constant tangential electromagnetic force in huge nuclear power turbo-generators. The front slot number of field winding bear large constant forces as seen in Fig. 6 and Fig. 7.

4 Conclusions

In this paper, the beneficial effect of nuclear power technology is discussed. And the factors that the nuclear power turbo-generators may encounter are displayed. Then the finite element model of a 4 pole nuclear power turbo-generator is established.

The damper winding and damper wedge show no influence on constant tangential electromagnetic force in huge nuclear power turbo-generators. The front slot number of field winding bear large forces.

References

1. Breeze P. The cost of electricity from nuclear power stations. In: Nuclear Power, pp. 95–99 (2017)
2. Alobaid, F., Mertens, N., Starkloff, R., et al.: Progress in dynamic simulation of thermal power plant. Prog. Energy Combust. Sci. **59**, 79–162 (2017)
3. Bertenshaw, D.R., Chan, T., Smith, A.C., et al.: Three-dimensional finite element analysis of large electrical machine stator core faults. IET Electr. Power Appl. **8**(2), 60–67 (2014)
4. Kaufhold, M., Aninger, H.: Electrical stress and failure mechanism of the winding insulation in PWM-inverter-fed low-voltage induction motors. IEEE Trans. Ind. Electron. **47**(2), 396–402 (2000)
5. Moore, B., Maughan, C.: Generator stator end winding resonance: problems and solutions. In: ASME Power Conference (2013)
6. Rajeevan, P.P., Sivakumar, K., Gopakumar, K., et al.: A nine-level inverter topology for medium-voltage induction motor drive with open-end stator winding. IEEE Trans. Ind. Electron. **60**(9), 3627–3636 (2013)

7. Alekseev, V.G., Leviush, A.I., Belozor, A.N., et al.: Failure of rotor protection due to unreliable contact of the relay brush and the shaft. Power Technol. Eng. **47**(2), 146–148 (2013)
8. Al-Ani, M.M.J., Barrans, S.M., Carter, J.: Rotor loss reduction using segmented inverter in surface-mounted permanent magnet drive. In: IEEE International Electric Machines and Drives Conference. IEEE (2017)
9. Chen, C.N., Han, J.T., Gong, W.P.: Heat transfer and hydraulic characteristics of cooling water in a flat plate heat sink for high heat flux IGBT. In: ASME International Mechanical Engineering Congress and Exposition (2016)

The Fuzzy Sliding Mode Variable Structure Control for Direct Flux Oriented Vector Control of Motor

Jie Zhao$^{(\boxtimes)}$, Peng Wang, and Zunhui Ge

Heilongjiang University of Science and Technology, Harbin, China
zhao_xxsc@163.com

Abstract. In this paper, a vector controller based on fuzzy sliding mode variable structure is designed for the influence of nonlinear and variable parameters on the performance of the system. On the basis of establishing the mathematical model of the asynchronous motor in the synchronous rotation orthogonal coordinate system oriented by the rotor flux, a sliding mode variable structure speed regulator using the exponential approach law is designed. The stability of the system is proved by the Lyapunov stability theorem, and the chattering of the system is reduced by the fuzzy algorithm. The simulation results show that compared with the traditional PI controller and the conventional equivalent sliding mode controller, the fuzzy sliding mode variable structure vector control system has better speed regulation performance, smaller torque ripple, and improves the robustness and anti-interference ability of the system.

Keywords: Vector control · Sliding mode variable structure · Fuzzy algorithm

1 Introduction

For many years, a lot of research has been done about flux observer and PI parameter regulation [1, 2] by the scholars and experts in the scientific research institutes and universities. In reference [3], multi-level inverter is used to attenuate the pulsations from the flux and torque. However the hardware structure of the system becomes complex. The speed observer and flux observer proposed in reference [4] can be applied to the control of the three-phase asynchronous motor direct torque. The references [5] can effectively attenuate of the torque, but PI controller is used in the torque modulate loop to slow down the torque response. Since Utkin applied sliding mode observer to motor control [6], sliding mode observer has attracted much attention [7–9] because of its strong parameter robustness and flexible design. Many scholars introduce a new control method into the motor drive system that has variable structure from the sliding mode and do a lot of research on it. In reference [10], sliding mode control is used to design speed loop and torque loop to reduce the response time and torque ripple of the motor. The application of high-order sliding mode is put forward in reference [11], but these are relatively complex. In reference [12, 13], the chattering phenomenon is effectively reduced by

X. Jiang and P. Li (Eds.): GreeNets 2020, LNICST 333, pp. 285–291, 2020.
https://doi.org/10.1007/978-3-030-62483-5_30

improving the traditional sliding mode approach law. However induction motor has the characters such as high-order, nonlinear, etc. The research of efficient control scheme of induction motor is still an important research topic because its strongly coupling and multivariable value.

In this research, the control method of sliding mode variable structure and vector technology are combined to reduce torque disturbance and keep switching frequency stable. On this basis, fuzzy control algorithm is used to change the speed of reaching the face of the sliding patterns, so as to weaken the high frequency chattering problem from the process of sliding pattern control. The system has better speed regulation performance, smaller torque ripple, and improves the robustness and anti-interference ability.

2   Dynamic Mathematical of Asynchronous Motor

In order to realize the vector control technology, the mathematical model of an asynchronous motor should be established firstly. In the case of neglecting the iron loss and magnetic circuit saturation, assmodel uming that the three-phase winding is symmetrical, the electrical angle difference in the space is 120°, the generated magnetomotive force with the air spacing conforms to the sinusoidal function distribution. And the universal model of the three-phase asynchronous motor is obtained. The altered model in the two-phase static coordinate of the asynchronous motor is obtained through 3/2 transformation. Then the mathematical model in d-q coordinate is taken by vector rotation transformation.

The voltage and current equations can be obtained by transforming flux equation and the voltage equation, as shown in Eq. (1):

$$
\begin{bmatrix} u_{sd} \\ u_{sq} \\ u_{rd} \\ u_{rq} \end{bmatrix} = \begin{bmatrix} R_s + L_s p & -\omega_{dqs} L_s & L_m p & -\omega_{dqs} L_s \\ \omega_{dqs} L_s & R_s + L_s p & \omega_{dqs} L_m & L_m p \\ L_m p & -\omega_{dqr} L_m & R_r + L_r p & -\omega_{dqr} L_m \\ \omega_{dqr} L_m & L_m p & \omega_{dqr} L_r & R_r + L_r p \end{bmatrix} \begin{bmatrix} i_{sd} \\ i_{sq} \\ i_{rd} \\ i_{rq} \end{bmatrix} \tag{1}
$$

Where i_{sd}, i_{sq} are stator current components; i_{rd}, i_{rq} are rotor current components; L_s is stator equivalent winding self inductance; L_r is rotor equivalent winding self inductance; L_m is the mutual inductance.

Where u_{sd}, u_{sq} are respectively the stator voltage components in d-q coordinate system; u_{rd}, u_{rq} are respectively the rotor voltage components in d-q coordinate system; ω_{dqs}, ω_{dqr} are respectively the stator and rotor angular velocity in d-q coordinate system; R_s, R_r are respectively the stator resistance and rotor resistance; p stands for differential operation.

The torque equation is derived and Eq. (2) is get

$$
T_e = n_g L_m \left(i_{sq} i_{rd} - i_{sd} i_{rq} \right) \tag{2}
$$

Where T_e is electromagnetic torque; n_g corresponds for polar logarithm.
The motion equation of motor references (3):

$$
T_e = T_L + \frac{J}{n_p} \times \frac{d\omega}{dt} \tag{3}
$$

Where J corresponds for moment of inertia; T_L is load torque; ω is turning angular speed of asynchronous motor.

3 Improved Sliding Mode Controller

3.1 The Sliding Mode Control

The whole motion can be divided into two stages. If the initial position of the controlled trajectory is not considered, the sliding mode control will force the trajectory to move towards the sliding manifold in the first stage. Then the trajectory is kept on sliding manifold by the control function and moves towards the desired equilibrium point.

For any sliding surface (switching surface) $s = s(x) = s(x_1, x_2, \ldots, x_n)$, the state space can be divided into two parts, namely s > 0 and s < 0. The control quantity u of the system is designed:

$$u = \begin{cases} u^+(x), \ s(s) > 0 \\ u^-(x), \ s(s) < 0 \end{cases} \tag{4}$$

where $u^+(x) \neq u^-(x), u^+(0) = u^-(0)$.

The selection of sliding surface is the key to ensure it has good dynamic characteristics and stable performance of sliding mode motion. Select the switch surface function as:

$$s = \omega_r - \omega_r^* \tag{5}$$

Where ω_r^* represents the set value of angular speed, and ω_r represents the actual value of angular velocity.

The setting of the reaching law determines the dynamic quality of the reaching motion. In the variable structure controller of sliding method, the main reason for chattering is that the reaching law outputs a high frequency signal with a large amplitude, and the state variable still has a certain speed when it approaches the sliding mode surface, which results in its motion path passing through the sliding method face. In the design of sliding method controller, the exponential approach law can reduce the approach speed from a large value to zero, which not only shortens the approach time, but also the speed slows down when the state track approaches the sliding mode face. Thus, the regulation process of the motor speed is accelerated and the chattering of the system is weakened.

The expression of approach law is shown in Eq. (6):

$$\dot{s} = -Cs - K\,sign(s) \tag{6}$$

In the above formula, C is the exponential approach rate, K is the coefficient, and C > 0, k > 0, so that the derivative between the two sliding surfaces is different and the system is stable. In formula (6), the size of C should be selected appropriately in the speed of the system's motion track approaching to the sliding surface, and K determines the motion quality.

In this system, the sliding mode controller is designed by exponential approach law, which can be obtained by combining formula (5) and formula (6)

$$\dot{s} = -C(\omega_r - \omega_r^*) - K\,sign(s) \tag{7}$$

From formula (7)

$$\dot{s} = \dot{\omega}_r = \frac{n_p \times (T_e^* - T_L)}{J} \tag{8}$$

Because the input of the speed regulator is $\Delta\omega = \omega_r^* - \omega_r$ and the output is T_e^*, so the equation of the speed control is shown.

$$T_e^* = \frac{J}{n_p}\left[-C(s) - K\,sign(s)\right] + T_L \tag{9}$$

3.2 Optimal Sliding Method Controller

In the optimal control system, a most important improvement of the controller design is made. In the process of designing the controller, there are two key links. They are building the fuzzy rule table and determining the structure of the fuzzy control algorithm. In the fuzzy controller structure of the optimal motor control system, input accurate value in the process of fuzzification, and then fuzzify it. Thus, the fuzzy value input logic decision-making function link is generated. Before this link, it is necessary to judge that the logic decision-making function can identify the fuzzy value. The fuzzy value to is processed clearly, so that the accurate value is generated.

In the sliding mode variable structure control, the coefficient K of the symbolic function in the switching control quantity determines the motion quality of the system in the sliding stage. Further optimization can reduce the chattering of the system in the sliding stage.

In the two-dimensional fuzzy controller, the derivative ds/dt and the sliding surface s are taken as the input. The value K is adjusted in real time by combining the fuzzy control rules. The obtained $\triangle K$ plus the initial value of K_0 is used to adjust the value of K in real time, attenuated the vibration of the variable structure motor drive system, and prepare for the movement quality of the system.

The input s, ds/dt and output $\triangle K$ in the domain of fuzzy controller are defined in the following forms:

The output and input fuzzy subsets are $\{-B, -M, -S, ZO, +S, +M, +B\}$.

The fuzzy subset is divided into seven segments. $-B, -M, -S$ stand for negative big, negative middle, and negative small separately. ZO is the symbol for zero. $+S, +M$ and $+B$ are similar, except they are positive. The continuity domains of input s, ds/dt and output $\triangle K$ are all defined in $[-3, 3]$, and the triangle membership function is adopted in the input and output of the controller. Its expression is shown in formula (12)

$$f(x, m, n, k) = \begin{cases} 0 & x \le m \\ \frac{x-m}{n-m} & m \le x \le n \\ \frac{k-x}{k-n} & n \le x \le k \\ 0 & x \ge k \end{cases} \tag{10}$$

The system uses the barycenter method to calculate the output variables according to "If A and B then C" the rule statements. Based on the distance between status position and the sliding face in the process of movement, a fuzzy rule table is constructed. The corresponding rule table of the fuzzy control system established in this system is shown in Table 1. In the actual operation of the system, if the movement status track get close to the sliding face, then the output value ΔK has a smaller value. when the distance away the switching face is far, the output value ΔK has a larger value.

Table 1. Rule table

ΔK	S						
	$-B$	$-M$	$-S$	ZO	$+S$	$+M$	$+B$
ds/dt							
$-B$	$-B$	$-B$	$-B$	$-M$	$-M$	$-S$	0
$-M$	$-B$	$-B$	$-M$	$-S$	$-S$	0	0
$-S$	$-M$	$-M$	$-S$	0	0	$+S$	$+S$
0	$-S$	$-S$	$-S$	0	0	$+S$	$+S$
$-S$	$-S$	$-S$	0	$+S$	$+M$	$+M$	$+M$
$+M$	$-S$	0	0	$+M$	$+M$	$+B$	$+B$
$+B$	0	0	$+S$	$+M$	$+B$	$+B$	$+B$

4 System Simulation Analysis

This design simulation model of sliding method optimized by fuzzy algorithm is established in Matlab environment. In this design, traditional PI controller and fuzzy sliding method optimization controller are applied to the speed regulator of the asynchronous motor control system, and the simulation dynamic response curve is obtained. Specific system simulation parameters are as follows: $U_n = 1140$ V, $f = 50$ Hz, $P_n = 40$ kW, $R_s = 0.435\ \Omega$, $L_s = 0.2$ mH, $R_r = 0.816\ \Omega$, $L_r = 0.2$ mH, $L_m = 0.194$ mH, $J = 0.19$ kg.m^2.

The simulation result of velocity and torque running curve of traditional PI control is shown.

When the motor is started, $T_L = 35$ N·m, and the given speed of the system $n = 500$ r/min. At 0.2 s, 0.4 s and 0.6 s, the system simulation response curves of torque and speed under different control strategies are obtained under sudden load and sudden load reduction.

The simulation result of velocity and torque running curve of the fuzzy sliding method is in Fig. 2.

Compared with Fig. 1 and Fig. 2, the fuzzy sliding mode control system takes less time from the start to the rated speed stability stage, and the overshoot of the system is significantly reduced. At 0.2 s and 0.6 s, the system increases and decreases the load respectively, the response curve of PI control is more fluctuant than that of optimization controller, and the recovery time is slightly shorter, but the regulation of optimization controller is relatively smooth. When the system increases the load again and the speed decreases in 0.6 s, the overshoot of fuzzy sliding mode control is about 10 r/min less than that of PI control, and the regulating time of fuzzy sliding method control is a little less. When the system reduces the load torque and the system increases the speed in 0.8 s, there is almost no overshoot in fuzzy sliding mode control, while the overshoot of PI control is about 5 r/min.

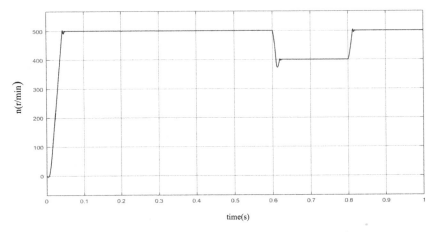

Fig. 1. Speed response curve of traditional PI control

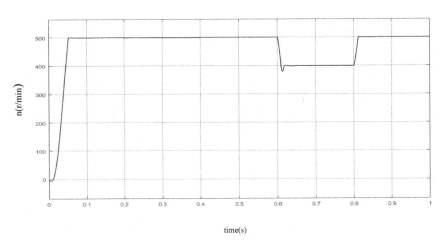

Fig. 2. Optimization controller

5 Conclusion

In this paper, through the theoretical analysis of the advantages and disadvantages of the asynchronous motor control system, combined with the asynchronous motor mathematical model, fuzzy and sliding method control theory, the simulation model of the frequency conversion speed control system of the shearer's traction unit based in fuzzy sliding mode control is established. Matlab tools are used to simulate the actual running of asynchronous motor, and then the simulation curves are analyzed theoretically. The results show that the fuzzy sliding mode control system is faster than the traditional PI algorithm system, which proves the correctness and feasibility of the design.

Acknowledgments. This work is supported by National Nature Science Foundation under Grant 51304075

References

1. Finch, J.W., Giaouris, D.: Controlled AC electrical drives. IEEE Trans. Ind. Electron. **55**(2), 481–491 (2008)
2. Zhifei, W., Xinghua, Z., Zhenxing, S., et al.: Direct torque control of induction motor based on adaptive sliding mode observer. Micromotor **41**(10), 64–69 (2013)
3. Guangzhao, L., Fei, L., Nanfang, Y., et al.: Study on direct torque control of permanent magnet synchronous motor powered by three-level inverter. J. Northwest. Univ. Technol. **30**(1), 22–26 (2012)
4. Yongheng, L., Xiaoyun, F., Zhen, W.: Space vector modulation direct torque control of induction motor based on stator flux sliding mode observer. Chin. J. Electr. Eng. **32**(18), 88–97 (2012)
5. Tang, L., Zhong, L., Rahman, M.F., et al.: A novel direct torque controlled interior permanent magnet synchronous machine drive with low ripple in flux and torque and fixed switching frequency. IEEE Trans. Power Electron. **19**(2), 346–354 (2004)
6. Utkin, V.I.: Sliding mode control design principles and applications to electric drives. IEEE Trans. Ind. Electron. **40**(1), 23–36 (1993)
7. Barambones, O., Alikorta, P.: Position control of the induction motor using an adaptive sliding-mode controller and observers. IEEE Trans. Ind. Electron. **61**(12), 6556–6565 (2014)
8. Shi, H., Feng, Y.: High-order terminal sliding mode flux observer for induction motors. Acta Automatica Sinica **2**, 288–294 (2012)
9. Vieira, R.P., Gastaldini, C.C., Azzolin, R.Z., Gründling, H.A.: Sensorless sliding-mode rotor speed observer of induction machines based on magnetizing current estimation. IEEE Trans. Ind. Electron. **61**(9), 4573–4582 (2014)
10. Hong, F., Quanshi, C., Guangyu, T.: Research on direct torque control of vehicle drive motor based on sliding mode. J. Automot. Saf. Energy Saving **1**(1), 59–66 (2010)
11. Di Gennaro, S., Rivera Domínguez, J., Meza, M.A.: Sensorless high order sliding mode control of induction motors with core loss. IEEE Trans. Industr. Electron. **61**(6), 26–28 (2014)
12. Lu Tao, Yu., Haisheng, S.B., et al.: Adaptive sliding mode maximum torque/current control of PMSM Servo System. Control Theor. Appl. **32**(2), 251 (2015)
13. Zhang, W., Miao, Z., Yu, X., et al.: Fractional calculus sliding mode control of permanent magnet synchronous motor based on improved sliding mode observer. Mot. Control Appl. **45**(7), 8 (2018)

The Influence of Word Attribute Information and Word Frequency Information on the Concreteness Effect of Words

Sun Fang[1] and Sui Xue[2(✉)]

[1] The School of Computer and Information Technology, Liaoning Normal University,
Dalian 116029, China
[2] The School of Psychology, Liaoning Normal University, Dalian 116029, China
Suixue@lnnu.edu.cn

Abstract. Using ERP recording technology our purpose is to explore the neural mechanism of the effect of word attribute information and word frequency information on concreteness. Experiment 1: under the lexical judgment task, the relationship between the two variables was investigated. The results showed that in N2 and P3 time window, there were differences between noun and verb processing. In P3 time window, concrete words and abstract words appear separate. In the process of N400, there are differences in the processing of nouns and verbs, concrete words and abstract words. In experiment 2, under the vocabulary judgment task, frequency and vocabulary type were taken as independent variables. The results showed that in N2 time window, high frequency vocabulary and low frequency vocabulary processing were separated. In P3 time window, noun and verb processing differences, concrete and abstract words began to appear separate. In N400, there are differences in the processing of nouns and verbs, concrete words and abstract words. The results suggest that word attributes and word types affect concreteness effect; the concreteness effect occurs in low-frequency words. The processing of concreteness effect of Chinese two-character words supports a single semantic processing model.

Keywords: Concreteness effect · Word attributes · Word frequency · Chinese two-character word

1 Introduction

Speech comprehension is a social phenomenon [1, 2]. Word recognition is the first stage of speech understanding. Word concreteness information and word frequency information affect word recognition. Concrete words refer to words that express specific concepts, such as table, bread and others. Abstract words refer to thoughts, feelings and other words that represent abstract concepts. In many experimental tasks, concrete words are processed faster than abstract words, that is, concrete effect.

Dual coding theory holds that there are two independent but interrelated processing systems in the human brain. They are semantic - based processing system (verbal system)

X. Jiang and P. Li (Eds.): GreeNets 2020, LNICST 333, pp. 292–305, 2020.
https://doi.org/10.1007/978-3-030-62483-5_31

and image - based processing system (non-verbal system). Concrete words are processed in verbal and non-verbal systems, while abstract words are mainly processed in the verbal system [3]. The Context Availability Model holds that when concrete and abstract words are processed in a system. The comprehension of verbal information depends heavily on the information of "context". It is more difficult to retrieve the "context" information of abstract words than that of concrete words when the words are presented separately. However, if appropriate context is provided, there is no significant difference in the processing of the two types of words, that is, the concreteness effect disappears [4].

Holcomb et al. investigated the concreteness effect of words under three different context conditions (appropriate, abnormal and neutral), and found that in abnormal and neutral sentences, concrete words produced larger N400 components than abstract words [5]. The Kounios and Holcomb studies also found that concrete words elicit larger N400 components than abstract words; In addition, in the right anterior brain region, concrete words and abstract words appear more separate, which indicates that concrete words have the advantage of right brain processing. But the study also found that the repetition of words makes the difference between concrete and abstract words weaken or even disappear [6]. Roberta and Alice took Italian as the experimental material and adopted the lexical judgment task, which resulted in significant concreteness effect. Different from the past, there was no N400 component. Concrete words in the middle occipital lobe produced larger N365 component than abstract words, and P3 component was found in the prefrontal lobe (the amplitude of abstract words was larger than that of concrete words) [7]

Using the PET technique and the vocabulary judgment task, Perani, Cappa, et al. found that the processing of abstract words was related to the selective activation of right temporal pole, right amygdala and bilateral inferior prefrontal cortex, while the processing of specific words did not find brain regions more active than that of abstract words [8]. Kiehl et al. [9] used the same experimental task as Perani et al. [19], and combined with fMRI technology, obtained consistent results. Fiebach and Friederici [10] followed the experimental task technique of Kiehl [9], but obtained different results. The processing of abstract words was related to the left inferior frontal triangle, while the processing of specific words was related to the left temporal lobe base. The results show that the processing of concrete words and abstract words is different, but they all have left brain dominance. However, Binder et al. [11] concluded another conclusion that the processing of abstract words activates the left inferior frontal lobe, while the processing of specific words activates bilateral angular gyrus and bilateral prefrontal cortex back.

Zhang qin [12] investigated the concreteness effect of Chinese words by manipulating word frequency, and found that there was a significant concreteness effect in the recognition of Chinese double-character words, but this advantage of specific words only existed in the range of low-frequency words. They also used repetition priming paradigm to investigate the influence of word context on the concreteness effect of two-character words, and found that repetition priming could not eliminate the concreteness effect.

Chen baoguo [13] used the vocabulary judgment task to investigate the influence of concreteness effect on vocabulary recognition. The results showed that concreteness and frequency had an interaction, but the concreteness effect only existed in high-frequency words. Zhang qin et al. [14] used ERP technology to study the processing difference between nouns and verbs and the relationship between concrete word and word attribute. The results showed that: in the 200–300 ms and 300–400 ms time Windows after the stimulus presentation, the ERP component induced by specific nouns in the whole brain was larger than that of specific verbs. Zhang, Guo, Ding, and Wang [15] used ERP technology to investigate the concrete effect of words by using word judgment task manipulation word frequency, and added concrete verbs and abstract verbs. The results show that, regardless of word frequency, concrete nouns produce more negative components between 200 ms–300 ms and 300 ms–500 ms than abstract nouns. They believe that concreteness and frequency of words may be two independent factors. The brain regions of concrete and abstract words were different in the 300 ms–500 ms time period, which supported the dual coding theory. Finally, the concreteness between verbs was smaller than that between nouns, and the difference was found in the middle left parietal lobe.

It can be seen that there is still controversy about the results of specific vocabulary processing. The first innovation of this study is to investigate the specificity effect by controlling word frequency and part of speech variables. The second innovation was the use of electroencephalography. The third is the Chinese double word as the experimental material

2   Experiment 1 the Influence of Word Attribute in Concreteness Effect

2.1   Method

Participations. The subjects were 22 college students of Liaoning normal university, aged 18–20 years old. All the subjects had not participated in similar experiments recently, and were healthy with normal vision or corrected vision. All the students who participated in the experiment received a nice gift and credits as a reward.

Materials. Material preparation and evaluation are as follows: (1) to select double-character concrete words (concrete nouns and concrete verbs) and abstract words (abstract nouns and abstract verbs) from the modern Chinese frequency dictionary. (2) these two-character words were randomly arranged and 20 students who did not participate in the formal experiment rated concreteness on a seven-point scale. 7 represents the strongest concreteness and 1 represents the strongest abstraction. (3) control the frequency of double-character words, the number of strokes of the first character, the number of strokes of the last character and the total number of strokes, and avoid the use of emotional words. (4) finally, 60 concrete nouns, concrete verbs, abstract nouns and abstract verbs were obtained respectively. (5) any word in a true word is modified to become a false word. Control the influence of the number of strokes of false words on the experimental results, and try to avoid false words with the same pronunciation, similar or ambiguous. (6) statistical information of 240 selected double-character words is shown in Table 1.

Experimental Apparatus. The experimental program was developed with e-prime 1.0, and the experimental materials were presented with a 15-in. monitor. EEG was recorded

Table 1. Statistics of experimental materials

C-A	CR	AWF	ASIC	ASLC	OMS
C–noun	6.44	67.73	7.77	7.62	15.38
	(0.33)	(0.01)	(2.64)	(2.46)	(3.84)
A–noun	1.93	64.93	8.10	8.45	16.55
	(0.23)	(0.01)	(2.90)	(2.35)	(3.76)
C–verb	5.39	72.29	8.53	7.75	16.28
	(0.45)	(0.01)	(2.38)	(2.81)	(3.80)
A–verb	2.52	60.79	8.60	7.13	15.73
	(0.25)	(0.00)	(2.39)	(2.13)	(3.02)

Note: frequency units are times/million. C-A: Concreteness - attribute. CR: Concreteness rating. AWF: The average word frequency. ASIC: Average stroke of initial character. ASLC: Average stroke of last character. OMS: Overall mean stroke

using EGI-128 channel recording system produced by EGI Corporation. The electroencephalogram was recorded by the international 10–20 system. With the Cz point as the reference electrode, electrodes were placed on the outside of the eyes, on the top and bottom of the eyes respectively to record the horizontal ophthalmic (HEOG) and vertical ophthalmic (VEOG). Sampling frequency is 250 Hz, and scalp electrode contact resistance is less than 50 kΩ.

Experimental Procedure. In this experiment, the experimental design is 2 (word type: concrete word, abstract word) × 2 (word attribute: noun, verb) completely repeat two factors design. The experiment adopted the word judgment task, and each subject was tested separately. Each subject completed the judgment task of 480 words.

After the electrode installation, the subject sat down facing the computer screen with a visual range of 80 cm. This experiment is divided into four blocks, each containing 120 trails. Firstly, the fixation point "+" is presented for 200 ms, followed by the target word, which is presented in black and white on the center of the computer screen. Each screen presents a word with a maximum view Angle of 2.5°. The subject pressed the key "1 or 4" to judge whether the word is the true or false word, and then the empty screen lasted for 500 ms. And then the next trial continues.

Data Analysis. The response time and error rate of the target words were analyzed statistically. Data whose response time of the target word is less than 250 ms and greater than 1500 ms are deleted. The subjects whose error rate was less than 80% were excluded. Finally, 20 valid subjects were obtained. SPSS16.0 statistical software was used to analyze the variance of repeated measures in response time

30 Hz low-pass filtering was performed off-line for the continuously recorded original EEG data. The analysis duration starts with the target word presenting, lasts 1000 ms, and the baseline is 200 ms before the target word presenting. After removing the data

segments containing eye movement artifacts, the data segments under the same conditions were averaged. After baseline correction and whole-brain averaging, the average waveforms of each subject in each experimental condition were obtained. According to the previous literatures [5–7] and the total average ERP waveform in this experiment, the average amplitude of N2 (150–280 ms), P3 (280–420 ms) and N400 (420–570 ms) were taken for analysis

2.2 Results

Twenty effective subjects were analyzed. The response time of true words and false words were 584.41 ms and 662.93 ms, respectively, and the response error rate was 0.06% and 0.05%, respectively. The variance analysis of response time shows that the main effect of lexical type is significant, subject analysis: $F(1,19) = 21.301$, $P = 0.001$; item analysis: $F(1,59) = 1.672$, $P = 0.201$. The main effect of lexical attributes is marginal significance, subject analysis: $F(1,19) = 0.413$, $P = 0.0528$; item analysis: $F(1,59) = 0.115$, $P = 0.0735$. There is no significant interaction between word types and lexical attributes, Subject analysis: $F(1,19) = 0.062$, $P = 0.807$; item analysis: $F(1,59) = 0.065$, $P = 0.800$. To identify the EEG sources of these differences, we further analyzed the ERP data. Repeated measurement ANOVA was performed on average amplitude, and Greenhouse-Geisser correction method was applied.

N2 (150–280 ms). Repeated measure ANOVA of 2 (concrete and abstract) × 2 (noun and verb) × 19 (electrode) was performed for the average amplitude of this period. The results showed that the electrode had a main effect, $F(18,324) = 38.429$, $P = 0. 001$. There was interaction between word attribute and electrode, $F(18,324) = 10.0267$, $P = 0.001$.

Further analysis of the interaction between word attribute and electrode showed that there was a main effect of word attribute on C4 electrode, $F(1,18) = 6.667$, $P = 0.019$; on the P4 electrode, $F(1,18) = 3.873$, $P = 0.065$; on the O2 electrode, $F(1,18) = 3.360$, $P = 0.083$ (marginal significant).

P3 (280–420 ms). Repeated measure ANOVA of 2 (concrete and abstract) × 2 (noun and verb) × 19 (electrode) was performed for the average amplitude of this period. The results show that the main effect of word type is marginal significant, $F(1,18) = 3.880$, $P = 0.064$. At the same time, the main effect of the word attribute of the target word is also near the marginal significant, $F(1,18) = 4.343$, $P = 0.052$. The main effect of electrode was also significant, $F(18,324) = 21.859$, $P = 0.001$. No interaction between the variables was found.

N400 (420–570 ms). Repeated measure ANOVA of 2 (concrete and abstract) × 2 (noun and verb) was performed separately for the average amplitude at each electrode. The results showed that there was a main effect of word type on C3 electrode, $F(1,18) = 6.559$, $P = 0.02$; On the PZ electrode, the main effect of word type was marginal significant, $F(1,18) = 3.913$, $P = 0.063$. On the O2 electrode, the main effect of word attributes was marginal significant, $F(1,18) = 3.379$, $P = 0.083$ (Fig. 1).

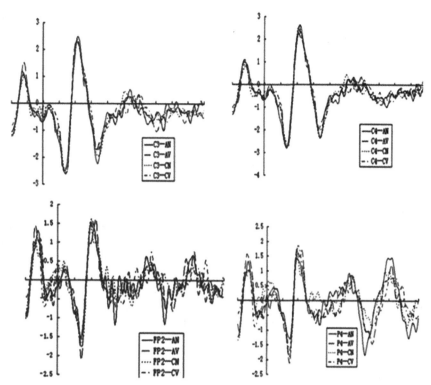

Fig. 1. Comparison of ERP waveform on some electrodes. Note: The abscissa represents time (-200 ms–1000 ms) ,and the ordinate represents the average amplitude (V). AN stands for abstract noun; AV stands for abstract verb; CN stands for concrete noun; CV stands for concrete verb.

2.3 Discussion

The behavioral data of experiment 1 show that there is a word superiority effect, that is, the real word reacts faster than the pseudoword word. The subjects were more likely to judge the correct word as wrong, but less likely to judge the wrong word as right. Moreover, the error rate of abstract words is lower than that of concrete words, perhaps because of the influence of the speed-accuracy tradeoff. But it hasn't been proven conclusively by other studies.

The study found that the difference between nouns and verbs began to appear about 200 ms after the target word was present. Statistical analysis of individual electrodes showed that EEG differences between nouns and verbs were significant at C4 (right center), P4 (right parietal), and O2 (right occipital) electrodes. The studies by Preissl etc. [16] and Pulvermller etc. [17] have shown that about 200 ms after the target word onset, nouns and verbs processing differences began to emerge. Pulvermller et al. [17] found that about 200 ms after stimulus presentation, EEG differences of content words and functional words began to emerge. In this study, the number of strokes, frequency and concreteness of experimental materials were strictly controlled. Therefore, the word attribute effect may reflect the semantic level difference of words with different properties. In this period, no main effect of word types was found, indicating that the processing of concrete words and abstract words had not been separated.

The difference between noun and verb processing continues to emerge in the next time window. Statistical analysis of individual electrode data shows that there is a significant difference in the processing of vocabularies of the two properties on FP2 (right frontal pole) and P4 (right parietal) electrodes. In this time window, the processing of concrete words and abstract words appeared different on the electrodes of P4 (right top), C3 (left center) and CZ (center point). In the 420–570 ms window, differences in the processing of nouns and verbs, concrete words and abstract words remain. The difference between noun and verb is shown by O2 (right occipital) and PZ (vertex) electrodes, while the difference between concrete and abstract words is shown by C3 (left center) and PZ (vertex) electrodes. N400 is closely related to language processing such as lexical access or semantic representation of words. In this study, it is believed that the presence of N400 components is more related to semantic representation. Whether there is word attribute effect or concreteness effect in the experiment, nouns are more related to perception, while verbs are more related to action procedures.

The difference between concrete words and abstract words is embodied in imaginability. Concrete words are more likely to make people produce representations. Thus, differences in semantic representation are likely to be reflected in electrical activity in the brain. However, the affirmation and clarification of this conclusion need further study.

3   Experiment 2 the Influence of Frequency on Concreteness Effect

3.1   Methods

Participation. The subjects were 24 college students of Liaoning normal university, aged 18–20. None of the participants had recently taken part in a similar experiment and were healthy with normal vision or corrected vision. All the participants received a nice gift and credits as a reward. None of the participants in experiment 2 participated in experiment 1

Materials. The material preparation and evaluation of this experiment includes the following steps: (1) Select concrete and abstract words from the dictionary of modern Chinese frequency. Among them, the low-frequency words are the words with a frequency below 10 times per million, and the high-frequency words are the words with a frequency above 90 times per million. (2) Print these two-character words in random order on a piece of paper. Twenty students who did not participate in the formal experiment rated the word's concreteness on seven-point scale. "7" means the word refers to very specific content, and "1" means the word refers to very abstract content. (3) Control the number of strokes of the first character, the number of strokes of the last character and the total number of strokes, and try to avoid the use of emotional words. (4) 400 double-character words were selected, including 100 high-frequency concrete words, 100 high-frequency abstract words, 100 low-frequency concrete words and 100 low-frequency abstract words. (5) The composition of the 400 false words was the same as in experiment 1. The statistical information of the 400 selected two-character words is shown in Table 2.

Table 2. Statistics of experimental materials

C–F	CR	AWF	ASIC	ASLC	OMS
C–HF	5.00 (0.92)	260.42 (0.02)	7.13 (2.44)	7.19 (2.83)	14.32 (3.58)
A–HF	2.42 (0.61)	4.02 (0.00)	7.36 (2.13)	7.67 (2.45)	15.03 (3.55)
C–LF	5.75 (0.78)	248.17 (0.01)	7.92 (2.25)	7.11 (2.43)	15.03 (3.17)
A–LF	2.58 (0.60)	4.11 (0.00)	7.67 (2.23)	7.41 (2.41)	15.16 (3.30)

Note:C-F: Concreteness - frequency. CR: Concreteness rating. AWF: The average word frequency. ASIC: Average stroke of initial character. ASLC: Average stroke of last character. OMS: Overall mean stroke. HF-high frequency, LF-low frequency, Frequency unit is in order per million

Apparatus. The experimental instrument is the same as experiment 1.

Design and Procedure. In this experiment, 2 (word type: concrete word, abstract word) × 2 (word frequency: high frequency, low frequency) two-factor measurement experiment design was adopted. Using the word judgment task, each subject was tested separately, and each subject judged 800 words. This experiment is divided into four blocks, each containing 200 trails. The rest of the experiment followed the same procedure as experiment 1.

Data Analysis. The response time and error rate of the target words were statistically analyzed. Data whose response time of the target word is less than 250 ms and greater than 1500 ms are deleted. The subjects whose overall accuracy rate was less than 80% were excluded. Finally, 21 subjects had valid data. A 30 Hz low-pass filter was performed offline for the recorded continuous original EEG data. The time epoch of analysis is the beginning of the target word presentation, lasting 1000 ms, and the baseline is 200 ms before the target word presentation.

After removing the data segments containing eye movement artifacts, the data segments under the same conditions were averaged. After baseline correction and brain average, the average waveform of each subject in each experimental condition was obtained. The average amplitude of N2 (150–280 ms), P3 (280–420 ms) and N400 (420–620 ms) were taken for analysis according to the previous literatures and the total ERP waveform in this experiment.

3.2 Results

The data of 21 effective subjects were analyzed. The response time of true words and false words were 603.84 ms and 674.48 ms, respectively, and the response error rate was 0.06% and 0.05%, respectively.

The variance analysis of response time data shows that the main effect of word types is significant, Subject analysis: $F(1, 20) = 19.653$, $P = 0.001$; item analysis: $F(1, 99) = 5.726$, $P = 0.019$. The main effect of word frequency was significant. Subject analysis: $F(1, 20) = 3.573$, $P = 0.001$. Item analysis: $F(1, 99) = 3.887$, $P = 0.001$. There is a significant interaction between word type and word frequency, Subject analysis: $F(1, 20) = 9.768$, $P = 0.005$; item analysis: $F(1, 99) = 3.813$, $P = 0.054$. The simple effect analysis of the interaction is found that the main effect of the word type is not significant, $P > 0.05$; When the target word is low-frequency, the main effect of word type is significant, subject analysis: $F(1, 20) = 22.333$, $P = 0.001$. item analysis: $F(1, 99) = 5.757$, $P = 0.018$.

In order to clarify the source of these differences, we analyzed the ERP data. Repeated ANOVA was performed on the average amplitude, and Greenhouse-Geisser correction method was applied.

N2 (150–280 ms). The average amplitude of this epoch was analyzed by 2 (concrete and abstract) × 2 (high and low frequency) × 19 (electrode) repeated measurement ANOVA. The results showed that the main effect of the target word frequency was significant, $F(1,21) = 10.614$, $P = 0.004$. The main effect of electrode was significant, $F(18,378) = 40.284$, $P = 0.001$. There was interaction between word frequency and electrode, $F(18,378) = 4.762$, $P = 0.008$.

Simple effect analysis shows that there is a main effect of target word frequency on FP1 (left frontal pole) electrode, $F(1,21) = 10.787$, $P = 0.004$; There is a main effect of target word frequency on the FP2 electrode, $F(1,21) = 9.072$, $P = 0.007$; There is a main effect of target word frequency on F4 (right frontal) electrode, $F(1,21) = 4.803$, $P = 0.04$; there was a main effect of target word frequency on the F8 (right anterior temporal) electrode, $F(1,21) = 4.534$, $P = 0.045$. In addition, there were significant frequency primary effects on T3 (left middle temporal), T5 (left posterior temporal), T6 (right posterior temporal), O2 (right occipital), FZ (frontal midpoint), and CZ (central point) electrodes.

P3 (280–420 ms). For the average amplitude of this period, 2 (concrete and abstract) × 2 (high and low frequency) × 19 (electrode) repeated measurement variance analysis was performed, and the results showed that the main effect of the electrode was significant, $F(18,378) = 23.175$, $P = 0.001$; There was interaction between word types and electrodes, $F(18,378) = 32.112$, $P = 0.001$; There was an interaction between the word frequency and the electrode, $F(18,378) = 5.651$, $P = 0.001$. A simple effect analysis of the interaction shows that there is a main effect of the target word type on the FP1 electrode, $F(1,21) = 4.282$, $P = 0.051$ (marginal significance); There is a main effect of target word frequency on F3 electrode, $F(1,21) = 10.517$, $P = 0.004$; There is a main effect of target word frequency on the C3 electrode, $F(1, 21) = 3.872$, $P = 0.062$ (marginal significance); There is a main effect of target word frequency on T3 electrode, $F(1, 21) = 9.506$, $P = 0.006$. In addition, there are also major effects of lexical types on T4, T5,

O1, O2, FZ and CZ electrodes. Moreover, there was a significant interaction between target word frequency and word type on the T6 electrode, $F(1, 21) = 14.874, P = 0.001$.

N400 (420–620 ms). Repeated measure ANOVA of 2 (concrete and abstract) × 2 (high and low frequency) was performed separately for the average amplitude of each electrode. The results showed that there was a main effect of word frequency on C4 electrode, $F(1, 21) = 6.130, P = 0.022$; on the P4 electrode, $F(1,21) = 5.613, P = 0.027$; on CZ electrode, $F(1,21) = 7.720, P = 0.011$; on the PZ electrode, $F(1,21) = 3.158, P = 0.090$ (edge significant). There was a main effect of lexical type on the P4 electrode, $F(1,21) = 10.764, P = 0.004$; on T6 electrode, $F(1,21) = 5.710, P = 0.026$; on O2 electrode, $F(1,21) = 6.033, P = 0.023$; on the PZ electrode, $F(1,21) = 4.798, P = 0.040$ (Fig. 2).

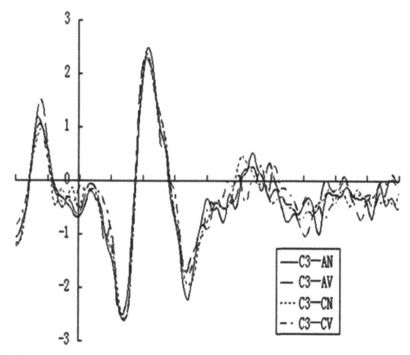

Fig. 2. Comparison of ERP waveform on some electrodes. Note: The abscissa represents time (−200 ms–1000 ms) ,and the ordinate represents the average amplitude (μV). HA stands for high frequency abstract words; HC stands for high frequency specific words; LA stands for low-frequency abstract words; LC stands for low frequency specific words.

3.3 Discussion

Experiment 2 also found the word superiority effect, the error rate of false words is lower than that of true words, and the error rate of abstract words is lower than that of concrete words. Behavioral data show that there is a significant effect of word frequency and word type, and there is an interaction between word frequency and word type. In the case of low frequency, the main effect of word type is significant, and the specific effect only exists in the range of low frequency words. It is common sense that the concreteness effect exists in low-frequency word. Whether abstract or concrete, the difference in processing disappears after the words are used quite a lot. The processing of high frequency abstract words is close to or equal to the processing of high frequency concrete words. However, for low-frequency word, the processing of abstract word and concrete word is still carried out in the original way. From this point of view, if a low-frequency abstract word has been used for quite a few times, it will not have a concrete effect. Therefore, faced with a new problem, what frequency range, word specificity effect exists; what frequency range, where does the concreteness effect disappear, and where is the critical point of that frequency? It deserves the attention of future research.

The results show that the main effect of the target word frequency is significant in 150–280 ms time window, and there is an interaction between the frequency and the electrode. From the beginning of word processing, word frequency plays a role, and the brain source of processing is also different. This result is consistent with the result of experiment 1. In experiment 1, the separation of noun and verb processing also occurs in this period. It can be speculated that this kind of word attribute effect may reflect the difference in semantic processing level of words with different properties.

In experiment 2, it was also observed that the separation of high frequency words and low frequency words with different properties also appeared in the 150–280 ms time window. Thus the results reinforced our hypothesis that differences on N2 waves indicate differences in the semantic level of words of different properties. In 280–420 ms, the processing of concrete and abstract words begins to separate. High and low frequency words continue to be processed differently.

In the process of 420–620 ms, the processing of different frequency words and the processing of different word types continue to be separated. However, there was no significant interaction between the two variables, and the N400 component was not affected by the frequency of the target word. In the study of Kounios and Holcomb [18, 19], the inconsistency between this response and N400 results also appeared. They argue that response time and EEG cannot be reduced to the same cognitive operation. Zhang [32] also showed such inconsistencies in the lexical judgment task. They explained that the measurement method of response time may be more sensitive to lexical tasks than lexical integration.

4 General Discussion

This study used the word judgment task, combined with ERP technology, to investigate the factors influencing the concreteness effect of Chinese two-character words. N2 is typically found in tasks such as describing response stimulus recognition, attention switching, suppressing motor responses, overcoming stereotypical responses, or conflict detection. Naatanen [20] proposed the following characteristics of N2 component: it was induced by deviant stimulus, even with only a very small deviation; Its enhancement results from an increase or decrease in event deviations. In this study, the separation of nouns and verbs, high frequency words and low frequency words occurred in the period of N2 component. This suggests that after processing the physical features of words, differences in brain processing begin to emerge. Word attributes and word frequency play an important role in the early stage of word processing.

P3 is regarded as a measurement index of resource allocation [21, 22]. The P3 amplitude increased as the subjects put more effort into the task. In our two experiments, the separation of the first processing of concrete words and abstract words occurred in this period. The processing of concrete words activates more brain regions than that of abstract words. This is consistent with the dual coding model we mentioned earlier. But the dual coding model emphasized that this difference occurred only in the right brain. Since the processing of specific words is performed by the verbal semantic processing system and the non-verbal (image) processing system, while abstract words are processed by the verbal semantic processing system, the difference between them is embodied in the non-verbal (image) system. This system is thought to exist in the right brain. The differences between the two kinds of words in this study are not limited to the right brain.

According to the extended dual coding model, the difference between concrete and abstract words can be expressed in speech semantic processing system and non-verbal (image) processing system. It is therefore difficult to say in which hemisphere the amplitude of the processing of concrete words is greater than that of abstract words. The results of this study found that the difference between concrete words and abstract words appeared not only in the left brain, but also in the right brain, so it was more consistent with the extended double-coding model. The contextual validity model holds that the processing of concrete words and abstract words is reflected in different degrees of processing in the same system. Therefore, this experiment does not support dual coding model.

From a certain point of view, the extended dual coding model has something in common with the context validity model. The extended dual coding model is the synthesis and extension of the dual coding model and the context validity model. More importantly, some viewpoints of dual coding model were removed and context validity model was added, which made it extremely difficult to distinguish the extended dual coding model from context validity model, not only from response time, latency of brain waves, amplitude or distribution of brain regions. More research is needed.

It is generally believed that the larger the amplitude of N400, the smaller the context priming amount of the target word. The smaller the amplitude is, the greater the amount of context priming of the target word is. N400 is also related to the successful completion of vocabulary integration. In the N400 time window, the difference in the processing of concrete and abstract words in the two experiments also did not exist in the unilateral brain region, thus further supporting our view that the dual coding model was not supported.

The nature of nouns and verbs is different and has been confirmed in many studies. In this study, when examining the influence of word attributes on the concreteness effect, it is found that word attributes are not the factors that affect the concreteness effect. Zhang et al. [13] used high frequency nouns, low frequency nouns and low frequency verbs to investigate the concrete effect. The study showed that the difference between verbs in ERP was smaller than that between nouns, but did not analyze the difference between nouns and verbs. Meanwhile, Tsai [23]'s study on the word judgment task showed that there was a significant interaction between word attributes and the cerebral hemisphere, and the processing of verbs was weaker than that of nouns in right brain.

The two studies also failed to show the interaction between word attribute and word type. Therefore, it is reasonable to believe that although nouns and verbs are words with different properties, such differences have been offset or ignored by the differences in the processing of concrete words and abstract words when making lexical judgments, and the specific reasons need to be further studied.

Word frequency is an indispensable factor in the study of word. This study shows that there is a significant interaction between word type and word frequency, that is, word frequency affects the specific effect of words. Further analysis shows that the specific effect occurs in the range of low-frequency vocabulary. The reason may be that, within the range of high frequency words, the specificity effect is not sensitive to the way in which it is measured. It is worth noting that word frequency and word type may be two independent variables. But further research is needed. The critical point of occurrence and disappearance of concreteness effect is also a main direction of future research.

References

1. Liu, Y., Liu, D.: Morphological awareness and orthographic awareness link Chinese writing to reading comprehension. Read. Writ. **33**(7), 1701–1720 (2020). https://doi.org/10.1007/s11145-019-10009-0
2. Caramazza, A., Laudanna, A., Romani, C.: Lexical access and inflectional morphology. Cognition **28**(3), 297–332 (1988)
3. Li, D., Song, D., Wang, T.: Concreteness and imageability and their influences on Chinese two-character word recognition. Read. Writ. **33**(6), 1443–1476 (2020). https://doi.org/10.1007/s11145-020-10016-6
4. Schwanenflugel, P.J.: Why are abstract concepts hard to understand? In: Schwanenfugel, P.J. (ed.) The Psychology of Word Meanings, pp. 223–250. N J Lawrence Erlbaum Associates, Hillsale (1991)
5. Holcomb, P.J., Kounios, J., Anderson, J.E., West, W.C.: Dual-coding, context-availability, and concreteness effects in sentence comprehension: an electrophysiological investigation. J. Exp. Psychol. Learn. Mem. Cogn. **25**(3), 721–742 (1999)
6. Kounios, J., Holcomb, P.J.: Concreteness effects in semantic processing: ERP evidence supporting dual-coding theory. J. Exp. Psychol. Learn. Mem. Cogn. **20**(4), 804–823 (1994)
7. Adorni, R., Proverbio, A.M.: The neural manifestation of the word concreteness effect: an electrical neuroimaging study. Neuropsychologia **50**(5), 880–891 (2012)
8. Perani, D., Schnur, T., Tettamanti, M., Gorno-Tempini, M., Cappa, S.F., Fazio, F.: Word and picture matching: a pet study of semantic category effects. Neuropsychologia **37**(3), 293–306 (1999)
9. Kiehl, K. A., Liddle, P. F., Smith, A. M., Mendrek, A., Forster, B.B., Hare, R.D.: Neural pathways involved in the processing of concrete and abstract words. Hum. Brain Mapp. **7**, 225–233 (1999)
10. Fiebach, C.J., Friederici, A.D.: Processing concrete words: fMRI evidence against a specific right-hemisphere involvement. Neuropsychologia **42**(1), 62–70 (2004)
11. Binder, J.R., Desai, R.H., Graves, W.W., Conant, L.L.: Where is the semantic system? A critical review and meta-analysis of 120 functional neuroimaging studies. Cereb. Cortex **19**(12), 2767–2796 (2009)

12. Chen, B.G., Peng, D.L.: The influence of concreteness of words on word recognition. Acta Psychologica Sinica **30**(4), 387–393 (1998)
13. Zhang, Q., Zhang, B.Y.: Concreteness effects of two – character chinese words. Acta Psychologica Sinica **29**(2), 216–224 (1997)
14. Zhang, Q., Ding, J.H., Guo, C.Y.: ERP difference between processing of nouns and verbs. Acta Psychologica Sinica **35**(6), 753–760 (2003)
15. Zhang, Q., Guo, C.Y., Ding, J.H., Wang, Z.Y.: Concreteness effects in the processing of Chinese words. Brain Lang. **96**(1), 59–68 (2006)
16. Preissl, H., Pulvermüller, F., Lutzenberger, W., Birbaumer, N.: Evoked potentials distinguish between nouns and verbs. Neurosci. Lett. **197**(1), 81–83 (1995)
17. Pulvermller, F., Lutzenberger, W., Birbaumer, N.: Electrocortical distinction of vocabulary types. Electroencephalogr. Clin. Neurophysiol. **94**(5), 357–370 (1995)
18. Kounios, J., Holcomb, P.J.: Structure and process in semantic memory: Evidence from event-related brain potentials and reaction times. J. Exp. Psychol. General **121**, 459–479 (1992)
19. Holcomb, P.J.: Automatic and attentional processing: An event-related brain potential analysis of semantic priming. Brain Lang. **35**(1), 66–85 (1998)
20. Näätänen, R., Picton, T.W.: N2 and automatic versus controlled processes. Electroencephalogr. Clin. Neurophysiol. **38**, 69–186 (1986)
21. Isreal, J.B., Chesney, G.L., Wickens, C.D., Donchin, E.: P300 and tracking difficulty: Evidence for multiple resources in dual-task performance. Psychophysiology **17**, 259–273 (1980)
22. Tian, J., Wang, J., Xia, T., Zhao, W., Xu, Q., He, W.: The influence of spatial frequency content on facial expression processing: an ERP study using rapid serial visual presentation. Sci. Rep. **8**(1), 2383 (2018). https://doi.org/10.1038/s41598-018-20467-1
23. Tsai, P.-S., Yu, B.H.Y., Lee, C.-Y., Tzeng, O.J.L., Hung, D.L., Wu, D.H.: An event-related potential study of the concreteness effect between Chinese nouns and verbs. Brain Res. **1253**, 149–160 (2009)

High-Precision Harmonic Analysis Algorithm Based on Five-Term MSD Second-Order Self-convolution Window Four-Spectrum Line Interpolation

Yang Qingjiang$^{(\boxtimes)}$ and Qu Xiangxiang

School of Electronics and Information Engineering, Heilongjiang
University of Science and Technology, Harbin 150022, China
1962578365@qq.com

Abstract. For the frequency spectrum leakage and fence effect generated in Fast Fourier Transform (FFT) during asynchronous sampling and integral period truncation affect the precision of harmonic detection, a new algorithm for harmonic analysis based on four-spectrum-line interpolation FFT with second-order self-convolution window with five-term Maximum-Sidelobe-Decay (MSD) was proposed and the polynomial fitting method was used to construct the four-spectrum-line interpolation correction formula. The simulation results showed that this algorithm could improve the detection accuracy of amplitude, phase and frequency by 1-2 orders of magnitude compared with other commonly used windowed interpolation algorithms.

Keywords: Harmonic analysis · Five-term MSD self-convolution window · Four-spectrum line interpolation · Polynomial fitting

1 Introduction

With the development of industry, attention is turning to the harmonic problem caused by the widely used nonlinear load. In order to reduce harmonic pollution, accurate detection of harmonics becomes the primary issue [1]. The fast Fourier transform (FFT) algorithm has become a typical method of harmonic detection due to its characteristics of reliability, simplicity, fast operation speed, and easy implementation in hardware circuits [2]. However, when FFT is used for harmonic analysis, the asynchronous sampling and non-integral period truncation of the signal will cause spectrum leakage and fence effect, which will seriously affect the accuracy of harmonic detection [3]. In order to suppress the effects of spectral leakage and fence effects, a windowed interpolation correction algorithm can be used. At present, the commonly used window functions include Hanning window [4], Blackman window [5], Blackman-Harris window [6], Nuttall window [7], etc., all of which suppress the effect of spectrum leakage to a certain extent. The more commonly used interpolation algorithms are multimodal spectral line

X. Jiang and P. Li (Eds.): GreeNets 2020, LNICST 333, pp. 306–319, 2020.
https://doi.org/10.1007/978-3-030-62483-5_32

interpolation [8, 9], cubic spline interpolation [10–12], Lagrange interpolation [13–15], etc.

There are various FFT window interpolation algorithms nowadays [16], and the traditional single detection method cannot adapt to the increasingly complex harmonic detection requirements. At present, many researchers at home and abroad have conducted in-depth research on the harmonic detection algorithm where a new method is introduced on the basis of the original algorithm to realize the optimization of some calculation processes [17–19], or several methods are used to overcome the defects of a single detection algorithm [20–22]. Liu Kaipei and others have derived correction formulas for four typical window functions, verifying that the four-spectrum line interpolation algorithm has higher accuracy than the commonly used bispectral and trispectral lines [23]. Xie Qiang et al. Combined the five-term MSD window with the three-spectrum line interpolation algorithm, and compared with other commonly used window function interpolation algorithms, it has relatively higher accuracy [24]. Nan Xue et al. Proposed a four-spectrum line interpolation FFT harmonic analysis algorithm based on five MSD-Rife windows, and concluded that the five MSD-Rife windows with MSD/Rife weight ratio $\alpha/\beta = 0.6/0.4$ have better suppression of spectrum leakage Function [25]. In order to further improve the precision of windowed interpolation algorithm, the self-convolution of five-term MSD window was conducted in this paper to generate five-term MSD second-order self-convolution window. The information of bilateral symmetry lines of the peak frequency was taken into full consideration meanwhile in the harmonic analysis of four-spectrum-line interpolation FFT. Finally, a new algorithm for harmonic analysis was put forward based on four-spectrum-line interpolation FFT with second-order self-convolution window with five-term MSD.

2 Five-Term MSD Second-Order Self-convolution Window

2.1 Five-Term MSD Window Characteristics

In the case of asynchronous sampling, in order to reduce the influence of spectrum leakage on the harmonic measurement, a window function with a low sidelobe peak level and a high sidelobe attenuation rate is used for windowing the signal. The pentad MSD window is a cosine combination window [26], and its general form is

$$w(n) = \sum_{m=0}^{M-1} (-1)^m a_m \cos(\frac{2\pi mn}{N}) \tag{1}$$

In the formula: N is the number of sampling points; M is the number of terms of the window function; $m = 0, 1, 2, \ldots, M-1$; $n = 0, 1, 2, \ldots, N-1$; the coefficient of a_m should be meet constrains $\sum_{m=0}^{M-1} (-1)^m a_m = 0$ and $\sum_{m=0}^{M-1} (-1)^m a_m = 0$.

In the five-term maximum sidelobe attenuation window, the five coefficients of a_m are taken as $a_0 = 0.2734375$, $a_1 = 0.4375$, $a_2 = 0.221875$, $a_3 = 0.0625$, $a_4 = 0.0078125$, respectively.

According to its time domain information, the normalized logarithmic spectrum and the normalized logarithmic spectrum of several commonly used window functions are drawn in the same graph, and the comparison graph is shown in Fig. 1. The specific sidelobe characteristics of the four window functions in Fig. 1 are shown in Table 1.

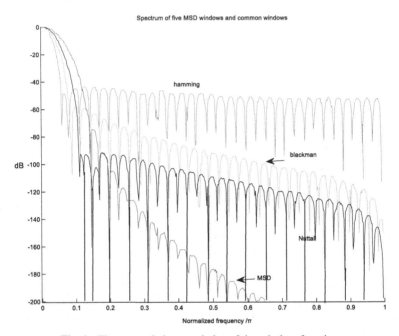

Fig. 1. The spectral characteristics of the window function.

Table 1. Sidelobe characteristics of each window function.

Window function name	Peak sidelobe/dB	Sidelobe decay rate/(dB/oct)
Hamming	−43	6
Blackman	−58	18
Nuttall	−83	24
Five-Term MSD	−75	54

With reference to Fig. 1 and Table 1, it can be seen that the five MSD window sidelobe peaks are −75 dB, and the sidelobe attenuation rate is 54 dB/oct. There is a clear advantage in the sidelobe attenuation rate, which is far greater than the other mentioned above. The window function has a very good attenuation effect on the side lobes, which can well suppress the influence of spectral leakage on other frequency points, and make the calculation result of the algorithm more accurate.

2.2 Five-Term MSD Self-Convolution Window

The self-convolution of the window function can improve the window function's ability to suppress sidelobe leakage and enhance the performance of the window function. In order to further improve the window function performance of the five-term maximum sidelobe attenuation window, the five-term MSD window is self-convolved And analyze the results of self-convolution.

Time-domain self-convolution of the five-term MSD windows yields

$$w_{MSD-p}(M) = \underbrace{w_{MSD}(N) * w_{MSD}(N) * \ldots * w_{MSD}(N)}_{p} \tag{2}$$

In the formula: p is the number of five MSD window functions participating in the self-convolution operation, which is also called the order of the self-convolution operation.

According to the convolution property, when two five-term MSD window sequences of length N are convolved, a sequence of length $2N - 1$ can be obtained. Perform a zero-padding operation at the beginning or end of the sequence to obtain a sequence of $2N$ in length. Similarly, by performing $p - 1$ convolution operations on five MSD window sequences of length N, a sequence of length $pN - p + 1$ can be obtained. By adding $p - 1$ zeros at the beginning or end, we can obtain Sequence of length pN.

From the discrete sequence Fourier transform, the frequency domain expression of the five-term MSD window is

$$W_{MSD}(w) = \sum_{m=0}^{M-1} (-1)^m \frac{a_m}{2} \left[W_R(w - \frac{2m\pi}{N}) + W_R(w + \frac{2m\pi}{N}) \right] \tag{3}$$

According to the nature of the convolution, the convolution in the time domain of the window function is equivalent to the product in the frequency domain, so the spectrum of the p-order five-term MSD self-convolution window is

$$W_{MSD-p}(w) = [W_{MSD}(w)]^p \tag{4}$$

The main lobe width of the p-order pentad MSD self-convolution window, that is $W_{MSD-p}(w)$, the point at which 0 is taken for the first time, which $|W_{MSD}(w)|$ is 0 at this time.

According to formula (3),

$$\begin{cases} \frac{N}{2} \left(w \pm \frac{2\pi m}{N} \right) = d\pi \\ \frac{1}{2} \left(w \pm \frac{2\pi m}{N} \right) \neq d\pi \end{cases} \quad d = 0, \pm 1, \pm 2 \cdots \tag{5}$$

Since the main lobe of the five MSD self convolution windows is $W_{MSD}(w)$ the point of the first 0, at this time, $d = \pm 1$, the main lobe width of the five MSD self convolution windows is

$$B_{MSD-p} = \frac{20\pi}{N} = \frac{20\pi p}{M} \tag{6}$$

From Eq. (6), the main lobe width of the p-order pentad MSD self-convolution window is inversely proportional to the length of the mother window. When the length M of the p-order pentad MSD self-convolution window is constant, the main lobe width of the p-order pentagonal MSD self-convolution window depends only on the convolution order p. The higher the convolution order, the larger the main lobe width, The lower the frequency resolution.

From Eq. (4), we construct first-order, second-order, and fourth-order pentad MSD self-convolution windows, and the corresponding spectrum curves are shown in Fig. 2.

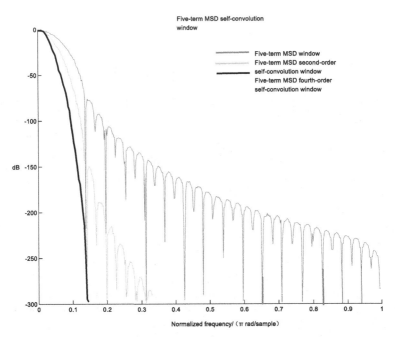

Fig. 2. Spectrum of five MSD self-convolution windows

It can be seen from Fig. 2 that with the increase of the self-convolution order of the five-term cosine self-convolution window, the sidelobe peak value gradually decreases, and the sidelobe attenuation gradually increases. The sidelobe peak of the second-order self-convolution window is -151.56 dB, the sidelobe peak of the fourth-order self-convolution window is -303.12 dB, and the sidelobe attenuation speed of the second-order self-convolution window is 108 dB/oct, and the fourth-order is 216 dB./oct. However, as the order p increases, the length of the convolution window will increase to p times the mother window. If the length of the sampling window is maintained, according to formula (6), the main lobe width will increase to the original p times. The main lobe width of the second-order self-convolution window is $40\pi/M$, but the fourth-order self-convolution window will reach $80\pi/M$. Although the rapid attenuation of the side-lobe peaks is conducive to the improvement of the calculation accuracy, too large the main lobe width will have a great impact on the resolution. Large, so we choose the five-term MSD second-order self-convolution window as the window function in harmonic detection.

3 Four-Spectrum Interpolation Algorithm

Taking a single frequency signal as an example for analysis, let $x(t)$ take the discrete time signal sampled uniformly at the sampling frequency f_s as:

$$x(n) = A_0 \sin(2\pi \frac{f_0}{f_s} n + \phi_0) \tag{7}$$

Where: A_0 is the amplitude of the signal; f_0 is the frequency of the signal; ϕ_0 is the initial phase angle of the signal; f_s is the sampling frequency, $n = 0, 1, 2, \cdots, N-1$, where N is the number of sampling points.

Windowing truncation of the signal $x(n)$ using the p-order pentad maximum sidelobe attenuation self-convolution window will obtain $x_w(n) = x(n) \times w(n)$, x_w is calculated in the frequency domain as

$$X_w(e^{jw}) = \sum_{n=0}^{N-1} x(n)w(n)e^{-jwn}$$

$$= \frac{A_0}{2j} \left[e^{j\phi_0} W(\frac{2\pi(f-f_0)}{f_s}) - e^{-j\phi_0} W(\frac{2\pi(f+f_0)}{f_s}) \right] \tag{8}$$

Ignoring the effect of the long-spectrum leakage at the frequency $-f_0$, the DFT transform of $x_w(n)$'s DTFT transform $X_w(k)$ of $X_w(e^{jw})$ is sampled at equal intervals $\Delta w = \frac{2\pi}{N}$:

$$X_w(k) = \frac{A_0}{2j} e^{j\phi_0} W(k - \frac{f_0}{\Delta f}) \tag{9}$$

Where $\Delta f = \frac{f_s}{N_c}$, $k = 0, 1, 2, \cdots, N-1$, $W(k)$ is the spectral function of the p-order pentad MSD self-convolution window. Because of the sampling point $N \rangle\rangle 1$, the expression of $W(k)$ is:

$$W(k) = \left\{ \frac{N}{\pi} e^{-j\frac{k\pi}{p}} \sin(\frac{k\pi}{p}) \left[\sum_{m=0}^{M-1} (-1)^m \frac{a_m k}{(k)^2 - p^2 m^2} \right] \right\}^p \tag{10}$$

In asynchronous sampling, the signal frequency $f = k\Delta f$ is difficult to be located at the sampling frequency point, that is, k is generally not an integer, and the FFT generates a spectrum leakage at this time. As shown in Fig. 3, the four spectral lines near the peak frequency k are k_1, k_2, k_3, k_4 and $k_1 < k_2 = k_1 + 1 < k_3 = k_2 + 1 < k_4 = k_3 + 1$, respectively. Let $\varepsilon = k - k_2 - 0.5$, because of $0 \leq k - k_2 \leq 1$, so $\varepsilon \in [-0.5, 0.5]$, find ε is a key step to accurately estimate the harmonic parameters.

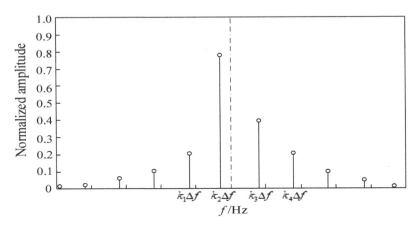

Fig. 3. FFT spectrum under asynchronous sampling

Record the amplitudes of the four spectral lines, respectively: $y_1 = |X_w(k_1)|$, $y_2 = |X_w(k_2)|$, $y_3 = |X_w(k_3)|$, $y_4 = |X_w(k_4)|$. Let $R = (2y_3 + y_4)$, $S = (2y_2 + y_1)$. Then according to formula (9) and bringing in ε, we get:

$$r = 2|W(-\varepsilon + 0.5)| + |W(-\varepsilon + 1.5)| \tag{11}$$

$$s = 2|W(-\varepsilon - 0.5)| + |W(-\varepsilon - 1.5)| \tag{12}$$

Let

$$\gamma = \frac{y_3 + y_4 - y_1 - y_2}{y_3 + y_4 + y_1 + y_2} \tag{13}$$

Then

$$\gamma = \frac{r - s}{r + s} \tag{14}$$

It can be seen from Eq. (14) that γ is a function of ε and $\gamma = g(\varepsilon)$, and the inverse function $\varepsilon = g^{-1}(\gamma)$ can be obtained to find the offset ε. Polynomial approximation can be used to calculate $\varepsilon = g^{-1}(\gamma)$. Use Matlab's polyfit function to fit the inverse function. If you fit $2q + 1$ times, you get:

$$\varepsilon \approx L(\gamma) = a_1\gamma + a_3\gamma^3 + \cdots + a_{2q+1}\gamma^{2q+1} \tag{15}$$

Where $a_1, a_3, \cdots, a_{2q+1}$ is an odd-order coefficient.
After finding ε, the frequency correction formula of the available signal is:

$$f = k\Delta f = (k_2 + 0.5 + \varepsilon)f_s/N \tag{16}$$

The phase correction formula of the signal is:

$$\phi_0 = \arg[X_w(k_2)] + \frac{\pi}{2} - \arg[W(-\varepsilon - 0.5)] \tag{17}$$

It can be seen from Fig. 3 that the position of k_2,k_3 is closer to k, so the two spectral lines are given greater weight, that is, the weights of k_1,k_2, k_3,k_4,and are added 1, 2, 2, and 1, respectively. The amplitude estimation formula is:

$$A_0 = \frac{2(y_1 + 2y_2 + 2y_3 + y_4)}{r + s} \tag{18}$$

When N is 512 or 1024, Eq. (18) can be simplified as $A_0 = N^{-P}(y_1 + 2y_2 + 2y_3 + y_4)u(\varepsilon)$,Where $u(\cdot)$ is an even function, so the approximation polynomial of the magnitude is:

$$A_0 = N^{-P}(y_1 + 2y_2 + 2y_3 + y_4) \cdot (a_0 + a_2\varepsilon^2 + \cdots + a_{2q}\varepsilon^{2q}) \tag{19}$$

In the formula, $a_0, a_2, \cdots, a_{2q}$ is an even term coefficient.
The frequency domain expression of the P-order pentad MSD self-convolution window can be simplified by formula (10), where M = 5.

$$|W(k)| = \left[\frac{N}{\pi} \left| \sin(\frac{k\pi}{p}) \sum_{m=0}^{4}(-1)^m \frac{a_m k}{k^2 - p^2 m^2} \right| \right]^p \tag{20}$$

Substituting Eq. (20) into Eq. (14), randomly taking a set of ε from $[-0.5, 0.5]$ and substituting it will get a corresponding set of γ values. The number of ε cannot be too small, otherwise it will affect the fitting accuracy.$\varepsilon \approx L(r)$ is obtained by a polynomial fitting function polyfit(γ, ε, i), where i represents the degree of fitting (i usually takes 5 or 7 times).

Take the fitting number $i = 7$, and the corresponding ε of the second-order pentad MSD self-convolution window is:

$$\varepsilon \approx L(\gamma) = 0.3000500\gamma^7 + 0.4966542\gamma^5 + 1.0904182\gamma^3 + 5.7181015\gamma \tag{21}$$

The Eq. (18) is combined with the Eq. (20), and the coefficient of $u(\varepsilon)$ is fitted by polyfit($\varepsilon, u(\varepsilon), i$).

$$u(\varepsilon) = 0.002803784\varepsilon^6 + 0.084853287\varepsilon^4$$
$$+1.794230810\varepsilon^2 + 19.639067884 \tag{22}$$

4 Simulation Experiment Analysis

In order to verify the accuracy of the algorithm in this paper, a 21st harmonic simulation analysis is performed. The signal model of the simulation sampling is:

$$x(n) = \sum_{i=1}^{21} A_i \sin(2\pi in\frac{f_1}{f_s} + \theta_i) \tag{23}$$

The amplitudes A_i of each harmonic and the phase θ_i of each harmonic are shown in Table 2. The fundamental frequency f_1 of the signal is 50.5, the sampling frequency f_s is 2520, and the number of sampling points N is 512.

Table 2. Fundamental and harmonic parameters of harmonic signals

Harmonic order	1	2	3	4	5	6	7
Amplitude/V	220	4.4	10	3	6	2.1	3.2
Phase/(o)	0.05	39	60.5	123	−52.7	146	97
Harmonic order	8	9	10	11	12	13	14
Amplitude/V	1.9	2.3	0.8	1.1	0.7	0.85	0.1
Phase/(o)	56	43.1	−19	4.1	40	10.5	115
Harmonic order	15	16	17	18	19	20	21
Amplitude/V	1	0.06	0.4	0.04	0.3	0.005	0.01
Phase/(o)	25	53.1	−132	85	0.8	53	−72

Window the signal model given by Eq. (23), and then perform simulation experiments according to the flow of Fig. 4.

Hanning window and Blackman-Harris (B-H) window four-line interpolation uses the correction formula given in Literature [23], five-term MSD window three-line interpolation uses the correction formula given in Literature[24], and five-term MSD window four-line interpolation uses literature The correction formula given in [25]. The five-term MSD second-order self-convolution window four-spectrum line interpolation uses the correction formula of this paper. Through comparison of simulation experiments, the relative errors of amplitude, phase and frequency are shown in Figs. 5, 6 and 7.

It can be seen from the simulation results that:①under the same conditions, compared with Hanning window and Blackman Harris window, the harmonic detection accuracy of the five MSD windows is higher, and the relative error of amplitude, phase and frequency is increased by about 3–4 orders of magnitude; ② compared to the five-term MSD window of the five-term MSD second-order self-convolution window, the relative errors in amplitude, phase, and frequency have increased by about 1–2 orders of magnitude; ③ on the premise of adding phase and window function, the accuracy of harmonic detection can be improved by four spectral line interpolation compared with three spectral line interpolation, which can be increased by 1–2 orders of magnitude.

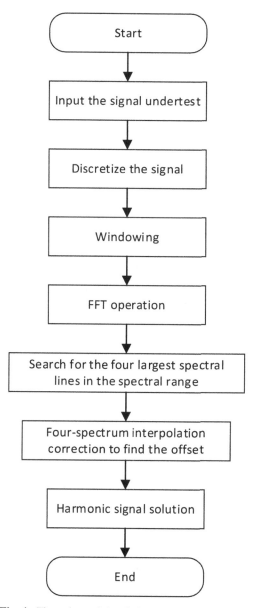

Fig. 4. Flow chart of simulation experiment program

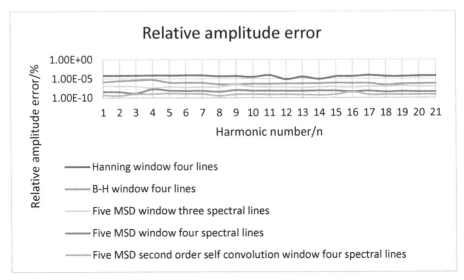

Fig. 5. Relative error of amplitude for different windowed interpolation algorithms

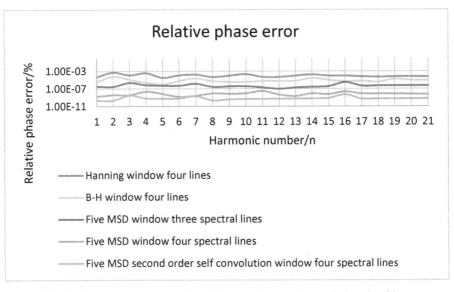

Fig. 6. Relative error of phase for different windowed interpolation algorithms

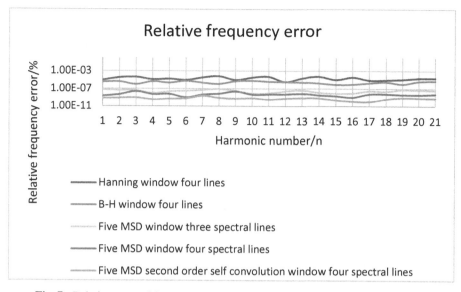

Fig. 7. Relative error of frequency for different windowed interpolation algorithms

5 Conclusion

In this paper, five-term MSD window with superior sidelobe performance were selected to construct five-term MSD self-convolution window. By analyzing their spectrum characteristics, it was concluded that the sidelobe performance of the five-term MSD self-convolution window was getting better with the increase of the convolution order, and the width of the main lobe also increased gradually. On this basis, a new algorithm for harmonic analysis based on four-spectrum-line interpolation FFT with second-order self-convolution window with five-term MSD was proposed, and the correction formulas of various parameters were calculated to realize the high-precision detection of amplitude, frequency and phase of each harmonic in the signal. Compared with other algorithms, this algorithm greatly improved the accuracy of harmonic detection and could be applied to the occasions where the accuracy of harmonic analysis were demanding.

References

1. Wen, H., Teng, Z.S., Guo, S.Y.: Triangular self-convolution window with desirable side lobe behaviors for harmonic analysis of power system. IEEE Trans. Instrum. Meas. **59**(3), 543–552 (2010)
2. Zeng, B.: Parameter estimation of power system signals based on cosine self-convolution window with desirable side-lobe behaviors. IEEE Trans. Power Delivery **26**(1), 250–257 (2011)
3. Nuttall, A.: Some windows with very good sidelobe behavior. IEEE Trans. Acoustics, Speech, Signal Process. **29**(1), 84–91 (2003)
4. Wen, H., et al.: Hanning self-convolution window and its application to harmonic analysis. Sci. China Ser. E: Technol. Sci. **52**(2), 467 (2009)

5. Kumar, A.: A comparative study of performance of blackman window family for designing cosine-modulated filter bank. In: (IACSIT—International Association of Computer Science and Information Technology) Proceedings of International Conference on Circuits, System and Simulation (ICCSS 2011) (IACSIT—International Association of Computer Science and Information Technology), pp. 315–321 (2011)

6. Agre, D.: Interpolation in the frequency domain to improve phase measurement. Measurement **41**(2), 151–159 (2008)

7. Zhu, Y., Wang, Y., Lin, T., Feng, C., Chen, J., Gao, Y.: Harmonic analysis of power system based on nuttall self-convolution window triple-spectral-line interpolation FFT. In: Proceedings of the 2017 2nd International Conference on Control, Automation and Artificial Intelligence (CAAI 2017) (2017)

8. Agrez, D.: Weighted multipoint interpolated DFT to improve amplitude estimation of multi-frequency signal. IEEE Trans. Instrum. Meas. **51**(2), 287–292 (2002)

9. Song, X., Li, D., Li, Z., Guo, W.: Harmonic analysis and simulation study using triple-spectrum-line interpolation fft algorithm. In: Proceedings of the 2015 International Conference on Materials Engineering and Information Technology Applications (2015)

10. Bogdanov, V.V., Volkov, Y.S.: Shape-preservation conditions for cubic spline interpolation. Siberian Adv. Math. **29**(1), 231 (2019)

11. Idais, H., Yasin, M., Pasadas, M., González, P.: Optimal knots allocation in the cubic and bicubic spline interpolation problems. Math. Comput. Simul. **164**, 131 (2019)

12. Li, H., Li, L., Di Zhao, : An improved EMD method with modified envelope algorithm based on C2 piecewise rational cubic spline interpolation for EMI signal decomposition. Appl. Math. Comput. **335**, 112 (2018)

13. de Camargo, A.P.: On the numerical stability of Newton's formula for Lagrange interpolation. J. Comput. Appl. Math. **365**, 112369 (2019)

14. Kobayashi, K., Tsuchiya, T.: Error analysis of Lagrange interpolation on tetrahedrons. J. Approximation Theory **249**, 105302 (2020)

15. Keller, W., Borkowski, A.: Thin plate spline interpolation. J. Geodesy **93**(9), 1251 (2019)

16. Li, P., Wei, Z., Lei, C., Songling, H.: Research on the performance comparison of harmonic analysis algorithms for power grid signals. Electr. Meas. Instrum. **57**(01), 1–20 (2020)

17. Wen, H., Teng, Z., Wang, Y., et al.: Optimized trapezoid convolution windows for harmonic analysis. IEEE Trans. Instrum. Meas. **62**(9), 2609–2612 (2013)

18. Junmin, Z., Kaipei, L., Li, W., et al.: An algorithm for harmonic analysis based on multiplication window function. Power Syst. Prot. Control **44**(13), 1–5 (2016)

19. Wang Ling, X., Baiyu, S.C., et al.: An approach for harmonic analysis based on a new type of cosine combination window interpolation FFT. Wuhan Univ. J. **47**(2), 250–254 (2014)

20. Candan, Ç.: A method for fine resolution frequency estimation from three DFT samples. IEEE Signal Process. Lett. **18**(6), 351–354 (2011)

21. Duda, K.: DFT interpolation algorithm for Kaiser-Bessel and Dolph-Chebyshev windows. IEEE Trans. Instrum. Meas. **60**(3), 784–790 (2011)

22. Li, Y.F., Chen, K.F.: Eliminating the picket fence effect of the fast fourier transform. Comput. Phys. Commun. **178**(7), 486–491 (2008)

23. Junmin, Z., Kaipei, L., Li, W., Wenjuan, C.: A rapid algorithm for harmonic analysis based on four-spectrum-line interpolation FFT. Power Syst. Prot. Control **45**(01), 139–145 (2017)

24. Shi, L., Xie, Q., Ma, X.: High accuracy analysis of harmonic algorithm based on 5-term maximum-sidelobe-decay window and triple-spectrum-line interpolation. Power Syst. Prot. Control **45**(07), 108–113 (2017)

25. Xue, N., Shulian, Y., Tianze, L., Jiazhen, Q.: Electrical harmonic analysis based on five-term MSD-rife window and four-spectrum-line interpolation FFT. Hydropower Energy Sci. **36**(10), 177–180 (2018)
26. Song, S., Ma, H., Xu, G., Wang, F., Wang, J.: Power harmonic analysis based on 5-term maximum-sidelobe-decay window interpolation. Autom. Electr. Power Syst., **39**(22), 83–89 + 103 (2015)

Optimal Configuration
of Micro-Energy-Network of Cogeneration Type
in Near Land Island

Weiguang Zhao, Zehao Ling[✉], Jingqiang Zhao, Ying Yang, Ze Xie, and Ze Chen

Heilongjiang University of Science and Technology, Harbin 150022, China
1033525946@qq.com

Abstract. In order to effectively solve the problem that the offshore island energy supply system can not be applied to the near land island, the micro energy network optimization configuration model under the typical scenes of the near land island is constructed, and the specific load data is used to configure the capacity of the main equipment in the scheme. The model satisfies the problems of electricity load, heat load and desalination. The objective function of the model is the minimum operating cost and the maximum customer satisfaction. Considering the constraints of various types of power generation and energy storage, thermal power balance and desalination, particle swarm optimization algorithm is used to solve the problem. Taking a micro energy network of an island as an example, this paper simulates the micro energy network of cogeneration type, and makes a comparative analysis with the common micro energy network of the island. The results show that the optimized configuration model of the micro energy network of the near land island studied in this paper can effectively provide the power load and heat load for the users, improve the satisfaction of the users, and thus improve the economy of the micro energy network.

Keywords: Near land island · Micro-energy-network · Energy supply system · Optimal allocation

1 Introduction

China has a large number of islands and a long coastline. How to ensure the safety and stability of energy supply is the top priority for island residents. The island is rich in renewable energy such as wind energy and wave energy. The island micro energy network provides energy for residents by combining multiple renewable energy sources, which is an effective way to solve the island energy supply problem [1–4]. In reference [5], According to the water demand of the island and the characteristics of the desalination system, and considering the economic and environmental benefits of the system operation, the paper puts forward the power distribution strategy of coordinating the

This work was financially supported by the fundamental research project of Heilongjiang Province (Hkdxp201905).

X. Jiang and P. Li (Eds.): GreeNets 2020, LNICST 333, pp. 320–329, 2020.
https://doi.org/10.1007/978-3-030-62483-5_33

desalination load, battery and diesel generator operation. In reference [6], a scheme of island micro grid power supply is proposed, in which the sea water pumped storage power station is used as energy storage equipment, and the wind farm and diesel generator are operated. Based on the modeling of wind farm, sea water pumping and storage power station and diesel generator respectively, considering their operation constraints and island load requirements, the scheme of optimal scheduling of island micro-grid is proposed. All of the above documents have optimized the energy storage system of the micro energy network on the basis of satisfying the power load of the residents, and they are all off grid islands. When the islands are close to the mainland, this mode cannot be applied. Therefore, this paper proposes a micro energy network system of land island combined heat and power supply, which takes the grid connected micro energy network system as the object, and operates the micro energy network in the scheduling cycle according to the time-sharing electricity price strategy based on the principle of minimum cost and maximum customer satisfaction, a micro energy network optimization model including each output unit, energy storage unit and cogeneration system is established and solved by particle swarm optimization. Taking an island micro energy network as an example, this paper compares the micro energy network system of cogeneration with that of common Island micro energy network, and verifies the validity and feasibility of the optimal configuration model and solution method of the near land island micro energy network studied in this paper.

2 Island Micro-Energy-Network Structure and Equipment Model

The micro energy network system studied in this paper is mainly composed of electricity, heat and residential water. Next, the mathematical model of micro energy network system equipment is analyzed (Fig. 1).

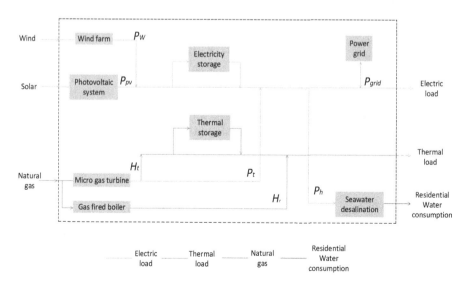

Fig. 1. Basic framework of micro-energy-network

2.1 Renewable Energy Power Generation Equipment

The mathematical models of wind power and photovoltaic power generation are as follows [7, 8]:

$$
P_w = \begin{cases} P_r & V_r \leq v \leq V_{co} \\ P_r(A + Bv + Cv^2) & V_{ci} \leq v < V_r \\ 0 & v < V_{ci} \; or \; v > V_{co} \end{cases} \tag{1}
$$

$$
\begin{cases} A = \frac{1}{(V_{ci}-V_r)^2}[V_{ci}(V_{ci}+V_r) - 4V_{ci}V_r(\frac{V_{ci}+V_r}{2V_r})^3] \\ B = \frac{1}{(V_{ci}-V_r)^2}[4(V_{ci}+V_r)(\frac{V_{ci}+V_r}{2V_r})^3 - (3V_{ci}+V_r)] \\ C = \frac{1}{(V_{ci}-V_r)^2}[2 - 4(\frac{V_{ci}+V_r}{2V_r})^3] \end{cases} \tag{2}
$$

$$
P_{PV} = f_{loss}P_{STC}\frac{G_{AC}}{G_{STC}}(1 + k(T_c - T_r)) \tag{3}
$$

Where, cut in wind speed $V_{ci} = 3$ m/s, rated wind speed $V_r = 12$ m/s, cut out wind speed $V_{co} = 22$ m/s; P_{STC} is the maximum output power of solar panel; G_{AC} is the average solar radiation intensity; G_{STC} is the solar radiation intensity under the standard test conditions, with a value of 1000 W/m^2; k is the power temperature coefficient, with a value of $-0.0047/°C$; T_r is the ambient temperature under the standard test conditions, It is 25 °C.

2.2 Seawater Desalination System

In order to meet the water demand of island residents, the desalination system is composed of a reservoir and several desalination units. The system model is as follows:

$$
\begin{cases} P_h = NP_N \\ N = \begin{cases} \left\lceil \frac{R_n(t)-R_x(t)+R_d(t)}{P_N} \right\rceil & R_n(t) > R_x(t) - R_d \\ 0 & R_n(t) \leq R_x(t) - R_d \end{cases} \end{cases} \tag{4}
$$

Where, N is the number of desalination units to be opened; P_N is the rated power of a single desalination equipment; $R_n(t)$, $R_x(t)$ and R_d are the water load required by the user at time t, the storage capacity and the minimum storage capacity of the reservoir at time t; the initial water volume of the reservoir is 50 t; the minimum water demand is 10 t.

2.3 CHP System

The mathematical models of micro gas turbine and gas boiler are as follows:

$$
\begin{cases} \eta_t(t) = 0.4166(\frac{P_t(t)}{200})^3 - 1.0134(\frac{P_t(t)}{200})^2 \\ \qquad +0.8365(\frac{P_t(t)}{200}) + 0.0926 \\ H_t(t) = \frac{P_t(t)(1-\eta_t(t)-\eta_l)}{\eta_t(t)} \\ F_{MT} = \frac{J_{NG}\sum P_t(t)\Delta t}{\eta_t(t)L_{NG}} \end{cases} \tag{5}
$$

$$\begin{cases} H_r = R_{GB}\eta_{GB} \\ F_{GB} = \dfrac{J_{NG}\sum H_r(t)\Delta t}{\eta_{GB}L_{NG}} \end{cases} \qquad (6)$$

Where, $\eta_t(t)$ is the generation efficiency of micro gas turbine at time t; $P_t(t)$ is the power of micro gas turbine at time t; $H_t(t)$ is the heat recovery of micro gas turbine at time t; η_l is the heat loss coefficient, taking 0.03; F_{MT} is the cost of purchasing natural gas for micro gas turbine; J_{NG}, L_{NG} is the price of natural gas and the calorific value of natural gas, taking 2 CNY/m^3 and 9.78 kwh/m^3 respectively. η_{GB} is the energy conversion efficiency of gas boiler, taking 0.91.

2.4 Energy Storage System

The energy storage system adopts battery power storage and heat storage tank. The mathematical model is as follows [9]:

$$\begin{cases} E_H(t+1) = E_H(t)(1-\sigma_E) + P_{H,c}(t)\eta_{H,c}\Delta t - \dfrac{P_{H,d}(t)\Delta t}{\eta_{H,d}} \\ E_G(t+1) = E_G(t)(1-\sigma_G) + P_{G,c}(t)\eta_{G,c}\Delta t - \dfrac{P_{G,d}(t)\Delta t}{\eta_{G,d}} \end{cases} \qquad (7)$$

In the formula, $P_{H,c}(t)$, $P_{H,d}(t)$, $P_{G,c}(t)$, $P_{G,d}(t)$ are the power of charging and discharging at time t, and the power of heat storage and heat release respectively; σ_E, σ_G are the self discharging and heat release rate of the battery and the heat storage tank respectively; $\eta_{H,c}$, $\eta_{H,d}$, $\eta_{G,c}$, $\eta_{G,d}$ are the charging and discharge efficiency of the battery and the heat storage and heat release efficiency of the heat storage tank respectively; the initial electric quantity and heat quantity of the battery and the heat storage tank are 50 KW; the minimum electric quantity is 40 kW.

3 Objective Function and Constraints

3.1 Objective Function

In this paper, the operation of micro energy network is to consider the operation cost and user satisfaction of micro energy network in a one-day time scale. This paper studies the optimization of micro energy network, including the minimum operation cost and the maximum user satisfaction.

1) Micro energy network economy

$$\min price = \min(F_{grid} + F_{MT} + F_{GB}) \qquad (8)$$

$$\begin{cases} F_{grid} = \sum_{i=1}^{24} P_{grid}(t)F_{grid}(t) \\ F_{MT} = \sum_{i=1}^{24} F_{MT}(t) \\ F_{GB} = \sum_{i=1}^{24} F_{GB}(t) \end{cases} \qquad (9)$$

Where, $F_{grid}(t)$ is hourly price; $P_{grid}(t)$, $F_{GB}(t)$, $F_{MT}(t)$ are the exchange value of power between the system and the grid at time t, the purchase cost of natural gas for gas turbine at time t and the purchase cost of natural gas at time t respectively.

2) User satisfaction

Customer satisfaction consists of power supply satisfaction and heating satisfaction. The satisfaction of power supply requires that the less the amount of purchased electric energy is, the higher the satisfaction of power supply is. The satisfaction of heat supply is the proportion of other heat energy in the heat load except the complete heat supply equipment. Customer satisfaction:

$$S(t) = \frac{1}{2}S_d(t) + \frac{1}{2}S_r(t) \tag{10}$$

$$\begin{cases} S_d(t) = \frac{P(t) - F_d(t)}{P(t)} \\ S_r(t) = \begin{cases} 0.6 & Q_r(t) = 0 \\ \frac{Q_r(t)}{Q(t)} & Q_r(t) > 0 \end{cases} \end{cases} \tag{11}$$

Where, $S_d(t)$ is the satisfaction degree of power supply; $S_r(t)$ is the satisfaction degree of heat supply; $P(t)$, $F_d(t)$, $Q(t)$ are respectively the power load required by the user, the total cost of power purchase and gas purchase, and the heat load required by the user at time t.

3.2 Constraint Condition

1) Heat balance constraint:

$$H_r(t) + H_t(t) + H_x(t) \geq Q_{load}(t) \tag{12}$$

Where, $Q_{load}(t)$ is the required heat load at time t; $H_x(t)$ is the difference between the heat stored in the heat storage tank at time t and the minimum heat stored.

2) Electrical balance constraints:

$$P_w(t) + P_{pv}(t) + P_x(t) + P_t(t) + P_{grid}(t) \geq P_{load}(t) + P_b(t) + P_h(t) \tag{13}$$

Where, $P_{load}(t)$ it is the required electric load at time t; $P_{grid}(t)$ is the power exchange value between the system and the external power grid at time t; $P_x(t)$ is the difference between the stored energy of the battery and the minimum stored energy at time t, and the required electric energy of the heat pump.

3) Micro gas turbine operation constraints:

$$P_t^{min} \leq P_t(t) \leq P_t^{max} \tag{14}$$

Where, P_t^{min} and P_t^{max} are the minimum and maximum output power of micro gas turbine.

4) Desalination constraints:

$$0 \leq N \leq N\mathrm{max} \tag{15}$$

Where, Nmax is the maximum number of desalination units opened.

4 Simulation Example

In order to verify the validity and correctness of the micro energy network optimization model of cogeneration in this paper, the comparison between the model and the island common micro energy network energy supply system is made and the conclusion is drawn. Diesel generator is the main power supply equipment and heat pump is the heat supply equipment. Select an island as the verification point of the micro energy network, and simulate the micro energy network. The main equipment parameters of the micro energy network are shown in Table 1, the purchase/sale price is shown in Table 2, the predicted output curve of wind power, photovoltaic power generation and electric heating load is shown in Fig. 2, and the residential water load is shown in Fig. 3.

Table 1. Device parameters

Parameter	Numerical value
Wind farm/kW	500
Photovoltaic system/kW	250
Micro gas turbine/kW	200
Gas fired boiler/kW	200
Diesel generator/kW	200
Heating coefficient of heat pump	3
Power of single desalination plant/kW	25
Water yield of single desalination plant/t/h	5

Table 2. Online purchase/sale price

	Time	Power purchase/(kW·h/CNY)	Selling electricity/(kW·h/CNY)
Peak time	21:00–4:00	0.82	0.6
Peacetime	5:00–6:00 20:00 10:00–12:00 14:00–17:00	0.65	0.6
Valley time	7:00–9:00 13:00–14:00 18:00–19:00	0.25	0.6

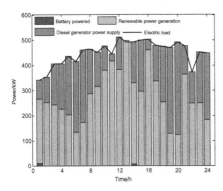

Fig. 2. Predicted output curve of renewable energy and electric heating load

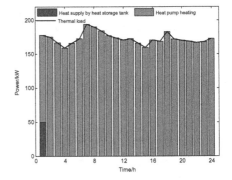

Fig. 3. Residential water load

The electric and thermal power dispatching results of the common energy supply system and the optimized configuration system of the island micro energy network are shown in Fig. 4, 5, 6 and 7 respectively.

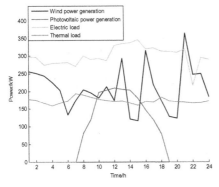

Fig. 4. Electric load dispatching results of common systems

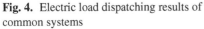

Fig. 5. Thermal load scheduling results of common systems

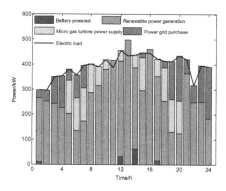

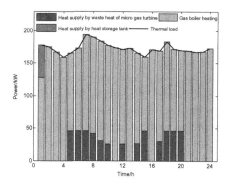

Fig. 6. The result of electric power dispatch in this paper

Fig. 7. The results of thermal power scheduling in this paper

It can be seen from the comparison between Fig. 6 and Fig. 7, Fig. 4 and Fig. 5 that the power supply system of common Island micro energy network will increase in the power load required by users every hour due to the heat pump heating. The power supply system of common Island micro energy network has only 13 h of slight power storage, and most of the time, the renewable energy can not meet the demand of users' power load, so diesel generator is needed for power generation Electricity causes the loss of cost and the aggravation of pollution, and heat supply is all completed by heat pump. The system studied in this paper can better meet the needs of users. It can store electricity at 11, 13 and 16 h, and provide power load for users in the following time. When the power supply of renewable energy is insufficient, it can choose the side with better economy to draw power from the grid and micro gas turbine, which can not only meet the demand of power load, but also reduce the expense. The heating system is composed of heat storage The waste heat of tank, gas boiler and micro gas turbine is no longer supplied by a single heat pump, which realizes the diversity of heating system.

Figure 8 and Fig. 9 show the operating costs and user satisfaction of the common island micro energy network energy supply system and the optimized configuration system in this paper.

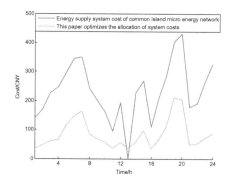

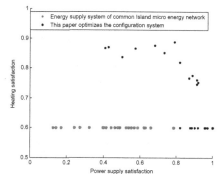

Fig. 8. Operating expenses

Fig. 9. User satisfaction

It can be seen from Fig. 8 that in one day, the hourly operation cost of the micro energy network system in this paper is about 100, which is far lower than that of the common micro energy network energy supply system. Only at 13:00, the cost is 0, which is lower than that of the system in this paper. This is because the thermal energy provided by the common micro energy network energy supply system is generated by electric energy conversion. When the renewable energy generation capacity is greater than the electric load, the battery is charged without using Diesel generator generates electricity, so the cost is 0. In this paper, the heating system of micro energy network is not completely affected by electric energy, so it needs to buy natural gas for heating, so the cost is slightly higher at 13:00.

It can be seen from Fig. 9 that the power supply satisfaction of the micro energy network studied in this paper is about 0.8 on average, the heating satisfaction is about 0.75 on average and the maximum is 0.9, and the two energy supply systems of the common micro energy network are about 0.4 and 0.6 respectively. Therefore, in this paper, the efficiency and customer satisfaction of converting cost into electricity are improved.

5   Conclusion

According to the characteristics of near land island, this paper proposes a micro energy network architecture of near land island cogeneration, which configures the capacity of the main equipment in the system, models a variety of energy supply and storage equipment in the micro energy network, and constructs two indicators that can reflect the operating cost and user satisfaction. According to the daily load demand, particle swarm optimization algorithm is used to solve the micro energy network, which is compared with the common Island micro energy network energy supply system. The example analysis shows that the micro energy network energy supply system studied in this paper can effectively reduce operating costs and improve user satisfaction.

References

1. Lei, J., Bai, H., Ma, X., et al.: Summary and prospect of multi-energy complementary island microgrid developed by china southern power grid. Southern Power Syst. Technol. **12**(3), 27–34 + 73 (2018)
2. Wang, J.: Modeling and simulation of island microgrid base on tidal current generation system. Zhejiang University (2018)
3. Yu, P., Liu, X., Sun, S., et al.: Study on operation control of island microgrid with high renewable energy penetration. Power Syst. Technol. **42**(3), 779–788 (2018)
4. Yao, Y., Ye, L., Qu, X., et al.: Exergy analysis model of wind-solar-hydro multi-energy generation power system. Electric Power Automation Equipment **39**(10), 55–60 (2019)
5. Zhang, J., Yu, L., Liu, N., et al.: Capacity configuration optimization for island microgrid with wind/photovoltaic/diesel/storage and seawater desalination load. Trans. Actions China Electrotech. Soc. **29**(2), 102–112 (2014)
6. Fan, L., Wang, K., Li, G., et al.: A optimization dispatch study of micro grid with seawater pumped storage plant in isolated islands. Power Syst. Technol. **40**(2), 382–386 (2016)

7. Sun, C., Wang, Z., Jiang, X., et al.: Coordinated optimization strategy for grid-connected microgrid considering demand response. Proc. CSU-EPSA **30**(01), 30–37 (2018)
8. Fang, B.: Multi-energy complementarity and integrated optimal allocation for hybrid renewable energy system. Hunan University (2017)
9. Li, Y.: Research on operation of the combined cooling, heating and power micro energy grid. Beijing Jiaotong University (2018)

Research on Sales Forecast of Electronic Products Based on BP Neural Network Algorithm

Linan Sun[✉], Guanghua Yu, and Zhuo Zhang

Heihe University, Heihe 164300, China
Sunlinan666666@163.com

Abstract. In order to solve the problem that the production volume and sales volume of electronic products cannot be matched in time, it is necessary to predict the order and sales volume, and then effectively control the production volume of manufacturers. This article first introduces the basic steps of implementing the BP neural network algorithm, and then uses MATLAB software to fit the original data based on the BP neural network algorithm to predict the sales volume of the latest generation of products sold by customers to customers in the next 20 weeks and the latest generation of products in different sales. The region's order volume in the next 20 weeks, and according to the forecast results to provide enterprises with production decisions to achieve timely matching of production volume and sales volume.

Keywords: BP neural network · Production volume · Sales volume

1 Introduction

With the rapid development of science and technology, electronic products have entered millions of households and become necessities for people's daily work and life. The rapid replacement of electronic products often makes product output and sales volume not match in time, resulting in product backlog or Deficit affects earnings. For this reason, the electronics industry urgently needs an effective product sales forecasting method to provide scientific decision support for its production plan. Orders and sales volume are important basis for forecasting production volume. Therefore, this article will build a sales forecast model and order forecast model for electronic products, and predict the sales volume of the latest generation of products sold by sellers to customers in the next 20 weeks based on the existing sales data of a certain brand of electronic products. Order volume for the next 20 weeks in different sales regions.

X. Jiang and P. Li (Eds.): GreeNets 2020, LNICST 333, pp. 330–342, 2020.
https://doi.org/10.1007/978-3-030-62483-5_34

2   BP Neural Network Algorithm

Among the neural networks, the most representative and widely used is the BP (Bake-Propagation) neural network (multi-layer feed forward error) proposed by Rumel Hart and McClelland of the University of California in 1985. Back Propagation Neural Network). This model is a supervised learning model with a strong self-organizing and adaptive ability. It can grasp the essential characteristics of the research system after learning and training on a representative sample, and has a simple structure and strong operability. Can simulate arbitrary non-linear input-output relationships. The specific implementation steps of the BP neural network algorithm are as follows:

The first step is to select the structure of the BP neural network based on the data signs. The BP neural network model used in this article The number of network layers is 2 and the number of hidden layer neurons is 10. The hidden layer and output layer neuron functions are selected as tansig function and purelin function, the network training method uses gradient descent method, gradient descent method with momentum and adaptive should lr the gradient descent method; The second step is to normalize the input data and output data; The third step is to construct a neural network using the function newff(); In the fourth step, before training the neural network, first set related parameters, such as the maximum number of training times, the accuracy required for training, and the learning rate; The fifth step is to train the BP neural network; The sixth step is to repeat the training until the requirements are met; The seventh step is to save the trained neural network and use the trained neural network to make predictions; In the eighth step, the predicted value is compared with the actual output value to analyze the stability of the model.

3   Sales Forecast

This section uses MATLAB software to fit the original data using the BP neural network algorithm, and finally completes the forecast of the sales volume of the latest series of each product. The specific program code is as follows:

```
p=(1:n);
t=[t1 t2 ......tn];
%Data normalization
[pn,minp,maxp,tn,mint,maxt]=premnmx(p,t);
%BP network training
net=newff(minmax(pn),[32,1],{'tansig','tansig','purelin'},'traingdx');
net.trainParam.show=1000;
net.trainParam.Lr=0.1;
net.trainParam.epochs=5000;
net.trainParam.goal=1e-2;
net=train(net,pn,tn);
%Simulation of raw data
an=sim(net,pn);
a=postmnmx(an,mint,maxt);
%Compared with actual data
x=1:n;
newk=a(1,:);
figure;
plot(x,newk,'r-o',x,t,'b--+');
legend('预测值','实际值');
xlabel('周数');
ylabel('销售量');
%Make predictions on new data
pnew=[n+1:n+20];
pnewn=tramnmx(pnew,minp,maxp);
anewn=sim(net,pnew);
anew=postmnmx(anewn,mint,maxt)
```

Enter the original data into the above code, and obtain the fitting curve of the sales volume of the a-3 generation products (see Fig. 1).

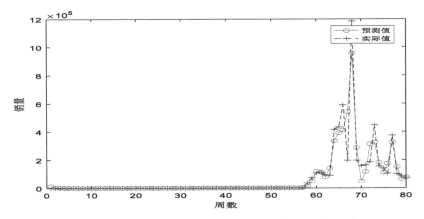

Fig. 1. Actual value and forecast value of sales of a-3 generation product sellers to customers.

After error analysis, the model is suitable for forecasting the sales volume of the product, and the original program is further improved to predict the sales volume of the a-3 generation products to the customers in the next 20 weeks(see Table 1).

Table 1. Sales value of a-3 generation product sellers to customers next 20 weeks.

Week number	1	2	3	4	5
Sales	143359	172831	176172	176463	176488
Week number	6	7	8	9	10
Sales	176490	176579	178659	175433	187642
Week number	11	12	13	14	15
Sales	196321	176440	176466	178762	176490
Week number	16	17	18	19	20
Sales	176490	163252	153621	14668	133695

The same processing method can predict the sales volume of the b-2 and c-1 products from the seller to the customer in the next 20 weeks. The fitting graphs of the actual and predicted values (see Fig. 2 and 3). Sales forecast values (see Table 2 and 3).

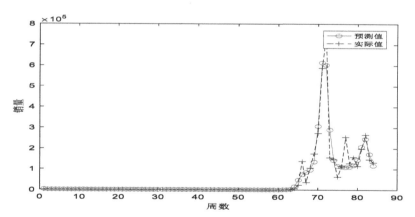

Fig. 2. The actual and predicted sales volume of the b2 generation product sellers to customers.

Through the above analysis, it is found that the BP neural network algorithm has a good fitting degree, the relative error between the predicted value and the actual value is small, the prediction result is more accurate, and it can be used for short-term and medium-term prediction.

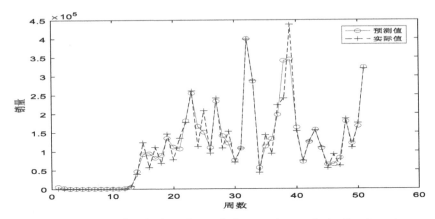

Fig. 3. Actual and predicted sales volume of c1 generation product sellers to customers.

Table 2. The sales value of the b-2 generation product sellers to customers next 20 weeks.

Week number	1	2	3	4	5
Sales	99359	94754	93600	93314	93243
Week number	6	7	8	9	10
Sales	93226	93222	93221	93762	99623
Week number	11	12	13	14	15
Sales	922168	95623	93256	93322	94563
Week number	16	17	18	19	20
Sales	977649	95321	93266	91237	91229

Table 3. Sales volume of c1 generation products sold to customers next 20 weeks.

Week number	1	2	3	4	5
Sales	336756	337216	3345662	339462	345982
Week number	6	7	8	9	10
Sales	345219	332346	337185	3373268	334232
Week number	11	12	13	14	15
Sales	335232	346259	363554	322496	332369
Week number	16	17	18	19	20
Sales	346813	345612	335694	321687	337432

4 Order Volume Forecast

This paper first uses EXCEL software to screen the d-3, e-2, and f-generation products by sales area and order type. According to the analysis of the characteristics of the filtered data, it is found that the prediction of the order volume can still use the BP neural network algorithm. The difference between the algorithm in this part and the second part lies in the selection of parameters and hidden layer neurons. The accuracy and training times of this part are higher than the second part. Another 30 hidden layer neurons are selected, which achieves a better simulation results. The effect. The results of fitting and prediction using MATLAB software are as follows: First of all, we carried out the order volume of the a-3, b-2, c-1 three-generation products that are not divided into sales areas in the next twenty weeks. Manufacturers received orders (type B + C) and sellers placed orders (type A + B). The specific prediction results are as follows (Fig. 4 and Table 4):

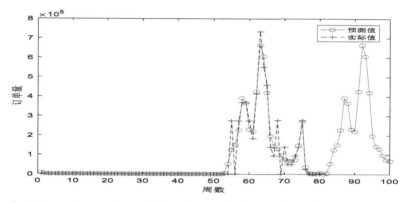

Fig. 4. A-3 generation product sellers' order quantity (A + B) actual data and forecast data.

Table 4. The forecast value of orders placed by a-3 generation product sellers next 20 weeks.

Week number	1	2	3	4	5
Order volume	1618	5331	52913	123943	150899
Week number	6	7	8	9	10
Order volume	227658	391698	365267	230814	220677
Week number	11	12	13	14	15
Order volume	425240	664172	607428	419209	197741
Week number	16	17	18	19	20
Order volume	145659	129577	95670	73021	68400

The forecast results show that in the next 20 weeks, the order volume of a-3 generation product sellers will first increase significantly, and will fall after reaching a certain number (Fig. 5 and Table 5).

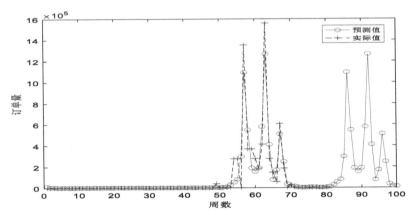

Fig. 5. Order data received by a-3 generation product manufacturers (B + C) actual data and forecast data.

Table 5. The forecast value of orders received by a-3 generation product manufacturers next 20 weeks

Week number	1	2	3	4	5
Order volume	9308	27117	55685	80166	296643
Week number	6	7	8	9	10
Order volume	1093835	546376	188619	156178	190653
Week number	11	12	13	14	15
Order volume	575802	1265769	409594	76835	171005
Week number	16	17	18	19	20
Order volume	510875	249303	33922	16593	12007

The forecast results show that the order volume of a-3 generation product manufacturers will have two peak periods in the next 20 weeks, and manufacturers can appropriately adjust their production strategies based on the forecast results (Fig. 6).

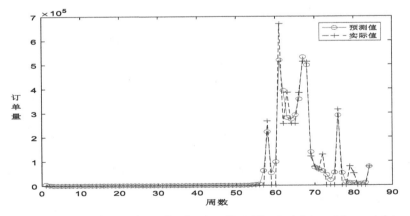

Fig. 6. B-2 generation product sellers' orders (A + B) actual data and forecast data.

It can be seen from the above figure that the fitting effect of the BP neural network model is better. After further error checking, it was found that the error was small, so the program was further improved for prediction. The predicted value.(see Table 6).

Table 6. B-2 generation product sellers' order volume forecast values for the next 20 weeks

Week number	1	2	3	4	5
Order volume	241864	293642	301244	302257	302390
Week number	6	7	8	9	10
Order volume	302408	302410	301345	302401	300411
Week number	11	12	13	14	15
Order volume	302345	312651	332461	313421	302011
Week number	16	17	18	19	20
Order volume	313541	245911	162111	92411	20411

The forecast results show that the order quantity of b-2 generation product sellers will pick up in the next 20 weeks, but the range is not large, and it will decline rapidly after the 17th week (Table 7 and Fig. 7).

Table 7. The forecast value of orders received by b-2 generation product manufacturers next 20 weeks.

Week number	1	2	3	4	5
Order volume	86486	90897	91573	91664	91676
Week number	6	7	8	9	10
Order volume	91677	91319	96678	91346	83461
Week number	11	12	13	14	15
Order volume	83497	76153	91699	91795	91278
Week number	16	17	18	19	20
Order volume	70325	66354	53196	63485	33467

The fitting and forecasting results show that the number of the two types of orders in the previous period of the product basically match, and the trend in the next 20 weeks is similar, but the number is different (Fig. 8 and Table 8).

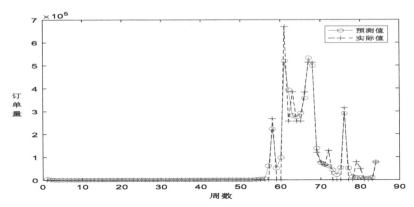

Fig. 7. The actual and forecast data of orders received by b-2 generation product manufacturers (type B + C).

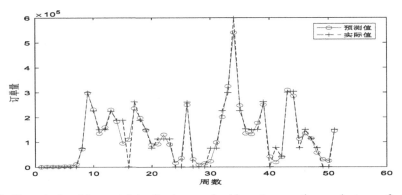

Fig. 8. The actual and forecast data of orders received by c-1 generation product manufacturers (type B + C).

Table 8. The forecast value of orders for c-1 generation product sellers in the next 20 weeks.

Week number	1	2	3	4	5
Order volume	194654	262736	302138	324758	316325
Week number	6	7	8	9	10
Order volume	315003	332587	69857	52867	42569
Week number	11	12	13	14	15
Order volume	158296	205432	241123	269158	85269
Week number	16	17	18	19	20
Order volume	52639	45891	234357	36985	25647

The forecast results show that the order volume of c-1 generation product sellers will fluctuate greatly in the next 20 weeks (Fig. 9 and Table 9).

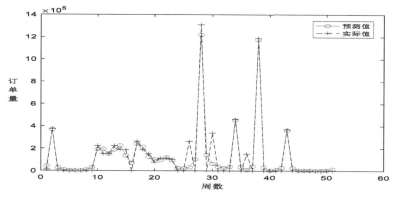

Fig. 9. The actual and forecast data of orders received by c-1 generation product manufacturers (type B + C).

Table 9. The forecast value of orders received by c-1 generation product manufacturers next 20 weeks.

Week number	1	2	3	4	5
Order volume	234873	321728	325511	325646	313549
Week number	6	7	8	9	10
Order volume	324637	346257	313425	315679	302467
Week number	11	12	13	14	15
Order volume	294637	237684	264359	334672	231467
Week number	16	17	18	19	20
Order volume	234678	297435	234357	224367	204367

The forecast results show that the orders received by c-1 generation product manufacturers in the next 20 weeks will remain high.

Secondly, according to the filtered data, predict the order volume of d-3, e-2, f products in different sales regions in the next 1 to 20 weeks. The specific fitting chart is as follows. The forecast results of the order volume in the next 20 weeks are not listed here. Now, readers can make their own predictions based on the program codes listed above, and the operation is simple and easy (Fig. 10).

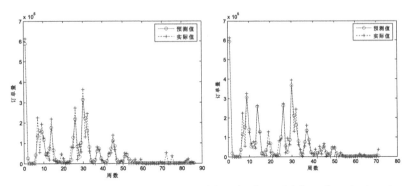

Fig. 10. (Left) Actual and predicted order quantities of sellers of d-3 products in the sales area a (Right) Actual and predicted orders for d-3 products in the sales area of the manufacturer.

The fitting and prediction results show that the two orders of d-3 products in the a sales region will enter a stable recession in the next 20 weeks, which indicates that these products will soon exit the market (Fig. 11).

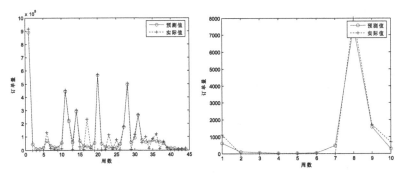

Fig. 11. (Left) Actual and predicted orders for d-3 products in c sales area; (Right) Actual value and forecast value of order quantity of sellers of d-3 products in b sales area.

The fitting and forecasting results show that the orders of manufacturers of d-3 products in the c sales region will gradually rise in the next 20 weeks. The orders of sellers of d-3 products in the b sales region will be smaller and should be increased in Product promotion in the region. The forecast results of the order quantity of the e-2 products in the sales region of a and the order quantity of the manufacturer show that the former has a certain degree of rebound after entering a long trough period. The main reason for the trough period is In the previous stage, the sellers in this sales area placed large orders; the latter placed large fluctuations in order quantities, which was mainly affected by changes in the order volume of C products (Figs. 12 and 13).

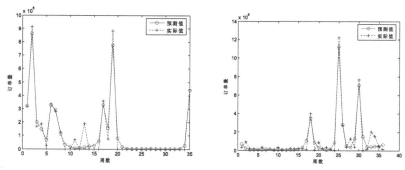

Fig. 12. (Left) Actual and predicted order quantities of sellers of e-2 products in sales area a (Right) Actual and predicted order quantities of manufacturers of e-2 products in a sales area.

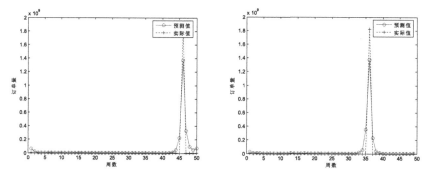

Fig. 13. (Left) Actual and predicted order quantities of sellers of product f in sales area a;(Right) Actual and predicted order quantities of producers of product f in sales region a.

The forecast results show that the order quantity of product f in sales area a has soared in a short period of time, but it has declined rapidly in the later period and continues to rise, indicating that the demand for product f in area a in the future is not high (Fig. 14).

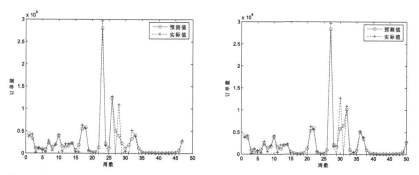

Fig. 14. (Left) Actual and predicted order quantities for manufacturers of product f in sales area c; (Right) Actual and predicted order quantities of sellers of product f in c sales area.

The fitting and forecasting results show that the two orders of f products in the c sales area have similar trends, indicating that sales of f products in this area are relatively stable.

Acknowledgement. 2019 Basic Scientific Research Operational Expenses Scientific Research Project of Hei-longjiang Provincial Department of Education, no. : 2019-KYYWF-0474.

References

1. Hu, Y.: Credit evaluation of personal housing loan repayment based on bp neural network. University of Science and Technology of China (2014)
2. Li, Q.: Fusion neural network model of BP neural network and its application in stock market forecasting. Jilin University of Finance and Economics (2013)
3. Deng, W.: MATLAB function all-around quick reference book. People's Posts and Telecommunications Press, Beijing (2012)
4. Li, B., Wu, L.: MATLAB data analysis method. Machinery Industry Press, Beijing (2012)
5. Ju, C.: Research and application of fuzzy neural network. University of Electronic Science and Technology Press (2012)
6. Deng, L., Li, B., Yang, G.: Analysis of matlab and financial model. Hefei University of Technology Press (2007)

A Multicast Routing Algorithm Under the Delay-Restricted Network Environment

Jinpeng Wang(✉), Xin Guan, Yang Zhou, Fan Cao, and Nianyu Zou

Information Science and Engineering College,
Dalian Polytechnic University, Dalian 116034, China
wangjp@dlpu.edu.cn

Abstract. Nowadays, the network is expected to service much more multimedia data with the improvement of the network communication bandwidth and development of the processing capacity. In the multicasting communication system, once every receiving end separately sends the data packets, the network resources intend to be wasted, and the calculating stress on the nodes is also going to be increased. A distributed delay-restricted multicast route heuristic method DMPH (delay-constrained minimal-cost path heuristic) is presented in this article. This algorithm proposed in the article can achieve a convergence speed, back up dynamic multicast, and offer the pretty good the network overhead performance. In order to gain the better performance of the network cost, the article firstly proposes a mathematical topological model of the problem over the delay-constrained multicast routing, and then presents a dynamic multicast delay-restricted multicast route algorithm which is called DMPH (Delay-Constrained Minimal-Cost Path Heuristic), fast-convergent and distributed. The computing simulation results suggest that the method pro-posed can gain the better performance of the network cost. In addition, the article just addressed the operation of the addition and quit of the nodes on the dynamic varying of the multicast group, but not concerned the over-head optimal. These problems mentioned above are all our future research work.

Keywords: Multicast · Delay-restricted multicast · Routing · Topology

1 Introduction

Nowadays, the network is expected to service much more multimedia data with the improvement of the network communication bandwidth and development of the processing capacity [1]. And the network is required that they should have perfect capacity on multicast by most of these services, especially for many multimedia services, just like the video/audio meeting online [2], the exchanging mock simulation, the game for multiplayer, and the distributed database [3] etc.

In the multicasting communication system, once every receiving end separately sends the data packets, the network resources intend to be wasted, and the calculating stress on the nodes is also going to be increased.

X. Jiang and P. Li (Eds.): GreeNets 2020, LNICST 333, pp. 343–349, 2020.
https://doi.org/10.1007/978-3-030-62483-5_35

In common meaning, it is pretty difficult to deal with the problem about the routing on QoS multicast [4]. The most important thing to a delay-constrained Steiner problem is to address the cost-optimization multicast tree under the delay-restricted environment [5]. Because the NP-complete is the key to address the optimization delay-restricted problem [6], the increase of the algorithm computing complexity is not able to be illustrated via the polynomials with the increase of the network ends. While the network is big, finding a Steiner tree which is unsuitable for network multicast applications intends to use a long time. Despite the heuristic methods are not able to achieve the optimal Steiner tree in most instances [6], they are also able to the quasi-Steiner tree with near-optimal cost in a relatively shorter time span. And it has the merits on the low complexity, easy application and is more reasonable in the multicast cost optimization.

A distributed delay-restricted multicast route heuristic method DMPH (delay-constrained minimal-cost path heuristic) is presented in this article. This algorithm proposed in the article can achieve a convergence speed, back up dynamic multicast, and offer the pretty good the network overhead performance.

2 The Mathematical Computing Model for the Delay-Restricted Multicasting Route Problem

Supposed a graph $G = (V, E)$, every edge $e \in E$ has its own two weight power functions: $C(e)$ and $D(e)$, inside $C(e)$ symbolize the positive real overhead of the link e, and $D(e)$ expresses the delay which is taken for transmitting message over the link e. For the graph G, supposed a source node $s \in V$ and a destination node set $D \subseteq V - \{s\}$. Then the delay-restricted Steiner tree T is a tree which is rooted in s and able to cover the all destination nodes. Once the condition which the delay constraint is met, the network overhead of the tree could be minimized as below:

$$\forall v \in D, \text{ while } \sum_{e \in P(s,v)} D(e) \prec \Delta \text{ is met, } \sum_{e \in T} C(e) \text{ is minimized} \tag{1}$$

In above, Δ is a positive real number, representing the boundaries of delay constraints, and $P(s, v)$ being the path from the source node s to the destination node v in the multicast tree.

3 Description of the Algorithm Steps

Based on the MPH arithmetic, DMPH algorithm is proposed in this paper, and its main idea is about: In the networks, there is not going to be many paths to d_i to meet the delay requirement in common if the shortest delay time from the source end s to the destination end d_i is somewhat close to the upper limit of Δ. However, once the shortest delay time to d_i is relatively long to the upper limit of Δ, then there will be more d_i paths to meet the delay requirements. Consequently, when we build a delay-restricted multicast tree, if we firstly add the destination ends with relatively big minimum delay to the multicast tree, then the destination ends with smaller minimum delay are connected to the multicast tree via sharing some paths of the current tree without violating the delay restrictions. These cases could come true, and the multicast tree also is optimized with much less overhead. On the contrary, once we firstly choose the ends with the smaller delay to connect to the tree, then the ends which meet the shortest time delay may give up the method of connecting to the multicast tree by sharing some paths of the current tree because of the limitation of delay constraint. It even needs to establish a path to connect itself to the group separately. So to plant trees is not easy to optimize the multicast tree cost.

Before describing the DMPH method, a tree to destination data structure T2D (shown in Table 1) and a parameter D_v that we need to use in the algorithm are defined firstly.

The table T2D has $|D|$ rows, every row has five columns: d_i, T2D $[d_i]$.cost, T2D $[d_i]$.trenode, T2D $[d_i]$.order and T2D $[d_i]$.tag. They represent respectively: the destination end di, the minimum overhead path $P(d_i, T)^1$ from node d_i to the current tree T under the condition of satisfying delay, the node which $P(d_i, T)$ can access in the tree T, the sequence number of destination node d_i plan to connect to the tree T, the flag whether d_i has been added to the tree T (yes: means Di has been added to the multicast tree; no: means Di has not been added to the multicast tree).

Table 1. Data structure of T2D

d_i	T2D[d_i].cost	T2D[d_i].trenode	T2D[d_i].order	T2D[d_i].tag
d_1				
......				
$d_{|D|}$				

Parametric D_v represents the path delay from source node s to node v in a multicast tree, and is kept as an information for each node v in the tree. The algorithm can be described below:

Step 1. To initialize the table T2D. Firstly, the tag of every destination node is set to be NO, and the trenode of each access end being s. Computing the minimum delay of every destination end D and the minimum cost which can meet the delay requirement, arranging these destination ends in non-ascending order withe the shortest delay they reach, but the nodes with the same minimum delay are arranged in a non-descending order of the minimum cost. The value of T2D[d_i].order is the location of the node Di in the sorted node set. The location also is the order in which it connects to the multicast tree, and supposing that the order of the sorted destination nodes is $d_1, d_2, \cdots, d_{|D|}$. T2D[$d_i$].cost is the minimum overhead which can meet the delay requirement from the source node to the destination node. It should be noted that if the shortest delay from one node to another is bigger than the limit Δ, there is no multicast tree that meets the delay requirement, so the process should be stopped.

---

[1] $P(d_i, T)$ It can be calculated from a delay-limited unicast routing algorithm, which does not mean absolute optimization.

Step 2. To assume $T = (\{s\}, \varnothing)$ and $i = 1$. Establishing a shortest overhead path $P(d_i,s)$ which meets the delay requirement from d_i to s, Then adding $P(d_i, s)$ into T, and updating T2D$[d_i]$.tag to yes. If $P(d_i, s)$ also pass through other destination ends, then T2D $[d_i]$.tag of this destination node also is updated to yes. Updating T and T2D.

Step 3. To assume that the access node from d_i to the tree is u_i, for every node on the newly added path $P(u_i, d_i)$. From the access end u_i to the destination node d_i, the following operations are performed step by step (taking one of the nodes v as an example):

To compute the smallest overhead path $P'(v, d_j)$ from the node v to every node d_j of the non-tree respectively under the condition of the delay limit $\Delta - D_v$, and if meeting the condition: $\text{cost}[P'(v, d_j)] \prec$ T2D$[d_j]$.cost, then

$$\begin{cases} \text{T2D}[d_j].\text{cost} = \text{cost}[P'(v, d_j)] \\ \text{T2D}[d_j].\text{trenode} = v \end{cases} \tag{2}$$

In the Eq. (2), $\text{cost}[P'(v, d_j)]$ being the overhead of the path $P'(v, d_j)$, and the node v transfer the T2D to the next node of the tree. Therefore, at destination node d_i, all non-tree destination nodes find the minimum overhead path and access node that can satisfy the delay requirement from the current tree, and then T2D stays at d_i.

Step 4. Let $i = i+1$, **and start from step 4 If the T2D [di].** To let tag is yes. Otherwise, d_{i-1} sends the message inform[2] to the node T2D $[d_i]$.trenode. Once T2D $[d_i]$.trenode receives the information, the smallest overhead path $P(\text{T2D }[d_i].\text{trenode}, d_i)$ to d_i which meets the delay requirement is constructed. Then this path will be accessed into the tree T, updating the tree T, and T2D $[d_i]$.tag to yes. If the path $P(\text{T2D }[d_i].\text{trenode}, d_i)$ also pass through other destination nodes, the T2D $[d_i]$.tag of the destination end is updated to yes.

Step 5. To keep the Operation. If $i \neq |D|$, then go to the step 3; the algorithm will not stop until $i = |D|$. T2D is established at the source node, with the addition of new paths to the tree, T2D will go through every new joined tree node (including destination nodes and non-destination nodes) one by one, which makes it easy that every new joined tree node can get the delay of the path from the source node to its tree.

From the algorithm above, the method DMPH is completely distributed, the source node only takes the responsibility for adding 3 into the multicast tree, and the rest of routes will be done via other nodes. This way allow that the two steps (route[3] 和connection configuration[4]) to establish the multicast connection could be done at the same time,

---

[2] Inform contains T2D and a message informing T2D $[d_i]$. tree node to establish a minimum overhead path to node d_i that satisfies the delay.

[3] Route: to look for a multicast routing tree initiated from a source node that covers all destination nodes;

[4] Connection Configuration: New connections are configured for each node on the tree, including reserving network resources and registering a new connections in a switching table.

it is much easier. However, in the centralized algorithm, when the source node (or a central node) calculates the routing, a connection configuration stage also is needed to be done separately. In addition, the centralized method is much more complex than the distributed algorithm. In the whole convergence time of this algorithm, DMPH intends to take at most $|D|$ times to connect the destination nodes from the tree ends to the multicast trees, up to $|D| - 1$ inform packets are sent to the tree node to notify the establishment of a path to connect a destination node to the tree, and a complete packet is needed to inform the source node the message that the tree construction has been completed. All in all, DMPH only take at most $|D|$ times to transfer the packets in the whole convergence time of this algorithm.

4   Addition and Quitting of the Nodes

Many multicast applications require the supports for dynamic multicasting, because the members of multicast group often change, the participating network nodes can join or leave the multicast group at any time, and the communication members are dynamic, with the addition and quitting of the multicast members, this type of the application needs to change the current multicast tree. So the multicast problem is the dynamic multicast routing, and DMPH addresses the dynamic multicast routing problem via using the methods below:

4.1   Addition of the Nodes

When a node d_q requests to be added into the current multicast group, if d_q is the Steiner node of the multicast tree (except the source node and destination node), no additional operation is needed to be performed. But if d_q is not on the multicast tree, the following operations below intends to be done:

The node d_q sends a request packet to the source node s for joining. When s receives the request information from d_q, it creates a require group with three items of information (new-node, cost and trenode), and sets new-node, cost and trenode to d_q, minimum overhead under delay constraint from s to d_q, and s respectively. s Sends require packets to each of its downstream nodes separately, the downstream node compares the minimum overhead of require packets to d_q under the requirement of time delay. If the overhead is less than the cost of require packets, cost is set to the overhead, and tree node is set to the downstream node. This downstream node transmits the require packets which are processed by it to every its downstream ends for performing the operations above. Therefore, when the require packet reaches at the all leaf nodes, each leaf end transmits the processed packet to the source node s. After the source node s receives the all require packets feed backed, the node which is the closest to d_q under the delay restriction will be found. Then the source end s sends the create packet to the node which is the closest to d_q to construct a minimum overhead path under the delay constriction, and add d_q to the tree.

4.2 Quitting of the Nodes

The treatment for the quitting of the nodes is so easy. If the quitting node is not the leaf node, there is no operation. If the quitting node is the leaf node, then this node and its upstream node will be deleted. The treatment for the quitting of the nodes is so easy. If the quitting node is not the leaf node, there is no operation. If the quitting node is the leaf node, then this node and its upstream node will be deleted until it meets a multicast node or a node with more than $2°$.

5 Experimental Cases

The Fig. 1 shows a graph with 18 ends, the parameters of its each edge is (overhead, delay), the source node is1, and its destination nodes set is {7, 14, 16, 18}. Figure 2 and 3 are the calculating results under the delay constriction $\Delta = 9$ and $\Delta = 25$ respectively, the tree overhead is 86 and 65, and the maximum delay from the source node to the destination end is 8 and 12.

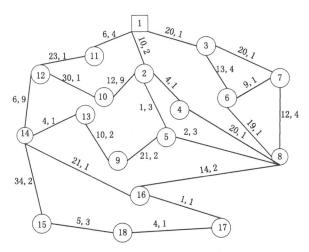

Fig. 1. A graph with 18 nodes

Now we could compare the algorithm to the minimum delay path tree method. When the algorithm starts, a shortest delay path from the source node to every destination nodes will be built, then these paths are combined, and combined result is exactly the final result.

The tree overheads are all 114 while the delay constriction $\Delta = 9$ and $\Delta = 25$. However, in the traditional method, the destination nodes which are the closest to the current tree could be linked to the tree, so that the overhead of the multicast tree equals 126 under $\Delta = 9$ and a circle is constructed. At the same time, the overhead of the multicast tree is 65 under $\Delta = 25$. Consequently, the overheads of the multicast tree which is computed via the DMPH algorithm proposed in this article are all not more than the traditional method.

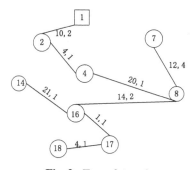

Fig. 2. Tree of $\Delta = 9$

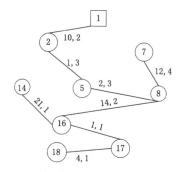

Fig. 3. Tree of $\Delta = 25$

6 Conclusion

This paper proposed delay-restricted multicast route heuristic method DMPH (delay-constrained minimal-cost path heuristic). The algorithm has the characteristics of distributed, fast convergence and dynamic multicast support, and achieves the good performance over the network overhead. In the article, we only talked about the algorithm under the delay constriction, and supposed the same delay-restricted limit for the all destination nodes. Virtually, there are different QoS limit, and each destination node has their own different QoS requirement. So a multicast routing algorithm with QoS limit intends to be more complicated. In addition, the article just addressed the operation of the addition and quit of the nodes on the dynamic varying of the multicast group, but not concerned the overhead optimal. If the optimal problem is involved in, the method will be more and more complicated. These problems mentioned above are all our future research work.

Acknowledgements. This research was financially supported via Project of the National Natural Science Foundation of China (61402069), the 2017 Project of the Natural Science Foundation of Liaoning province (20170540059), the General project of Liaoning education department in 2016 (2016J205).

References

1. Lee, S.-J., Su, W., Gerla, M.: On-demand multicast routing protocol in multi hop wireless mobile networks. Mob. Netw. Appl. **7**, 441–453 (2002)
2. Wang, J., Zhengpeng, Y., Gillbanks, J., Sanders, T.M., Zou, N.: A power control algorithm based on chicken game theory in multi-hop networks. Symmetry **11**(5), 718 (2019)
3. Sahasrabuddhe, L.H., Mukherjee, B.: Multicast routing algorithms and protocols: a tutorial. IEEE Netw. **1**(2), 90–102 (2000)
4. Lee, S.-J., Su, W., Gerla, M.: Ad hoc wireless multicast with mobility prediction.: Book title. In: Proceedings of IEEE ICCCN 1999, no. 7, pp. 4–9 (1999)
5. Wang, J., Zhang, S., Zhang, J.: Multi-hop maimal ratio combining (MHMRC) diversity based on virtual cellular network. Jilin Univ. **41**, 533–536 (2011)
6. Wang, J., Zou, N., Zhang, Y., Li, P.: Study on downlink performance of multiple access algorithm based on antenna diversity. ICIC Express Lett. **9**, 1221–1225 (2015)

The Research and Implementation of Image Style Conversion Algorithm Based on Deep Convolutional Neural Network

Huang Yaoqun[✉], Xia Hongyang, and Kang Hui

School of Electronics and Information Engineering, Heilongjiang University of Science and Technology, Harbin 150022, China
huangyaoqun@126.com

Abstract. With the development of deep learning and image technology, the deep convolutional neural network has been widely used to deal with image problems. In this paper, the pretrained vgg-19 convolutional network model is adopted to extract and define the loss function according to the image characteristics, and preset model parameters. The model training is completed through reverse propagation gradient descent and optimization iteration, finally, the artistic painting style conversion of photos is realized. At the same time, by adjusting the size of style weight and content weight, the output image is more inclined to the style picture, or more inclined to the content picture. Finally, the objective evaluation of the output image is has been completed by comparing the image style conversion results of the TensorFlow and the PyTorch. The results show that under the same iteration times, the PyTorch framework has relatively small computation, fast processing speed, and better image color retention effect, in contrast the TensorFlow frame retains more features of style images.

Keywords: Image style conversion · Deep learning · VGG-19 convolutional network model

1 Introduction

The image style conversion is a technology that extracts the image content contained in one image and the image style contained in the other model for synthesis, and then formed a new model, which can be widely used in animation production, advertising design, mobile image processing, and other fields.

Before the rise of deep learning, the image style conversion methods used by people were difficult to meet the actual needs, and the synthesized images were relatively rough. With the maturity of computer technology and the fierce development of deep learning, the convolutional neural network was adopted to image style conversion, have changed the concept of image style conversion, make the processing of image style conversion is no longer at the pixel level, still a global translation by extracting image features, in term of tonal, texture and spatial relations of image conversion [1], its basic principle is that

X. Jiang and P. Li (Eds.): GreeNets 2020, LNICST 333, pp. 350–359, 2020.
https://doi.org/10.1007/978-3-030-62483-5_36

by establishing loss function of making image between style image and loss function of generating image between content image, import the neural network model for training, finally produces vivid and better visual effect image through the optimization iteration [2].

2 VGG -19

Realizing image style conversion requires building, a new model should be built established on the deep convolutional neural network for training. The purpose of using a pre-trained convolutional neural network is to save time and space costs. The convolutional operation is a process in which the convolutional kernel makes a small range weighted sum on each position of the input data by using the sliding window. Therefore, the convolutional operation can be popularly understood as a process of "filtering." After the interaction between the convolutional kernel and the input data, the filtered image was obtained by extracting the image characteristic. As the convolutional layer gets deeper and deeper, the higher accuracy of image feature extraction, the pre-trained VGG-19 convolutional neural network model is adopted in this paper, and its underlying architecture has been shown in Fig. 1.

To determine the style layer and content layer in the convolutional network layer and establish the loss function, it is necessary to reconstruct the feature information extracted from each segment. As shown in Fig. 2, by comparing the reconstruction results of the first layer of convolution and the original image that almost the same, also the second layer, until the results of the third and fourth layer are relatively fuzzy, the fifth layer convolutional reconstruction results can distinguish is a monkey, until the figure fc7, relu7 layer the characteristics of the original image are indistinct, cannot identify completely, the neural network learned more and more general information [3].

Comparing the reconstruct effect of each layer, it can be seen that the reconstruct effect of the shallow layer is often better. The convolutional characteristics basically retains the shape, position, color, texture and other information in the selected original image. The deep corresponding restored image loses some color and texture information, but generally retains the shape and position of the object in the original image. By comparison, we can see that for content images, the effect is inversely proportional to the depth of the convolutional layer, and the shallower the convolutional layer, the better the content characteristics in the content image can be restored. However, for style images, the deeper the convolutional layer is, the better it can restore the style features in style images.

In this paper, the content loss function of the image is established in the slightly fuzzy fourth layer, which saves a lot of high-frequency components. The image style loss function is set from segment 1 to layer 5 to ensure the image style conversion results are smoother, and the style features are prominent.

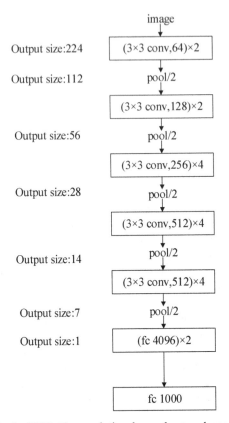

Fig. 1. VGG-19 convolutional neural network model

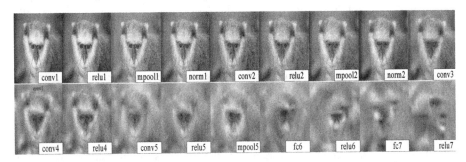

Fig. 2. The reconstruction of the convolutional layer feature information

3 Define the Loss Function

Firstly, the process of image characteristics extraction by VGG-19 convolutional network is compared with image style conversion. The purpose of the VGG-19 convolutional network is to extract the feature information from the input image and output the image category. But in the application of image style conversion, it is necessary to use the middle feature of the VGG-19 convolutional layer to restore the original image corresponding to these characteristices. In other words, the image style conversion is exactly the opposite of VGG-19. The input is these characteristics, and the output image corresponds to these characteristics.

3.1 Content Loss Function

The content loss Function is the difference of images content. Image characteristics are extracted from the pretraining network can be compared instead of the direct comparison between images. Features can be regarded as higher-dimensional pictures, so the better results can be achieved by using image features comparison [4].

The C_{nn} represents a pre-training network that only contains the extracts characteristics by the convolution training network before. The X represents an arbitrary input picture, then $Cnn\ (X)$ can stand for a set of collections of input image's characteristics of each layer which extracted through preliminary training network, each characteristics collection has a three-dimensional matrix, let the $F_{XL} \in C_{nn}(X)$ represents the characteristics collection which has been removed from the L layer of the network, its size is $h \times w \times d$, the matrix can be unfolded into a one-dimensional vector, regarded as this input picture's content of the L layer in the network is F_{XL}, If we need to compare the content differences between the two images, then the size of the two images should be the same. For example, Y is another picture. We can define the content loss function of the two photos in the L layer by using formula 1. In the equation, $F_{XL}(i)$ represents the elements of the expansion vector from the characteristics collection of the L layer.

$$D_C^L(X, Y) = \|F_{XL} - F_{YL}\|^2 = \sum_i (F_{XL}(i) - F_{YL}(i))^2 \qquad (1)$$

3.2 Style Loss Function

The definition of style loss function of the style image is not as intuitive as the definition of content loss, so we need to introduce a *Gram* matrix to represent the image style, and then calculate the style loss function. The size of the *Gram* matrix is determined by the thickness d of the characteristic graph. For each element in the *Gram* matrix, the i and j layers of the thick feature diagram are taken out first . In this way, two $h \times w$

matrices have been obtained, which are expressed as F_{XL}^i and F_{XL}^j respectively. Then, the corresponding elements of the two matrices are multiplied and summed, $Gram(i, j)$ have been obtained, shown in formula 2.

$$G_{XL}(i, j) = < F_{XL}^i, F_{XL}^j > = \sum_K F_{XL}^i(k) \cdot F_{XL}^i(k) \tag{2}$$

$$D_S^L(X, Y) = \|G_{XL} - G_{YL}\|^2 = \sum_{k,l} (F_{XL}(k, l) - F_{YL}(k, l))^2 \tag{3}$$

Gram matrix of each element are all related to the characteristic collection of the pattern in layer i and layer j, has been expressed as the association matrix, if the *Gram* (i, j) is defined as the picture in the output of the convolution L network layer style, the style difference of two images can be described by the difference of *Gram* matrices, style loss function has been shown in formula 3.

4   Model Establishment

After selecting the content layer and the style layer, the content loss function, and the style loss function need to be added to the VGG-19 convolutional network for pre-training. The specific implementation process has been divided into two parts.

In the first part, the pre-trained VGG-19 network segment is named for comparison with the content layer and the style layer. The specific steps are as follows: Set variable $i = 0$, and then through loop iterates of VGG-19 in each layer, if condition statement judgment the layer is Conv layer, made $i = i + 1$, and the layer named conv_i, if the coating is not Conv, Elif statement whether the sheet is used to activate the sheet, if the layer called *ReLU_i*, and the sheet is not enabled, then continue to use elif statement judge the sheet is the most prominent pooling layer, f the layer called *pool_i*, if the coating is not more than three layers of cycle judgment, otherwise, Add the unique layer to the original model.

The second part, take the content loss function and loss function in content layer and style layer, steps are as follows: if statement judges the layer named in the first part with the contents of the selected layer, if in the content layer, add content after the layer loss function, and add the layer to the content of network loss, in the same way judge the named layer in the layer style, if the loss function is added after the layer style. Loss of the network will be added to the layer style. After such a complete network structure is traversed, a new network model of image style conversion is constructed, it has been shown in Fig. 3.

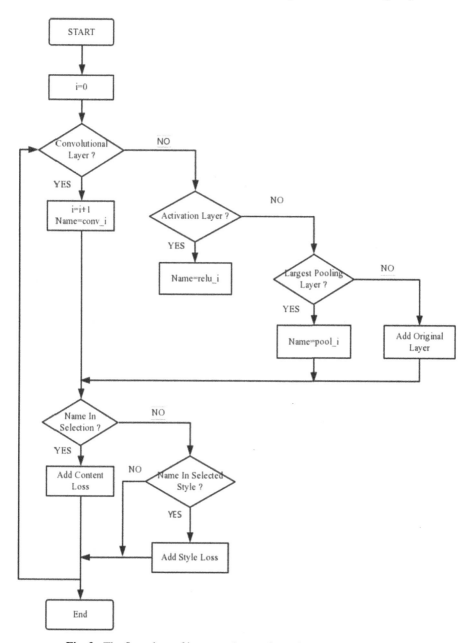

Fig. 3. The flow chart of image style transformation model construction

5 Model Training

The style conversion output image is obtained by model training. After the model training is completed, the style image and the content image are imported into the model, the style loss and the content loss are cyclically calculated, and the style loss and the content loss are weighted, and then the gradient is used. Decreasing the backpropagation update parameters to minimize the loss function, through optimization iterations, and finally output the style conversion image.

The block diagram of the back propagation gradient descent calculation loss parameter is shown in Fig. 4. First, after an image passes through the convolution layer, it is a three-dimensional array. After an image passes through the convolution layer, several three-dimensional arrays Ki are obtained. These arrays represent the content of the image. The multi-dimensional array can be used to obtain the Gram matrix operation to obtain the style of this image. After extracting the content and style of the image, the content loss and style loss are calculated**Error! Reference source not found.**. In the style transformation, content loss *DC (X,C)* and style loss *DS (X,C)* need to be minimized, so the gradient calculation of these two values needs to be carried out, the gradient calculation is shown in formula 4.

$$\nabla(X, S, C) = \sum_{L_c} w_{CL_c} \cdot \nabla^{L_c}(X, C) + \sum_{L_S} w_{SL_S} \cdot \nabla^{L_S}(X, S) \tag{4}$$

In the formula, L_c and L_s represent the layer output required by the content and style respectively. These parameters can be set arbitrarily according to the desired effect. The two w represent the weight assigned to the content and style, respectively which can be set arbitrarily. The size of the weight will affect the degree of style conversion image.

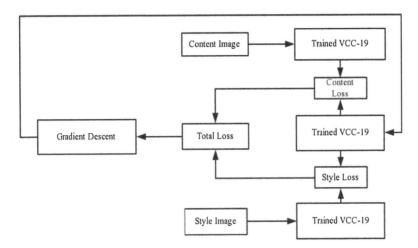

Fig. 4. The block diagram of reverse calculation of loss parameters

6   Results and Evaluation

In terms of the evaluation of the effect of image style conversion, there are two main evaluation methods, namely subjective evaluation and objective evaluation. In subjective evaluation, the main factors that lead to the effect of a picture are each person's personal preferences and evaluation methods, and can also be compared with other experimental results to analyze the quality of the experimental results. In objective evaluation, you can compare the output picture effect by comparing the size of content loss and style loss when the number of iterations is the same.

6.1   Subjective Evaluation of Different Weights

In this paper, the personal evaluation is obtained by adjusting the different weights in the loss function. Figure 5 is the style image, and Fig. 6 is the content image. When the weight of the loss is different, the degree of image information content and style conversion that have been retained can be controlled, so that more content or style oriented images can be obtained, shown in Fig. 7.

Fig. 5. The style image   **Fig. 6.** The content image

It can be seen from the comparison that when the content weight is larger, the content characteristics of the output image are closer to the content picture, on the contrary, when the style weight is larger, the content style of the output image is closer to the style image.

6.2   Objective Evaluation Under Different Deep Learning Frameworks

By using two different deep learning frameworks, TensorFlow and PyTorch, under the same number of iterations, the time used and the effect of the output image were compared to achieve the objective evaluation.

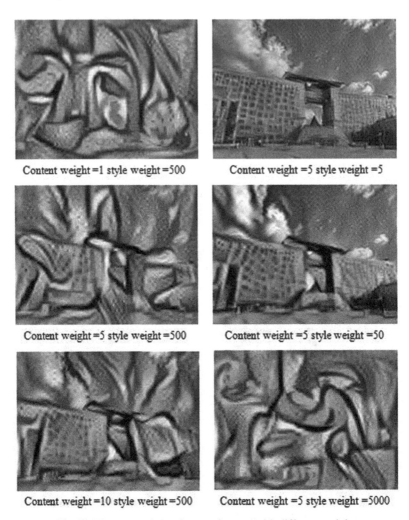

Content weight =1 style weight =500   Content weight =5 style weight =5

Content weight =5 style weight =500   Content weight =5 style weight =50

Content weight =10 style weight =500   Content weight =5 style weight =5000

Fig. 7. The comparison of output images with different weights

Figure 10 is generated by the TensorFlow deep learning framework, and Fig. 11 is made by the PyTorch deep learning framework, both of which have 200 iterations. From the comparison of the output image effect, it can be concluded that the image color retention effect of PyTorch deep learning box is better than that of the TensorFlow deep learning frame, which retains the feeling of style image swirl more. Meanwhile, the comparison of output efficiency under the two frameworks is shown in Table 1.

It can be seen that with the same number of iterations, the PyTorch framework has a relatively small amount of computation, so it has faster operation speed and a lower CPU proportion Fig. 8 and Fig. 9.

Fig. 8. The style image

Fig. 9. The Content image

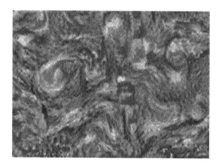

Fig. 10. TensorFlow output image

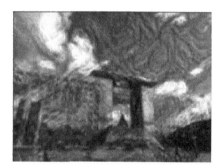

Fig. 11. PyTorch output image

Table 1. The comparison of test results

	Time	CPU Max
TensorFlow framework	313.84 s	98%
PyTorch framework	186.87 s	70%

References

1. Yaling, D., Wubin, Z., Renhuang, J.: Image style transfer technology based on convolutional neural network. Mod. Comput. **630**(30), 49–53 (2008)
2. Wuyang, L.: Shallow theory of image style transformation based on deep learning. Digit. Commun. World, (2) (2018)
3. Mahendran, A., Vedaldi, A.: Understanding deep image representations by inverting them. In: Proceedings of the IEEE Conference on Computer Vision and Pattern Recognition (2014)
4. Gatys, LA., Ecker, AS., Bethge, M.: Image style transfer using convolutional neural networks. In: 2016 IEEE Conference on Computer Vision and Pattern Recognition (CVPR). IEEE (2016)
5. Guohui, F.: Classification of small-scale images based on VGG model of convolutional neural network (2018)

Author Index

Printed in the United States
By Bookmasters